MEDICAL BIOCHEMISTRY

Books are to be returned on or before
the last date below.

LIBREX–

MEDICAL BIOCHEMISTRY
Human Metabolism in
Health and Disease

MIRIAM D. ROSENTHAL
ROBERT H. GLEW

WILEY

A JOHN WILEY & SONS, INC., PUBLICATION

For general information on our other products and services or for technical support, please contact our Customer Care Department within the United States at 877-762-2974, outside the United States at 317-572-3993 or fax 317-572-4002.

Wiley also publishes its books in a variety of electronic formats. Some content that appears in print may not be available in electronic format. For more information about Wiley products, visit our web site at www.wiley.com

Library of Congress Cataloging-in-Publication Data:

Rosenthal, Miriam D.
  Medical biochemistry : Human metabolism in health and disease /
Miriam D. Rosenthal and Robert H. Glew.
      p.   ;   cm.
  Includes index.
  ISBN 978-0-470-12237-2
1. Metabolism.   2. Metabolism–Disorders.   I. Glew, Robert H.   II. Title.
  [DNLM: 1. Metabolism.   2. Metabolic Diseases.   QU 120 R815t   2009]
  QP171.R65   2009
  612.3'9–dc22   2008029609

Printed in the United States of America

10 9 8 7 6 5 4 3 2 1

CONTENTS

PREFACE

Human metabolism is a key component of the basic science knowledge that underlies the practice of medicine and allied health professions. It is fundamental to understanding how the body adapts to physiologic stress, how defects in metabolism result in disease, and why data from the clinical chemistry laboratory are useful to diagnose disease and monitor the efficacy of treatment. Over the more than three decades that each of the authors has been teaching biochemistry to medical students, we have found students increasingly overwhelmed with details that tend to obscure rather than elucidate principles of human metabolism.

Our main aim in writing this book was to provide students in the health professions with a concise resource that will help them understand and appreciate the functions, constituent reactions, and regulatory aspects of the core pathways that constitute human metabolism and which are responsible for maintaining homeostasis and well-being in humans. We have tried to accomplish this by emphasizing function, regulation, and disease processes, while minimizing discussion of reaction mechanisms and details of enzyme structure.

Each chapter is organized in a consistent manner beginning with an explanation of the main functions of the pathway under discussion. Next comes a brief accounting of the cells, tissues, and organs in which the pathway is expressed and the conditions under which the normal function of the pathway is especially important. The bulk of each chapter is devoted to the reactions that account for the function of the pathway, with emphasis on key steps in the pathway. The next section of each chapter discusses the ways in which the activity of the pathway is regulated by hormones, genetic factors, or changes in the intracellular concentration of key metabolites. Each chapter concludes with a discussion of the more common and illustrative diseases that result from defects in or derangements of regulation of the pathway.

This volume is deliberately modest in size. Instead of providing exhaustive coverage of all the reactions that human cells and tissues are capable of executing, we have chosen examples that illustrate the physiologic and pathophysiologic significance of the topic. The authors' expectation is that each chapter will be read for comprehension rather than to provide abundant fact and detail. During their subsequent education and professional careers, the readers will undoubtably have need for more information on many topics discussed in this book. We hope that this book will provide them with the tools to comprehend and appreciate the detailed resources—both print and electronic—that contain the ever-expanding body of knowledge on human metabolism in health and disease.

MIRIAM D. ROSENTHAL
ROBERT H. GLEW

ACKNOWLEDGMENTS

We are grateful to our colleagues and friends who generously devoted time to reading selected chapters and provided the authors with invaluable feedback: William L. Anderson, Suzanne E. Barbour, Alakananda Basu, David G. Bear, Edward J. Behrman, Frank J. Castora, Anca D. Dobrian, Diane M. Duffy, Venkat Gopalan, Maurice Kogut, William Lennarz, Robert B. Loftfield, Gerald J. Pepe, Karl A. Schellenberg, David L. Vanderjagt, Dorothy J. Vanderjagt, and Howard D. White.

A special thanks to Mary H. Hahn and Charles D. Varnell, Jr., at Eastern Virginia Medical School, who provided the students' perspective of the book, for their insights on clarity and accessibility. We also appreciate the perceptive critiques provided by the University of New Mexico Medical School class of 2011.

The authors are indebted to Lucy Hunsaker, who drafted the figures. Her uncommon patience and good judgment in making the many revisions required to get the figures into final form are greatly appreciated.

We also thank the helpful people at John Wiley & Sons: Darla Henderson who championed our initial proposal, and Michael Foster, Rebekah Amos, Anita Lekhwani, and Rosalyn Farkas who shepherded the book all the way to publication.

THE AUTHORS

Miriam D. Rosenthal, Ph.D., is Professor of Biochemistry at Eastern Virginia Medical School. She received her B.A. in biology from Swarthmore College in 1963, followed by M.S. (1968) and Ph.D. (1974) degrees in biology from Brandeis University. Since 1977, Dr. Rosenthal has developed curricula, provided instruction, and conducted assessment of medical and other health professions students in biochemistry, molecular biology, cell biology, and human genetics. She has served as Course Director of Medical Biochemistry since 1997.

Robert H. Glew, Ph.D., is Emeritus Professor of Biochemistry and Molecular Biology at the University of New Mexico School of Medicine, where he was chair of the department from 1990 to 1998. He received a B.S. in food science from the University of Massachusetts, Amherst in 1962, M.S. in nutrition and food science from the Massachusetts Institute of Technology in 1964, and Ph.D. in biochemistry from the University of California, Davis in 1968. Dr. Glew has taught medical biochemistry at half a dozen medical schools and teaching hospitals in the United States and West Africa.

Drs. Rosenthal and Glew previously coedited *Clinical Studies in Medical Biochemistry* (3rd ed., 2006, Oxford University Press, New York). The book uses case presentations to develop the contextual basis of human metabolism, nutrition, and the molecular bases of disease.

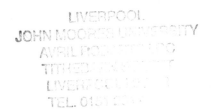

CHAPTER 1

INTRODUCTION TO METABOLISM

1.1 INTRODUCTION

Intermediary metabolism is the name given to the sequences of biochemical reactions that degrade, synthesize, or interconvert small molecules inside living cells. Knowledge of the core metabolic pathways and their interrelations is critical to understanding both normal function and the metabolic basis of most human diseases. Rational interpretation and application of data from the clinical chemistry laboratory also requires a sound grasp of the major metabolic pathways. Furthermore, knowledge of key biochemical reactions in the two dozen or so core metabolic pathways in humans is essential for an understanding of the molecular basis of drug action, drug interactions, and the many genetic diseases that are caused by the absence of the activity of a particular protein or enzyme.

1.1.1 Metabolic Pathways

Metabolism occurs in small discrete steps, each of which is catalyzed by an enzyme. The term *metabolic pathway* refers to a particular set of reactions that carries out a certain function or functions. The pathway of gluconeogenesis or glucose synthesis, for example, operates mainly during a period of fasting, and its primary function is to maintain the concentration of glucose in the circulation at levels that are required by glucose-dependent tissues such as the brain and red blood cells. Another example of a metabolic pathway is the tricarboxylic acid (TCA) cycle, which oxidizes the two

Medical Biochemistry: Human Metabolism in Health and Disease By Miriam D. Rosenthal and Robert H. Glew
Copyright © 2009 John Wiley & Sons, Inc.

carbons of acetyl-coenzyme A (acetyl-CoA) to CO_2 and water, thus completing the catabolism of carbohydrates, fats (fatty acids), and proteins (amino acids).

1.1.2 Metabolic Intermediates

Biochemical pathways are comprised of organic compounds called *metabolic intermediates*, all of which contain carbon, hydrogen, and oxygen. Some metabolic intermediates also contain nitrogen or sulfur. In most instances, these compounds themselves have no function. An exception would be a compound such as citric acid, which is both an intermediate in the TCA cycle and a key regulator of other pathways, including oxidation of glucose (glycolysis) and gluconeogenesis.

1.1.3 Homeostasis

Homeostasis refers to an organism's tendency or drive to maintain the normalcy of its internal environment, including maintaining the concentration of nutrients and metabolites within relatively strict limits. A good example is glucose homeostasis. In the face of widely varying physiological conditions, such as fasting or exercise, both of which tend to lower blood glucose, or following the consumption of a carbohydrate meal that raises the blood glucose concentration, the human body activates hormonal mechanisms that operate to maintain blood glucose within rather narrow limits, 80 to 100 mg/dL (Fig. 1-1). Hypoglycemia (low blood glucose) stimulates the release of gluconeogenic hormones such as glucagon and hydrocortisone, which promote the breakdown of liver glycogen and the synthesis of glucose in the liver (gluconeogenesis), followed by the release of glucose into the blood. On the other hand, hyperglycemia (elevated blood glucose) stimulates the release of insulin, which promotes the uptake of glucose and its utilization, storage as glycogen, and conversion to fat.

Maintenance of the blood calcium concentration between strict limits is another example of homeostasis. The normal total plasma calcium concentration is in the range 8.0 to 9.5 mg/dL. If the calcium concentration remains above the upper limit of normal for an extended period of time, calcium may deposit, with pathological consequences in soft tissues such as the heart and pancreas. Hypocalcemia (a.k.a. tetany) can result in muscle paralysis, convulsions, and even death; chronic hypocalcemia causes rickets in children and osteomalacia in adults. The body uses vitamin D and certain hormones (e.g., parathyroid hormone, calcitonin) to maintain calcium homeostasis.

1.2 WHAT DO METABOLIC PATHWAYS ACCOMPLISH?

1.2.1 Generation of Energy

The primary dietary fuels for human beings are carbohydrates and fats (triacylglycerols). The human body also obtains energy from dietary protein and—for some

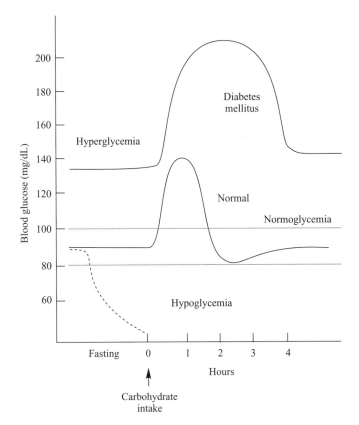

FIGURE 1-1 Changes that occur in the blood glucose concentration in a healthy adult, a person with type II diabetes mellitus, and a person experiencing fasting hypoglycemia. Following ingestion of a carbohydrate-containing meal, there are three features that distinguish the glucose vs. time curve for the person with type II diabetes relative to the healthy adult: (1) the initial blood glucose concentration is higher (approx. 135 vs. 90 mg/dL), (2) the rise in in the glucose level following the meal is greater; and (3) it takes longer for the glucose concentration to return to the fasting glucose level.

people—ethanol. Metabolism of these fuels generates energy, much of which is captured as the high-energy molecule adenosine triphosphate (ATP) (Fig. 1-2). The ATP can be used for biosynthetic processes (e.g., protein synthesis), muscle contraction, and active transport of ions and other solutes across membranes.

1.2.2 Degradation or Catabolism of Organic Molecules

Catabolic pathways usually involve cleavage of C—O, C—N, or C—C bonds. Most intracellular catabolic pathways are oxidative and involve transfer of reducing equivalents (hydrogen atoms) to nicotinamide-adenine dinucleotide (NAD^+) or flavine-adenine dinucleotide (FAD). The reducing equivalents in the resulting NADH or

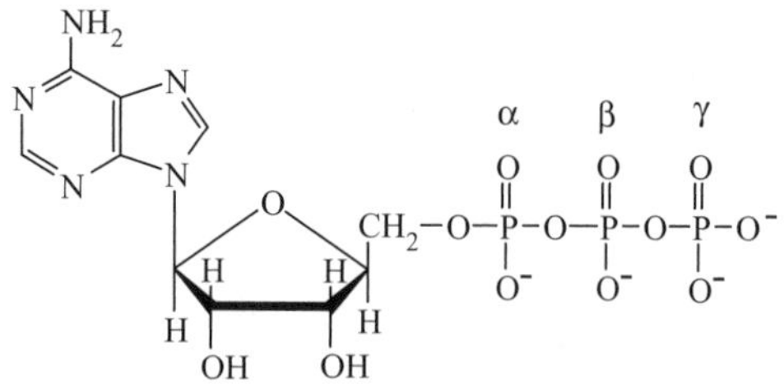

FIGURE 1-2 Structure of adenosine triphosphate.

$FADH_2$ can then be used in biosynthetic reactions or transferred to the mitochondrial electron-transport chain for generation of ATP.

1.2.2.1 Digestion. Before dietary fuels can be absorbed into the body, they must be broken down into simpler molecules. Thus, starch is hydrolyzed to yield glucose, and proteins are hydrolyzed to their constituent amino acids.

1.2.2.2 Glycolysis. Glycolysis is the oxidation of glucose into the three-carbon compound pyruvic acid.

1.2.2.3 Fatty Acid Oxidation. The major route of fatty acid degradation is β-oxidation, which accomplishes stepwise two-carbon cleavage of fatty acids into acetyl-CoA.

1.2.2.4 Amino Acid Catabolism. Breakdown of most of the 20 common amino acids is initiated by removal of the α-amino group of the amino acid via transamination. The resulting carbon skeletons are then further catabolized to generate energy or are used to synthesize other molecules (e.g., glucose, ketones). The nitrogen atoms of amino acids can be utilized for the synthesis of other nitrogenous compounds, such as heme, purines, and pyrimidines. Excess nitrogen is excreted in the form of urea.

1.2.3 Synthesis of Cellular Building Blocks and Precursors of Macromolecules

1.2.3.1 Gluconeogenesis: Synthesis of Glucose. This pathway produces glucose from glycerol, pyruvate, lactate, and the carbon skeletons of certain (glucogenic) amino acids. Gluconeogenesis is crucial to maintaining an adequate supply of glucose to the brain during fasting and starvation.

1.2.3.2 Synthesis of Fatty Acids. Excess dietary carbohydrates and the carbon skeletons of ketogenic amino acids are catabolized to acetyl-CoA, which is then utilized for the synthesis of long-chain (C16 and C18) fatty acids. Storage of these fatty acids as adipocyte triacylglycerols provides the major fuel source during the fasted state.

1.2.3.3 **Synthesis of Heme.** Heme is a component of the oxygen-binding proteins hemoglobin and myoglobin. Heme also functions as part of cytochromes, both in the mitochondrial electron transport chain involved in respiration-dependent ATP synthesis and in certain oxidation–reduction enzymes, such as the microsomal mixed-function oxygenases (e.g., cytochrome P450). Although most heme synthesis occurs in hemopoietic tissues (e.g., bone marrow), nearly all cells of the body synthesize heme for their own cytochromes and heme-containing enzymes.

1.2.4 Storage of Energy

Cells have only a modest ability to accumulate ATP, the major high-energy molecule in human metabolism. The human body can store energy in various forms, described below.

1.2.4.1 **Creatine Phosphate.** Most cells, especially muscle, can store a limited amount of energy in the form of creatine phosphate. This is accomplished by a reversible process catalyzed by creatine kinase:

$$\text{ATP} + \text{creatine} \rightleftharpoons \text{creatine phosphate} + \text{ADP}$$

When a cell's need for energy is at a minimum, the reaction tends toward the right. By contrast, when the cell requires ATP for mechanical work, ion pumping, or as substrate in one biosynthetic pathway or another, the reaction tends to the left, thereby making ATP available.

1.2.4.2 **Glycogen.** Glycogen is the polymeric, storage form of glucose. Nearly all of the body's glycogen is contained in muscle (approximately 600 g) and liver (approximately 300 g), with small amounts in brain and type II alveolar cells in the lung. Glycogen serves two very different functions in muscle and liver. Liver glycogen is utilized to maintain a constant supply of glucose in the blood. By contrast, muscle glycogen does not serve as a reservoir for blood glucose. Instead, muscle glycogen is broken down when that tissue requires energy, releasing glucose, which is subsequently oxidized to provide energy for muscle work.

1.2.4.3 **Fat or Triacylglycerol.** Whereas the body's capacity to store energy in the form of glycogen is limited, its capacity for fat storage is almost limitless. After a meal, excess dietary carbohydrates are metabolized to fatty acids in the liver. Whereas some of these endogenously synthesized fatty acids, as well as some of the fatty acids obtained through the digestion of dietary fat, are used directly as fuel by peripheral tissues, most of these fatty acids are stored in adipocytes in the form of triacylglycerols. When additional metabolic fuel is required during periods of fasting or exercise, the triacylglycerol stores in adipose are mobilized and the fatty acids are made available to tissues such as muscle and liver.

1.2.5   Excretion of Potentially Harmful Substances

1.2.5.1   Urea Cycle. This metabolic pathway takes place in the liver and synthesizes urea from the ammonia (ammonium ions) derived from the catabolism of amino acids and pyrimidines. Urea synthesis is one of the body's major mechanisms for detoxifying and excreting ammonia.

1.2.5.2   Bile Acid Synthesis. Metabolism of cholesterol to bile acids in the liver serves two purposes: (1) it provides the intestine with bile salts, whose emulsifying properties facilitate fat digestion and absorption, and (2) it is a mechanism for disposing of excess cholesterol. Humans cannot break open any of the four rings of cholesterol, nor can they oxidize cholesterol to carbon dioxide and water. Thus, biliary excretion of cholesterol—both as cholesterol per se and as bile salts—is the only mechanism the body has for disposing of significant quantities of cholesterol.

1.2.5.3   Heme Catabolism. When heme-containing proteins (e.g., hemoglobin, myoglobin) and enzymes (e.g., catalase) are turned over, the heme moiety is oxidized to bilirubin, which after conjugation with glucuronic acid is excreted via the hepatobiliary system.

1.2.6   Generation of Regulatory Substances

Metabolic pathways generate molecules that play key regulatory roles. As indicated above, citric acid (produced in the TCA cycle) plays a major role in coordinating the activities of the pathways of glycolysis and gluconeogenesis. Another example of a regulatory molecule is 2,3-bisphosphoglyceric acid, which is produced in a side reaction off the glycolytic pathway and modulates the affinity of hemoglobin for oxygen.

1.3   GENERAL PRINCIPLES COMMON TO METABOLIC PATHWAYS

1.3.1   ATP Provides Energy for Synthesis

Anabolic or synthetic pathways require input of energy in the form of the high-energy bonds of ATP, which is generated directly during some catabolic reactions (such as glycolysis) as well as during mitochondrial oxidative phosphorylation.

1.3.2   Many Metabolic Reactions Involve Oxidation or Reduction

During catalysis, oxidative reactions transfer reducing equivalents (hydrogen atoms) to cofactors such as NAD^+, $NADP^+$ (nicotinamide-adenine dinucleotide phosphate) or FAD. Reduced NADH and $FADH_2$ can then be used to generate ATP through oxidative phosphorylation in mitochondria. NADPH is the main source of reducing equivalents for anabolic, energy-requiring pathways such as fatty acid and cholesterol synthesis.

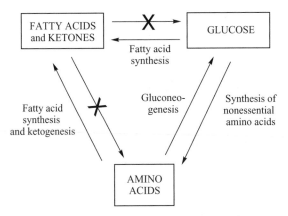

FIGURE 1-3 Possible interconversions of the three major metabolic fuels in humans. Note that glucose and amino acids cannot be synthesized from (even-carbon) fatty acids.

1.3.3 Only Certain Metabolic Reactions Occur in Human Metabolism

It is important to appreciate that although humans possess the machinery to interconvert many dietary components, not all interconversions are possible. Thus, humans can convert glucose into long-chain fatty acids, but they cannot convert even-carbon-numbered long-chain fatty acids into glucose (Fig. 1-3).

1.3.4 Some Organic Molecules Are Nutritionally Essential to Human Health

Certain key cellular components cannot be synthesized in the body and must therefore be provided preformed in the diet and are therefore designated as *essential*. These molecules include two polyunsaturated fatty acids (linoleic and α-linolenic) and the carbon skeletons of some of the amino acids. They also include the vitamins (such as thiamine and niacin), most of which serve as components of enzymatic cofactors. By contrast, other important compounds, such as glucose and palmitic acid, are not essential in the diet. Glucose, whose blood levels are crucial to homeostasis, can be synthesized from glycerol, lactate, pyruvate, and the carbon skeletons of glucogenic amino acids when dietary glucose is not available.

1.3.5 Some Metabolic Pathways Are Irreversible or Contain Irreversible Steps

One example of an irreversible pathway is glycolysis, the multistep catabolic pathway that oxidizes glucose to pyruvate or lactate. Gluconeogenesis is essentially the reverse of glycolysis and is the process by which pyruvate (or a number of other molecules such as lactate and the carbon skeleton of the amino acid alanine) can be used to synthesized glucose. Although glycolysis and gluconeogenesis share many enzymes, specific gluconeogenic enzymes are required to bypass the steps in glycolysis that are irreversible under physiological conditions.

1.3.6 Metabolic Pathways Are Interconnected

The initial step in glycolysis is the phosphorylation of glucose to form glucose 6-phosphate. Glucose 6-phosphate is also utilized in two other key metabolic pathways: glycogen synthesis and the pentose phosphate pathway (a.k.a. the hexose monophosphate shunt), which generates ribose 5-phosphate and NADPH.

1.3.7 Metabolic Pathways Are Not Necessarily Linear

Both the tricarboxylic acid (TCA) cycle and the urea cycle are circular pathways. In each case the pathway is initiated by addition of a small molecule to a key metabolic intermediate (oxaloacetate in the TCA cycle and ornithine in the urea cycle). At the end of one cycle, the key intermediate is regenerated and available to participate in another turn of the cycle. Although the TCA and urea cycles can be depicted as simple circular pathways, metabolites can enter into—or be removed from—the pathways at intermediate steps. For example, the amino acid glutamate can be used to generate α-ketoglutarate, a key intermediate in the TCA cycle.

1.3.8 Metabolic Pathways Are Localized to Specific Compartments Within the Cell

Many metabolic pathways occur within the mitochondria, including β-oxidation of fatty acids, the TCA cycle, and oxidative phosphorylation (Fig. 1-4). Other pathways are cytosolic, including glycolysis, the pentose phosphate pathway, and fatty acid synthesis. Still others, including the urea cycle and heme synthesis, utilize both mitochondrial and cytosolic enzymes at different points in the pathways.

1.3.9 A Different Repertoire of Pathways Occurs in Different Organs

All cells are capable of oxidizing glucose to pyruvate via glycolysis to generate ATP. However, since red blood cells lack mitochondria, they cannot further oxidize the resulting pyruvate to CO_2 and water via pyruvate dehydrogenase and the TCA cycle. Instead, the pyruvate is converted to lactate and released from the red blood cells.

Most cells and organs can also utilize fatty acids as fuels. Although neural cells do contain mitochondria, they do not oxidize fatty acids. The brain is therefore dependent on a constant supply of glucose to provide energy. The gluconeogenesis pathway that provides glucose for the brain occurs in the liver and to a lesser extent in the renal cortex.

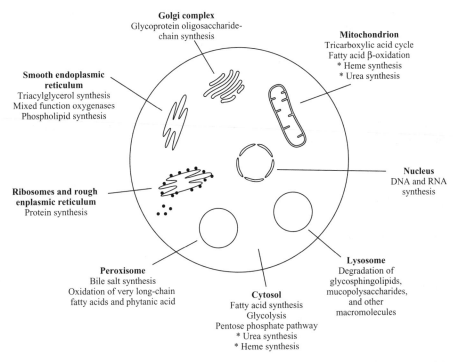

Golgi complex
Glycoprotein oligosaccharide-
chain synthesis

Mitochondrion
Tricarboxylic acid cycle
Fatty acid β-oxidation
* Heme synthesis
* Urea synthesis

**Smooth endoplasmic
reticulum**
Triacylglycerol synthesis
Mixed function oxygenases
Phospholipid synthesis

Nucleus
DNA and RNA
synthesis

**Ribosomes and rough
enplasmic reticulum**
Protein synthesis

Peroxisome
Bile salt synthesis
Oxidation of very long-chain
fatty acids and phytanic acid

Cytosol
Fatty acid synthesis
Glycolysis
Pentose phosphate pathway
* Urea synthesis
* Heme synthesis

Lysosome
Degradation of
glycosphingolipids,
mucopolysaccharides,
and other
macromolecules

FIGURE 1-4 A liver cell, showing where various metabolic pathways occur. An asterisk indicates a pathway, portions of which occur in more than one intracellular compartment.

1.3.10 Different Metabolic Processes Occur in the Fed State Than During Fasting

After a meal, metabolic pathways are utilized to process the digested foods and store metabolites for future utilization. Postprandially, glucose is plentiful and utilized both for energy generation and to replenish glycogen stores (primarily in muscle and liver). Excess glucose is metabolized to fatty acids in liver and fat cells and the resulting triacylglycerols are stored in adipocytes.

By contrast, when a person is fasting there is a need to generate energy from endogenous fuels. Consequently, the metabolic pathways involved in fuel metabolism are regulated in such a way as to promote the oxidation of stored fuels, including the fatty acids stored in adipose tissue in the form of triacylglycerols and, to a lesser extent, glycogen stored in liver and muscle. In fact, during a fast, most of the body's energy needs are satisfied by the oxidation of fatty acids.

1.3.11 Metabolic Pathways Are Regulated

All this specialization of organs and coordination of metabolism in the fed or fasted states is a highly regulated process with several levels of regulation. One level of regulation is gene transcription and translation, which determines which enzymes are actually present within a cell. A second level of control is substrate-level regulation, whereby concentrations of key metabolites activate or inhibit enzymatic reactions. A metabolite that acts to regulate several pathways is citrate, which both inhibits glycolysis and activates the first step in the pathway of fatty acid synthesis.

Hormones represent yet another level of control. Hormones act to coordinate processes between the organs of complex, multicellular organisms. For example, insulin, the main hormonal signal of the fed state, regulates both enzyme activity (at the level of enzyme dephosphorylation) and gene transcription.

1.4   WHAT IS THE BEST WAY TO COMPREHEND AND RETAIN A WORKING KNOWLEDGE OF INTERMEDIARY METABOLISM?

Before learning about the various enzyme-catalyzed reactions and intermediates that comprise a particular metabolic pathway, one should appreciate the major functions which that pathway serves in the body and how the pathway relates to other pathways. Particularly in the context of medical biochemistry, it is also important to understand how the pathway is regulated and how it affects (or is affected by) disease processes. As you go through this book you will find that each chapter is organized so as to answer the following questions:

1. Why does the pathway exist? That is, what are its functions?
2. Where does the pathway take place (i.e., what organ, tissue, cell, subcellular compartment, or organelle)?
3. When does the pathway operate, and when is it down-regulated: during the fasted state or the fed state; during rest or extreme physical activity; during a particular stage of development (e.g., the embryo, the neonate, old age)?
4. What are the actual steps of the pathway, and what cofactors does it require?
5. How is the pathway regulated?
6. What can go wrong? Problems can include hormonal dysregulation (e.g., diabetes mellitus), inborn errors of metabolism (e.g., adrenoleukodystrophy), and nutritional deficiencies (e.g., protein–calorie malnutrition, iron-deficiency anemia). Normal metabolic homeostasis is also profoundly altered by toxins and during infections.

CHAPTER 2

ENZYMES

2.1 THE NATURE OF ENZYMES

Enzymes are catalysts that greatly increase the rate of chemical reactions and thus make possible the numerous and diverse metabolic processes that occur in the human body. Catalysts increase the rate of a reaction without affecting its equilibrium. Enzymes can increase the rate of physiological reactions by as much as 10^{10}-fold. They accomplish this feat by decreasing the amount of energy required for activation of the initial reactants (substrates), thereby increasing the percentage of substrate molecules that have sufficient energy to react (Fig. 2-1).

With the exception of a few ribonucleic acid (RNA) molecules (ribozymes) that catalyze reactions involving nucleic acids, enzymes are proteins. Every enzyme has an active site that is composed of specific amino acid side chains which are brought into close proximity when the enzyme is folded into its active conformation. During the course of the reaction that it catalyzes, the enzyme's active site stabilizes the transition state, which is an intermediate conformation between substrates and products. The interaction between active site and substrate(s) is thus responsible for the catalytic efficiency of the enzyme as well as its substrate specificity. After the reaction occurs, the products are released from the enzyme and the active site is available to bind additional substrate molecules.

Medical Biochemistry: Human Metabolism in Health and Disease   By Miriam D. Rosenthal and Robert H. Glew
Copyright © 2009 John Wiley & Sons, Inc.

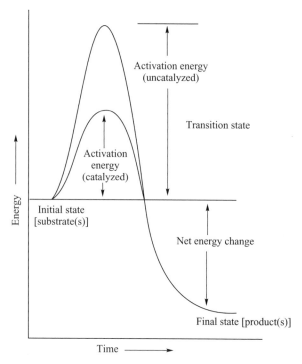

FIGURE 2-1 Activation energy of a chemical reaction.

2.2 TYPES OF ENZYMES

There are more than 2500 different enzymes in the human body. It is useful to group them into six major classes based on the type of reaction they catalyze.

2.2.1 Oxidoreductases

Oxidative reactions remove electrons, usually one or two electrons per molecule of substrate, while reductive reactions accomplish the converse. The substrate in an oxidation–reduction reaction may be a metal, as in the case of the one-electron oxidation of the ferrous ion of hemoglobin to the ferric ion of methemoglobin, or an organic compound as illustrated by the two-electron, reversible oxidation of lactate to pyruvate.

Oxidoreductases transfer electrons from one compound to another, thus changing the oxidation state of both substrates. Some oxidoreductases, such as lactate de-hydrogenase, catalyze the removal of two hydrogen atoms (two electrons plus two hydrogen ions) to an acceptor molecule such as nicotinamide-adenine dinucleotide (NAD^+) as illustrated by the lactate dehydrogenase reaction (Fig. 2-2):

$$\text{lactate} + NAD^+ \rightleftharpoons \text{pyruvate} + NADH + H^+$$

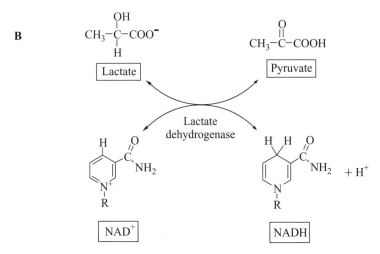

A

Oxidized nicotinamide adenine dinucleotide
(NAD$^+$)

B

Lactate

Pyruvate

Lactate
dehydrogenase

$+ H^+$

NAD$^+$

NADH

FIGURE 2-2 Lactate dehydrogenase is an oxidoreductase that uses the cofactor NAD$^+$ as a hydrogen acceptor: (A) structure of NAD$^+$; (B) lactate dehydrogenase reaction.

A second cofactor that serves as an acceptor of hydrogen atoms is flavin-adenine dinucleotide (FAD, Fig. 2-3):

$$\text{succinate} + \text{FAD} \rightarrow \text{fumarate} + \text{FADH}_2$$

In general, most oxidation–reduction (redox) reactions that oxidize oxygen-bearing carbons utilize NAD^+ (or the related cofactor NADP^+), whereas reductions or oxidations of carbon atoms that do not have oxygen attached utilize flavin mononucleotides (FMN or FAD).

Other oxidoreductases, such as 5-lipoxygenase (Fig. 2-4A), are dioxygenases, which catalyze the addition of both atoms of molecular oxygen into the substrate:

$$\text{arachidonic acid} + \text{O}_2 \rightarrow \text{5-hydroperoxyeicosatetraenoic acid}$$

Still other oxidoreductases are monooxygenases or mixed-function oxidases, which catalyze even more complex reactions. For example, phenylalanine hydroxylase (Fig. 2-4B) catalyzes the reaction

$$\text{phenylalanine} + \text{O}_2 + \text{BH}_4 \rightarrow \text{tyrosine} + \text{BH}_2 + \text{H}_2\text{O}$$

In this reaction, two organic substrates are oxidized: One atom of molecular oxygen is used to oxidize phenylalanine; the other combines with the two hydrogen atoms removed from tetrahydrobiopterin (BH_4), generating dihydrobiopterin (BH_2) and water.

2.2.2 Transferases

Transferases catalyze reactions in which a functional group is transferred from one compound to another. Transaminases, such as aspartate aminotransferase (Fig. 2-5A), catalyze the reversible transfer of an amino group from an amino acid to an α-ketoacid, thus generating a new amino acid and a new α-ketoacid:

$$\text{aspartate} + \alpha\text{-ketoglutarate} \rightleftharpoons \text{oxaloacetate} + \text{glutamate}$$

Similarly, kinases transfer phosphate groups from adenosine triphosphate (ATP) to acceptor molecules such as glucose in the reaction catalyzed by hexokinase or glucokinase (Fig. 2-5B):

$$\text{glucose} + \text{ATP} \rightarrow \text{glucose 6-phosphate} + \text{adenosine diphosphate (ADP)}$$

Unlike the aminotransferase reactions, which are reversible, most reactions catalyzed by kinases are irreversible under physiological conditions.

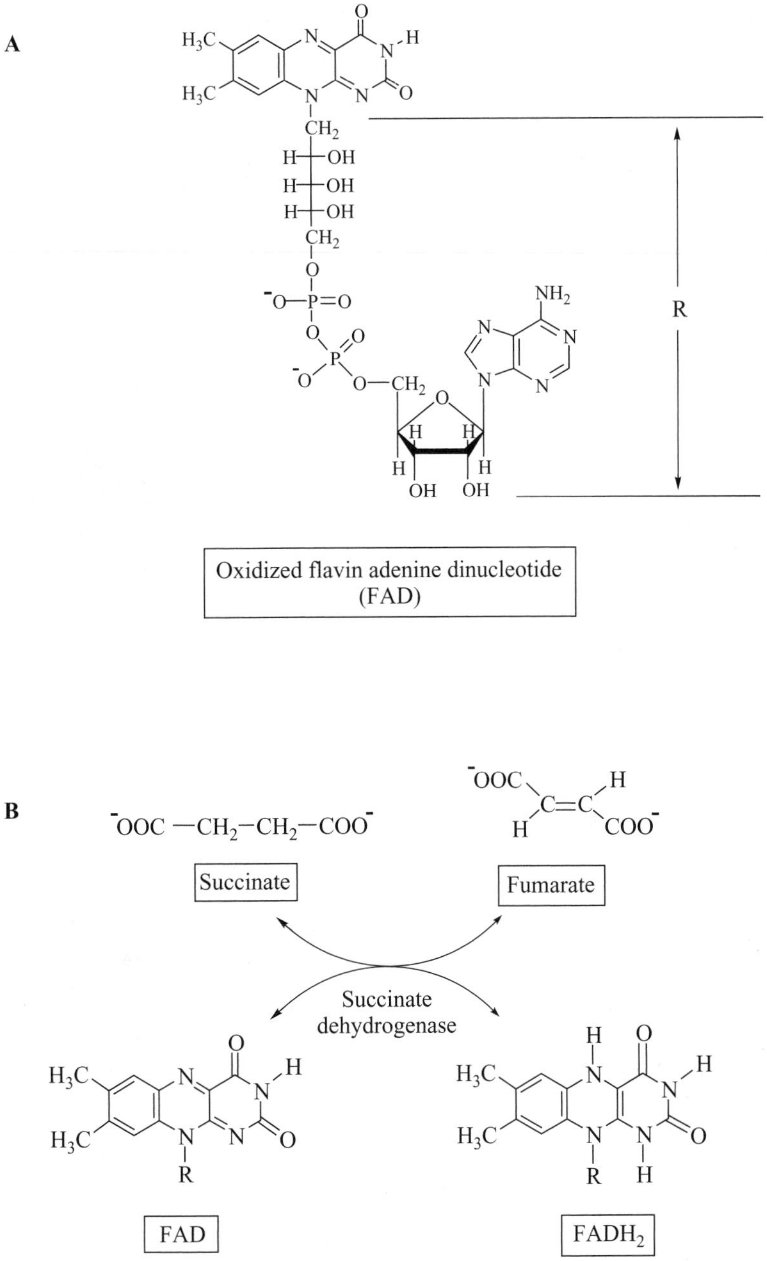

A

Oxidized flavin adenine dinucleotide
(FAD)

B

$^-OOC-CH_2-CH_2-COO^-$

Succinate

Fumarate

Succinate
dehydrogenase

FAD

FADH$_2$

FIGURE 2-3   Succinate dehydrogenase is an oxidoreductase that uses the cofactor FAD as a hydrogen acceptor: (A) structure of FAD; (B) succinate dehydrogenase reaction.

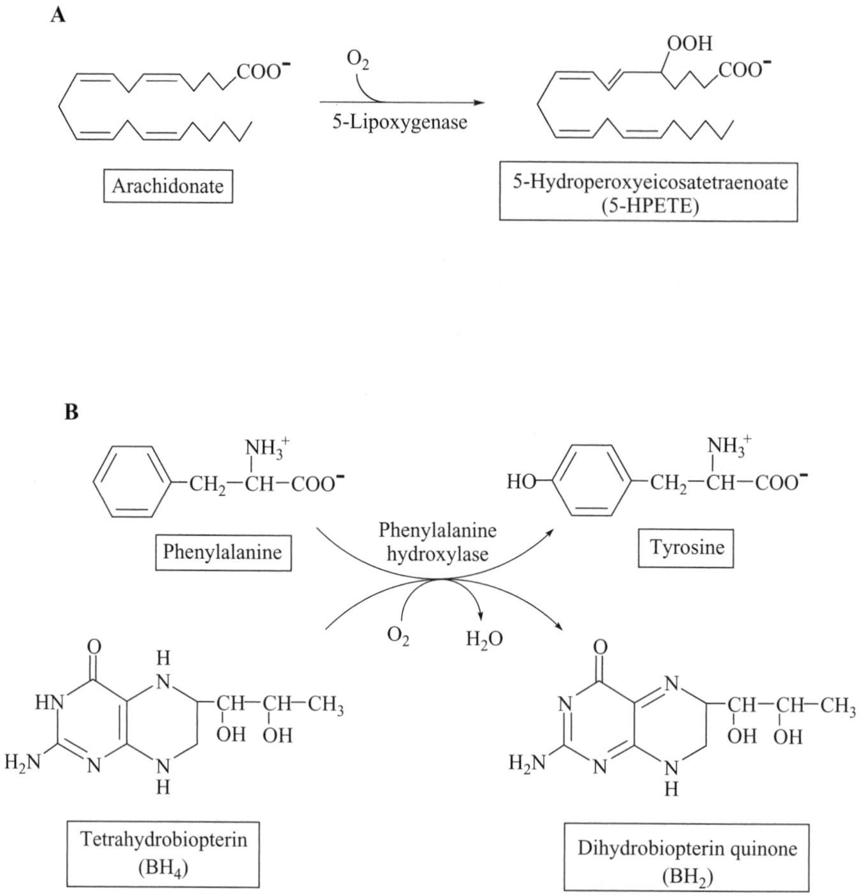

A

B

FIGURE 2-4 Oxidoreductase reactions utilizing molecular oxygen: (A) the reaction catalyzed by 5-lipoxygenase; (B) the reaction catalyzed by the sequential actions of phenylalanine hydroxylase and a subsequent dehydratase that removes water.

2.2.3 Hydrolases

Hydrolases cleave carbon–oxygen, carbon–nitrogen, or carbon–sulfur bonds by adding water across the bond. One example of a hydrolase is the digestive enzyme maltase, which hydrolyzes the glycosidic bond in the disaccharide maltose (Fig. 2-6):

$$\text{maltose} + H_2O \rightarrow 2 \text{ glucose}$$

2.2.4 Lyases

Lyases cleave carbon–oxygen, carbon–nitrogen, or carbon–sulfur bonds but do so without addition of water and without oxidizing or reducing the substrates. A good

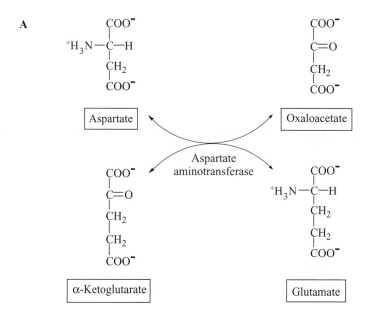

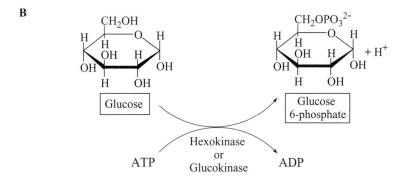

FIGURE 2-5 Transferase reactions: (A) aspartate aminotransferase uses pyridoxal phosphate (PLP) as a cofactor; (B) hexokinase and glucokinase do not utilize cofactors.

FIGURE 2-6 Maltase is a hydrolase.

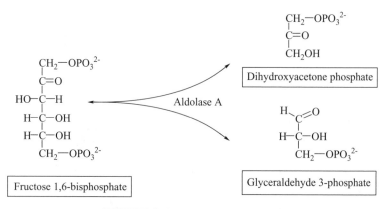

FIGURE 2-7   Aldolase A is a lyase.

example of a lyase is aldolase A (Fig. 2-7), which as an enzyme of the glyco-lytic pathway, catalyzes the reversible cleavage of the six-carbon sugar fructose 1,6-bisphosphate into two three-carbon sugar phosphates:

$$\text{fructose 1,6-bisphosphate} \rightleftharpoons \text{glyceraldehyde 3-phosphate}$$
$$+ \text{dihydroxyacetone phosphate}$$

Note that in the reverse reaction, aldolase A functions as a synthase, forming a new C−C bond.

2.2.5   Isomerases

Isomerases catalyze intramolecular rearrangements of functional groups that re-versibly interconvert optical or geometric isomers. One example is glucose 6-phosphate isomerase (Fig. 2-8A), which converts glucose 6-phosphate, an aldo-sugar phosphate, to the isomeric keto-sugar phosphate, fructose 6-phosphate:

$$\text{glucose 6-phosphate} \rightleftharpoons \text{fructose 6-phosphate}$$

When an isomerase catalyzes an intramolecular rearrangement involving movement of a functional group, it is called a *mutase*. For example, as part of the two metabolic pathways that synthesize and break down glycogen, phosphoglucomutase (Fig. 2-8B) catalyzes the reversible transfer of a phosphate group between the hydroxyl group on C1 (of the hemiacetyl ring form of glucose) and the C6 hydroxyl group of glucose:

$$\text{glucose 6-phosphate} \rightleftharpoons \text{glucose 1-phosphate}$$

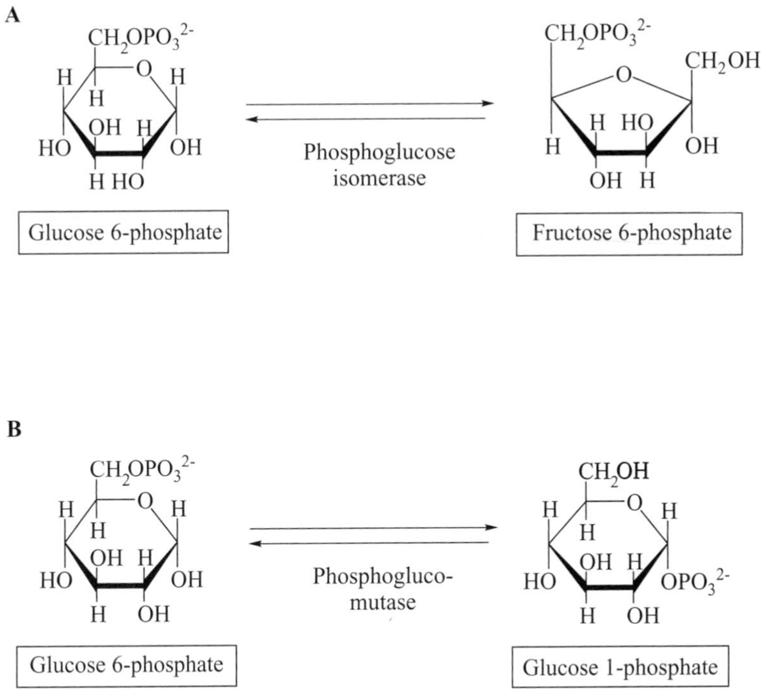

A

Glucose 6-phosphate

Phosphoglucose
isomerase

Fructose 6-phosphate

B

Glucose 6-phosphate

Phosphogluco-
mutase

Glucose 1-phosphate

FIGURE 2-8 Reactions catalyzed by isomerases. (A) Phosphoglucose isomerase; (B) phosphoglucomutase.

2.2.6 Ligases

Ligases catalyze biosynthetic reactions that form a covalent bond between two substrates. Ligases differ from lyases such as aldolase A (discussed above) in that they utilize the energy obtained from cleavage of a high-energy bond to drive the reaction. The molecule with the high-energy bond is usually ATP, which is concurrently converted to ADP with the release of inorganic phosphate. An example of a ligase is pyruvate carboxylase, which forms a new C−C bond by adding CO_2 from bicarbonate to pyruvate, the three-carbon end product of aerobic glycolysis (Fig. 2-9):

$$\text{pyruvate} + HCO_3^- + \text{ATP} \rightarrow \text{oxaloacetate} + \text{ADP} + P_i$$

Some ligases that catalyze synthetic reactions in which two substrates are joined and a nucleotide triphosphate (e.g., ATP) is hydrolyzed are designated by the term *synthetase*. In contrast, the term *synthase* is used to describe enzymes that catalyze reactions in which two substrates come together to form a product, but a nucleotide triphosphate is not involved in the reaction. An example of a synthase is citrate synthase, where the energy to drive the reaction is provided by the thioester of

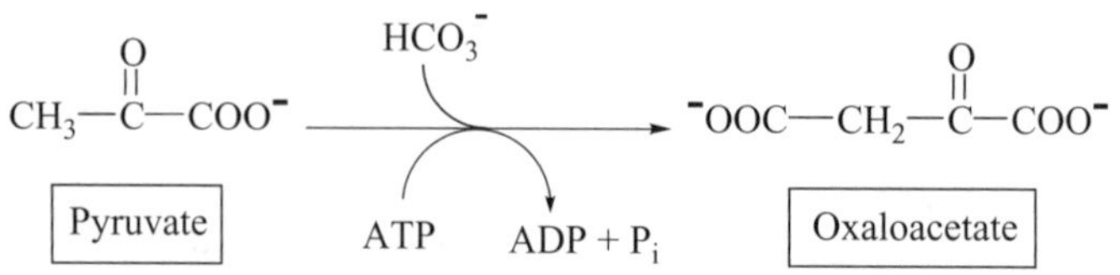

$$CH_3-\overset{\overset{\text{O}}{\|}}{C}-COO^- \xrightarrow{\hspace{3cm}} {}^-OOC-CH_2-\overset{\overset{\text{O}}{\|}}{C}-COO^-$$

Pyruvate   ATP   ADP + P$_i$ Oxaloacetate

HCO$_3^-$

FIGURE 2-9   Pyruvate carboxylase, a biotin-containing enzyme, catalyzes a ligase reaction.

acetyl-CoA:

$$\text{oxaloacetate} + \text{acetyl-CoA} \rightarrow \text{citrate} + \text{CoASH}$$

2.2.7 Nonenzymatic Reactions

Not all physiologically or pathophysiologically relevant reactions that take place in the body are catalyzed by enzymes. For example, the covalent attachment of glucose to hemoglobin to form glycated hemoglobin (HbA1c) occurs spontaneously and does not involve an enzyme. The fact that the extent of this glycation reaction in blood is determined solely by the plasma glucose concentration is the basis for the usefulness of the HbA1c measurement as a way to monitor glucose control. The high reactivity of glucose (as well as galactose and other monosaccharides) with proteins is attributable to the intrinsic affinity of aldehyde groups for the amino groups of proteins, resulting in protein adducts that can act as neoantigens. Similarly, the covalent attachment of acetaldehyde, an intermediate in ethanol metabolism, to a wide range of proteins may account for some of the pathology associated with excessive consumption of ethanol.

Another example of an important nonenzymatic reaction in humans is the autooxidation of oxyhemoglobin to methemoglobin, which generates the superoxide anion:

$$\text{hemoglobin (Fe}^{2+}) + O_2 \rightarrow \text{methemoglobin (Fe}^{3+}) + O_2^{\bullet-}$$

Methemoglobin does not bind oxygen and is a potent oxidizing agent that can damage the red cell membrane.

2.3 SMALL MOLECULES AND METAL IONS CAN CONTRIBUTE TO ENZYME-BASED CATALYSIS

2.3.1 Cofactors

Enzymatic catalysis often involves utilization of an additional small organic molecule called a *cofactor*. Certain cofactors, such as biotin and thiamine pyrophosphate, function only when they are attached covalently to their respective enzymes. In such cases the enzyme–coenzyme complex is called a *holoenzyme*, whereas the term *apoenzyme* refers to the protein component alone. In other cases, the cofactor acts more like a second substrate. A good example of this is NAD^+, which is converted to $NADH + H^+$ when it receives two hydrogen atoms (or two electrons plus protons) during the course of the redox reaction catalyzed by lactate dehydrogenase. The NADH molecule subsequently transfers the hydrogen atoms to another acceptor (e.g., FAD in the mitochondrial electron transport chain) and is thus available to participate in the catalytic dehydrogenation of another molecule of lactate. These NAD^+-utilizing enzymes are usually designated as dehydrogenases.

Most cofactors usually participate in the catalysis of many different reactions, often using a similar reaction mechanism. The cofactor does this by binding to the various enzymes, each of which has a particular active site whose structure and binding properties determine its unique substrate specificity. Thus, lactate dehydrogenase catalyzes the reaction

$$\text{lactate} + NAD^+ \rightleftharpoons \text{pyruvate} + NADH + H^+$$

whereas alcohol dehydrogenase catalyzes the reaction

$$\text{ethanol} + NAD^+ \rightleftharpoons \text{acetaldehyde} + NADH + H^+$$

2.3.2 Vitamins Are Components of Many Enzymatic Cofactors or Coenzymes

Vitamins are small organic molecules that are not synthesized in the body and are therefore essential dietary nutrients. Many of the vitamins are cofactors or components of cofactors. Because they play a catalytic role, they are required in the diet in only small amounts and are referred to as *micronutrients*. The vitamins that are cofactors or cofactor precursors include all the water-soluble B vitamins, vitamin C, and the fat-soluble vitamin K (Table 2-1).

2.3.2.1 *Thiamine (Vitamin B1).* Thiamine is utilized to synthesize thiamine pyrophosphate, which contributes to the transfer of active aldehyde intermediates during several reactions of carbohydrate metabolism. These include pyruvate dehydrogenase, the tricarboxylic acid cycle enzyme α-ketoglutarate dehydrogenase and transketolase, an enzyme that is a component of the pentose phosphate pathway.

2.3.2.2 *Riboflavin (Vitamin B2).* Riboflavin is a component of FAD (flavin-adenine dinucleotide, Fig. 2-3A) and FMN (flavin mononucleotide), which participate in numerous oxidation–reduction (redox) reactions and the process of ATP generation in mitochondria. FAD-linked dehydrogenases convert succinate to fumarate in the TCA cycle and fatty acyl-CoA to β-hydroxy fatty acyl-CoA during β-oxidation of fatty acids.

TABLE 2-1 Cofactor Roles of Vitamins

Vitamin	Coenzyme	Typical Reaction Type
Thiamine (B_1)	Thiamine pyrophosphate (TPP)	Oxidative decarboxylation of α-ketoacids
Riboflavin (B_2)	Flavin-adenine dinucleotide (FAD) Flavin-adenine mononucleotide (FMN)	Oxidation–reduction
Niacin (B_3) (nicotinate)	Nicotinamide-adenine dinucleotide (NAD^+) Nicotinamide-adenine dinucleotide phosphate (NAD^+)	Oxidation–reduction
Pantothenate (B_5)	Coenzyme A (CoASH) Acyl carrier protein (ACP)	Acyl transfer
Pyridoxine (B_6) Pyridoxal (B_6) Pyridoxamine (B_6)	Pyridoxal phosphate	Transamination and deamination of amino acids
Biotin (B_7)	N-Carboxybiotinyl lysine	Carboxylation
Folic acid	Tetrahydrofolate (TH_4)	One-carbon transfer
Cobalamin (B_{12})	Methylcobalamin Adenosyl cobalamin	Methylation of homocysteine to methionine Conversion of methylmalony-CoA to succinyl-CoA
Ascorbic acid (C)	Ascorbic acid	Hydroxylations in the synthesis of collagen, norepinephrine, and carnitine
Phylloquinone (K) Menaquinone (K)	Vitamin K hydroquinone (KH_2)	γ-Carboxylation of glutamate residues

2.3.2.3 *Niacin (Vitamin B₃)*. Niacin is a component of NAD^+ (nicotinamide-adenine dinucleotide) (Fig. 2-2A), and $NADP^+$ (nicotinamide-adenine dinucleotide phosphate), which participate in many redox reactions, such as those catalyzed by lactate dehydrogenase and fatty acyl-CoA dehydrogenase. $NADP^+$ differs from NAD^+ in that it has a phosphate group on C6 of the ribose moiety to which the adenosine moiety is attached. NADH, the reduced form of NAD^+, also donates electrons to the mitochondrial electron transport chain, which is a series of oxidation–reduction reactions that ultimately generate ATP. $NADP^+$ is a substrate or cofactor in the glucose 6-phosphate dehydrogenase reaction of the pentose phosphate pathway, and NADPH provides reducing equivalents for the synthesis of fatty acids and cholesterol.

2.3.2.4 *Pyridoxine, Pyridoxal, and Pyridoxamine*. These are forms of vitamin B_6 and precursors of pyridoxal phosphate (PLP). PLP is a cofactor for many enzymes that catalyze reactions involving amino acids, such as the various aminotransferases, amino acid decarboxylases, and the ligase enzyme δ-amino-levulinic acid (ALA) synthetase, which catalyzes the regulated step of heme synthesis.

2.3.2.5 Biotin. Biotin is active when it is attached covalently to enzymes. It binds CO_2 and transfers this one-carbon unit to organic acceptors (e.g., acetyl-CoA, pyruvate) as part of the catalytic mechanism of enzymes such as acetyl-CoA carboxylase and pyruvate carboxylase.

2.3.2.6 Folate. Folate is the precursor of tetrahydrofolate (THF), which is the cofactor involved in the transfer of one-carbon groups other than CO_2. THF plays a central role in the synthesis of purines, which are the building blocks for both deoxyribonucleic acid (DNA) and RNA.

2.3.2.7 Cobalamin (Vitamin B$_{12}$). Cobalamin is the cofactor that participates in the transfer of a methyl group in the regeneration of methionine from homo-cysteine. Cobalamin is also the precursor of deoxyadenosylcobalamin, which is the cofactor for methylmalonyl-CoA mutase, an enzyme involved in the metabolism of propionic acid.

2.3.2.8 Pantothenic Acid. Pantothenic acid is a component of coenzyme A (CoASH) and acyl carrier protein (ACP). The sulfhydryl group of CoASH forms thioester bonds with the carboxyl groups of acetate, long-chain fatty acids, and other organic acids. CoASH serves as a carrier for the activated forms of organic acids during many reactions, including those involved in the TCA cycle, fatty acid oxidation, the catabolism of the carbon skeletons of branched-chain amino acids, and the conjugation of bile salts with glycine or taurine. Acyl carrier protein is the carrier of acyl groups during the de novo synthesis of fatty acids.

2.3.2.9 Ascorbic Acid (Vitamin C). Ascorbic acid is a cofactor in hydroxyl-ation reactions, most prominently the hydroxylation of proline residues of collagen (Fig. 2-10) and the synthesis of norepinephrine from dopamine. Ascorbate is oxi-dized to dehydroascorbate during the course of these hydroxylation reactions and is regenerated by dehydroascorbate reductase, using reduced glutathione (GSH) as the source of reducing equivalents and generating oxidized glutathione (GSSG):

$$\text{dehydroascorboate} + 2\text{GSH} \rightarrow \text{ascorbic acid} + \text{GSSG}$$

2.3.2.10 Vitamin K. The two major dietary molecules with vitamin K activity are menaquinone, synthesized by bacteria, and phylloquinone, a product of green plants. Vitamin K is the cofactor for enzymes that γ-carboxylate specific glutamate residues of calcium-binding proteins (Fig. 2-11), such as prothrombin and other proteins of the blood-clotting cascade, and osteocalcin, a major bone protein. Vitamin K under-goes oxidation during γ-carboxylation reactions and is subsequently regenerated in

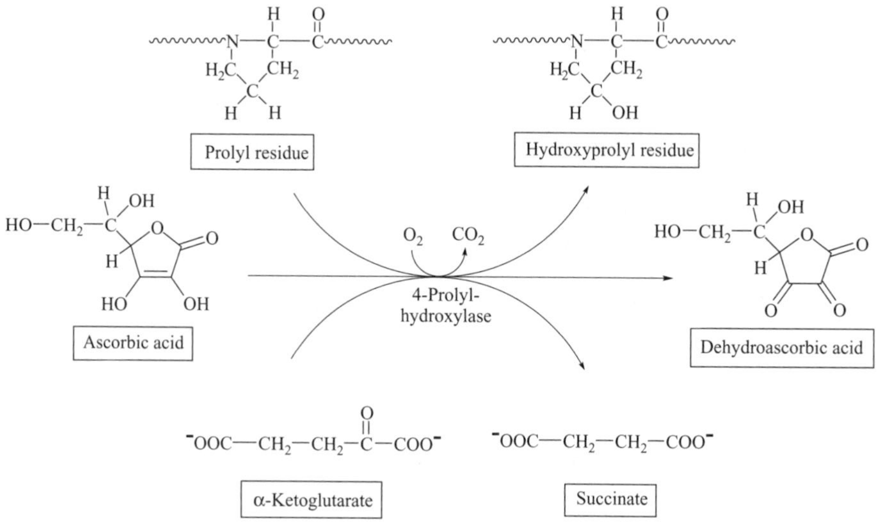

FIGURE 2-10 Role of ascorbic acid in the hydroxylation of a prolyl residue in collagen.

two reduction reactions catalyzed by vitamin K epoxide reductase and vitamin K reductase, respectively.

2.3.2.11 *Not All Cofactors Are Derived from Vitamins.* It is worth emphasizing that not all cofactors are synthesized from a vitamin. For example, since tetrahydrobiopterin (BH_4, Fig. 2-4B), the cofactor for phenylalanine hydroxylase and other enzymes that hydroxylate aromatic amino acids, is synthesized in the body from guanosine triphosphate (GTP), it is not a vitamin. Similarly, lipoic acid, which is one of several cofactors for the pyruvate dehydrogenase and α-ketoglutarate dehydrogenase complexes, is not a vitamin. It should also be noted that not all vitamins are precursors of cofactors. Indeed, vitamin K is the only one of the four fat-soluble vitamins that plays a direct catalytic role in an enzyme-catalyzed reaction in the body. Two other fat-soluble vitamins, retinol (vitamin A) and cholecalciferol (vitamin D), are actually precursors of hormones that regulate transcription of DNA, and thus gene expression. Retinol is also the precursor of 11-*cis*-retinal, which is an important constituent of rhodopsin, the visual pigment of the eye. α-Tocopherol (vitamin E), the fourth fat-soluble vitamin, is an antioxidant.

2.3.3 Many Enzymes Utilize Metal Ions as Part of Their Catalytic Mechanisms

Many enzymes utilize inorganic ions to bind the substrate and polarize critical functional groups. Examples of metal ions and the enzymes they function with include:

Zn^{2+}: alcohol dehydrogenase, carboxypeptidase

Mg^{2+}: ATP-dependent reactions such as hexokinase

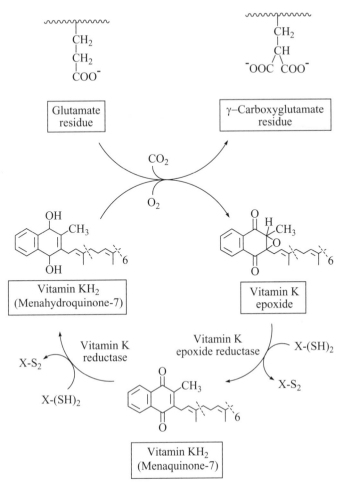

FIGURE 2-11 Role of vitamin K in the γ-carboxylation of glutamyl residues of proteins and the regeneration of reduced vitamin K. The figure shows menaquinone-7, which contains six additional isoprene units (portion between dashed lines); other menaquinones contain 6 to 13 isoprene units. Phylloquinone, obtained from plants, contains the same 2-methyl-1, 4-naphthoquinone ring, with a saturated rather than an unsaturated hydrocarbon tail. X designates the polypeptide cofactor thioredoxin, which is converted from the reduced state $[X-(SH)_2]$ to the oxidized state $(X-S_2)$ by both vitamin K reductase and vitamin K epoxide reductase.

Fe^{3+} and Cu^{2+}: components of the cytochrome oxidase complex, which catalyzes the last step in the electron transport chain in which the protons and electrons are transferred to molecular oxygen

Se^{2+}: glutathione peroxidase, which is involved in the cellular defense against free radicals

2.4   HOW DO ENZYMES WORK?

Biological catalysts increase the rate of a chemical reaction, permitting reactions to occur that would otherwise be so slow as to be incompatible with life. Mammalian enzymes have evolved to catalyze reactions under physiological conditions, that is, at 37°C and usually at a pH near neutrality. They commonly accelerate reactions by factors of 10^6 to 10^{10} and are usually highly specific for their substrates.

The active site of an enzyme is the pocket in the protein where the substrate or substrates are bound. Substrates are bound to enzymes in what is referred to an *enzyme–substrate* (ES) *complex* by multiple weak (usually noncovalent) interactions, particularly ionic and hydrogen bonds. Binding of substrates to the enzyme's active site stabilizes the reaction intermediate or transition state, thereby decreasing the amount of activation energy required for the reaction to occur (Fig. 2-1). Theoretically, all chemical reactions are reversible to some extent. Enzymes catalyze both the forward and reverse reactions.

2.4.1   What Determines the Direction in Which Reversible Reactions Proceed?

An example of a reversible reaction is the one catalyzed by lactate dehydrogenase:

$$\text{lactate} + \text{NAD}^+ \rightleftharpoons \text{pyruvate} + \text{NADH} + \text{H}^+$$

Whether one starts with the substrates (shown on the left) or the products (on the right), the lactate dehydrogenase reaction, like all reactions, will eventually reach an equilibrium or steady-state condition. At equilibrium, the relative proportion of reactants on the left and products on the right will be determined by the change in free energy of the reaction ($\Delta G^{\circ\prime}$); in other words, the reaction will proceed in the direction that releases energy ($\Delta G^{\circ\prime} < 0$) rather than one that requires a net input of energy.

For reversible reactions, the major factor that determines the rates of reactions in the forward and reverse directions is the relative concentration of substrates and products. For reactions like the one catalyzed by lactate dehydrogenase, the direction of the reaction is determined primarily by the NADH/NAD$^+$ ratio. Thus, when exercising muscle produces more NADH than can be utilized by the mitochondrial oxidative phosphorylation system, the buildup of NADH drives lactate dehydrogenase to produce lactate from pyruvate. Conversely, when hepatocytes are actively utilizing NADH for ATP production via oxidative phosphorylation, the NADH level falls and the concentration of NAD$^+$ increases, thereby causing lactate dehydrogenase to generate pyruvate from lactate.

2.4.2   Irreversible Reactions

There are many reactions that are essentially irreversible under physiological conditions. These irreversible reactions are *exergonic*, meaning that they give off significant energy. Biochemists consider a reaction to be irreversible when the free-energy change ($\Delta G^{\circ\prime}$) is -4 kcal/mol or more negative. An example of a physiologically irreversible reaction is that catalyzed by glucose 6-phosphatase:

$$\text{glucose 6-phosphate} + H_2O \rightarrow \text{glucose} + P_i$$

The reverse reaction, that is, the formation of glucose 6-phosphate, would require the input of significant energy. Neither glucokinase nor hexokinase, the two enzymes that catalyze the synthesis of glucose 6-phosphate from free glucose, can directly reverse the reaction catalyzed by glucose 6-phosphatase. Instead, both of these enzymes utilize the energy associated with one of the high-energy bonds of ATP to phosphorylate glucose:

$$\text{glucose} + ATP \rightarrow \text{glucose 6-phosphate} + ADP$$

Acetyl-CoA carboxylase is another enzyme that utilizes the high-energy γ-phosphate bond of ATP to drive a reaction that would be irreversible without the participation of ATP:

$$\text{acetyl-CoA} + CO_2 + ATP + H_2O \rightarrow \text{malonyl-CoA} + ADP + P_i$$

In this case, the terminal (γ) phosphate of ATP (Fig. 1-2) is not incorporated into the product of the reaction. Instead, two reactions (hydrolysis of ATP and carboxylation of acetyl-CoA) are coupled, with the favorable (energy-yielding or *exergonic*) hydrolysis of ATP being used to drive the unfavorable (energy-requiring or *endergonic*) carboxylation of acetyl-CoA.

2.4.3 Isozymes Are Different Proteins That Catalyze the Same Reaction

As described above, glucokinase and hexokinase both catalyze the synthesis of glucose 6-phosphate from glucose and ATP. However, the two enzymes differ with regard to both their catalytic properties and their protein structures, and are therefore called *isozymes* or *isoenzymes*. Hexokinase, the isozyme present in almost every cell of the body, has a high affinity for glucose and is therefore active even at relatively low concentrations of glucose (Fig. 2-12). By contrast, glucokinase, which is found primarily in liver, is relatively inactive at low concentrations of glucose. Glucokinase has a higher maximal activity than hexokinase and is able to respond to increased blood glucose concentrations by rapidly synthesizing glucose 6-phosphate. Biochemists quantify these differences by indicating that glucokinase has both a higher V_{max} (maximal reaction velocity) and a higher K_m (the substrate concentration required to support half-maximal activity) than hexokinase. As shown in Figure 2-12, the K_m of hexokinase for glucose is 0.01 mM, while the lower affinity of glucokinase for glucose is reflected by its higher K_m of 5 to 10 mM. This difference in K_m values between the two isozymes permits the liver to remove glucose rapidly from the blood

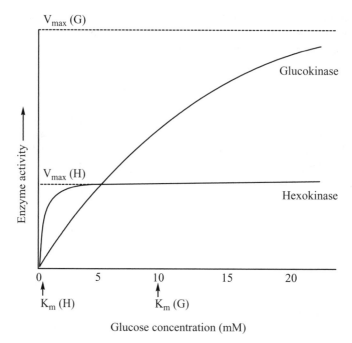

FIGURE 2-12   Glucokinase (G) and hexokinase (H) are isozymes with different kinetic properties.

when the glucose concentration is high, while leaving glucose available for glucose-dependent tissues (e.g., red blood cells, brain) when the body's glucose reserves are low.

2.5   HOW IS ENZYME ACTIVITY REGULATED?

Regulation of the functioning of the many enzymes in the body is central to co-ordinating the multiple pathways of metabolism and maintaining homeostasis. One key mechanism for regulating the level of activity of a particular enzyme in a cell is regulation of gene expression, since if an enzyme is not synthesized in the appropriate cell or at a particular time, the reaction it catalyzes will not occur.

The activities of existing enzymes are themselves also regulated, both by intracellular availability of metabolites and by covalent modifications (e.g., phosphorylation). In addition, many pharmaceutical agents act by inhibiting the activity of one or more enzymes. For example, patients who have elevated plasma cholesterol levels may be prescribed one of the statins, a class of drugs that inhibit HMG-CoA reductase, which catalyzes the rate-limiting step in cholesterol synthesis. The major mechanisms for regulation of enzymatic activity are described below.

2.5.1 Competitive Enzyme Inhibition

Competitive inhibition occurs when a molecule that is not a substrate for the enzyme in question, but which is structurally similar to the substrate, competes with the substrate and blocks its binding to the active site of the enzyme. Occupation of the active site by the inhibitor decreases the activity of the enzyme, particularly when the concentration of the substrate is low relative to that of the inhibitor.

Figure 2-13A depicts a Lineweaver–Burke double-reciprocal plot which illustrates how enzyme kinetics can be altered by the presence of a competitive inhibitor. For many enzymes, plotting $1/V$ (initial velocity) vs. $1/[S]$ (where $[S]$ is the substrate

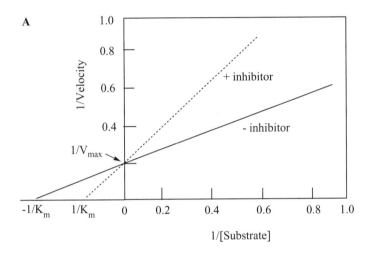

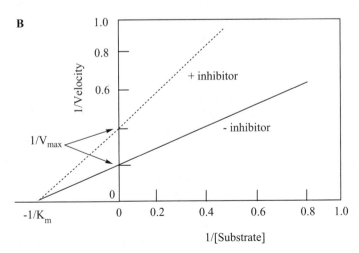

FIGURE 2-13 Enzyme inhibition illustrated with double-reciprocal Lineweaver–Burke plots: (A) competitive inhibition; (B) noncompetitive inhibition.

concentration) results in a linear graph. In this format, the X intercept of the line is $-1/K_m$ and the Y intercept is $1/V_{max}$. As shown in the figure, the competitive inhibitor increases the K_m but does not affect the V_{max} of the reaction.

Many pharmaceutical agents are competitive inhibitors of specific enzymes. For example, dicumarol inhibits catalysis involving vitamin K. Since many enzymes in the blood-clotting cascade are activated by γ-carboxylation, dicumarol acts as an anticoagulant that reduces the risk of thrombus formation.

In some cases, two different molecules may both be substrates for the same enzyme, with each acting as a competitive inhibitor of the metabolism of the other. One such example is alcohol dehydrogenase, which catalyzes the oxidation of both ethanol and methanol:

$$\text{ethanol} + \text{NAD}^+ \rightarrow \text{acetaldehyde} + \text{NADH} + \text{H}^+$$
$$\text{methanol} + \text{NAD}^+ \rightarrow \text{formaldehyde} + \text{NADH} + \text{H}^+$$

Although methanol itself is intoxicating, it is the metabolites of methanol (formaldehyde and formic acid) that are responsible for the blindness and death that result from methanol poisoning. One treatment for acute methanol poisoning involves intravenous administration of ethanol (plus glucose). Ethanol acts as a competitive inhibitor of the conversion of methanol to formaldehyde, thereby preventing accumulation of toxic metabolites until the methanol can be cleared by the kidneys. Glucose is administered to correct the hypoglycemia caused by ethanol.

2.5.2 Noncompetitive Enzyme Inhibition

A noncompetitive inhibitor binds to its target enzyme and cannot be displaced by excess substrate. Thus, this type of inhibitor diminishes the fraction of the enzyme pool that is catalytically competent. As illustrated in Figure 2-13B, noncompetitive inhibitors reduce the V_{max} of the reaction but do not change the enzyme's K_m.

Aspirin's action as a noncompetitive inhibitor is the major reason that it is a drug of choice for long-term therapy to decrease the risk of cardiovascular crises. Aspirin is a member of a class of drugs called *nonsteroidal anti-inflammatory agents* (NSAIDs) that inhibit the cyclooxygenase isozymes, thereby decreasing thromboxane production and platelet aggregation. Aspirin is an irreversible inhibitor since the molecule covalently acetylates a serine residue at the active site of the enzyme, inactivating cyclooxygenase permanently. By contrast, the inhibitory actions of other NSAIDs, such as ibuprofen, are attributable to reversible, noncovalent interactions between drug and enzyme. The reaction of aspirin with cyclooxygenase is particularly effective in platelets because platelets are incapable of synthesizing new enzyme protein.

Some of the deleterious effects of heavy metals, such as mercury and lead, result from their actions as noncompetitive enzyme inhibitors. For example, mercury inhibits glyceraldehyde 3-phosphate dehydrogenase, an enzyme in the glycolytic pathway, while lead inhibits heme synthesis.

2.5.3 Allosteric Regulation

Many enzymes contain regulatory sites that are physically separated from the active site. Binding of small molecules to one or more regulatory sites alters the three-dimensional structure of the enzyme, which increases or decreases its catalytic activity. *Allosteric regulation*, as this phenomenon is called, provides a mechanism by which enzymatic activities can be modulated by compounds that have little or no structural similarity to the substrate(s) but which instead, reflect the overall metabolic state or needs of the cell. Allosteric enzymes usually exhibit sigmoidal (S-shaped) kinetic curves (Fig. 2-14) rather than simple hyperbolic curves (Fig. 2-12). Activators of allosteric enzymes shift the V vs. S curve to the left, whereas allosteric inhibitors shift the curve to the right.

2.5.3.1 End-Product Inhibition. There are many instances in which the final endproduct of a multienzyme metabolic pathway is an allosteric inhibitor of an enzyme that catalyzes an early and irreversible step of the pathway. This form of allosteric regulation prevents accumulation of additional end product and of metabolic intermediates once a cell has sufficient supplies of that metabolic end product. Examples of this are seen in the pathways that generate heme, long-chain fatty acids, and cholesterol, where the end products inhibit δ-aminolevulinic acid synthase, acetyl-CoA carboxylase, and HMG-CoA reductase, respectively.

2.5.3.2 Regulation by Molecules That Signal the Availability of Precursors. Allosteric regulation provides a mechanism by which flux through a particular

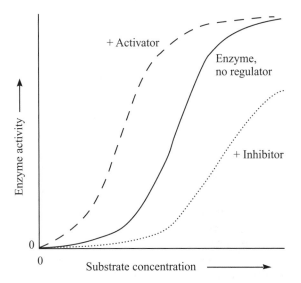

FIGURE 2-14 Sigmoidal kinetics of an allosteric enzyme and the effects of an activator and an inhibitor on the enzyme kinetics.

pathway can be rendered responsive to the overall nutritional state and needs of the cell. One such important small regulatory metabolite is citrate, an intermediate in the tricarboxylic acid cycle. Citrate allosterically stimulates liver cells to synthesize both fatty acids and glucose (gluconeogenesis) while inhibiting the breakdown of glucose by glycolysis.

2.5.3.3 *Regulation by the Energy Charge of the Cell.* Allosteric mechanisms also serve to regulate many metabolic pathways in response to a high ATP/ADP ratio, which is indicative of a plentiful supply of energy, or conversely, to high concentrations of ADP and adenosine 5′-monophosphate (AMP), which occur when ATP supplies have been depleted. An enzyme whose activity is regulated by the energy charge of the cell is the muscle isozyme of glycogen phosphorylase, which releases glucose (as glucose 1-phosphate) from glycogen stores. AMP is an allosteric activator of glycogen phosphorylase, and ATP is an allosteric inhibitor of the enzyme.

2.5.4 Regulation of Enzyme Activity by Covalent Modification

Many enzymes are activated or inhibited by covalent modification of their polypeptide structure. The modifications can be reversible or irreversible.

2.5.4.1 *Phosphorylation/Dephosphorylation.* The most common covalent modification utilized in regulating human metabolism is the reversible phosphorylation of enzyme proteins. In most cases, phosphorylation is the result of a hormone-stimulated signal-transduction cascade, thus providing a mechanism by which intracellular enzymatic activity can be modulated in response to intercellular signaling.

For example, glucagon and epinephrine both stimulate the activity of a serine–threonine protein kinase called *protein kinase A* (PKA). Briefly, this particular signaling pathway involves binding of glucagon or epinephrine to its respective transmembrane receptors, activation of a GTP-binding or G-protein, and activation of adenylyl cyclase, an enzyme that synthesizes cyclic AMP (cAMP) from ATP (Fig. 2-15):

$$ATP \rightarrow cAMP + PP_i$$

cAMP is an allosteric activator of PKA. Once activated, PKA uses ATP to phosphorylate specific serine and threonine residues on critical enzymes that regulate the flux of intermediates through key metabolic pathways. As discussed in more detail in Chapter 8, PKA-catalyzed phosphorylation activates enzymes involved with glycogen breakdown (glycogen phosphorylase). PKA-catalyzed phosphorylation also activates enzymes involved with mobilization of triacylglycerol stores and gluconeogenesis. Concurrently and conversely, PKA-catalyzed protein phosphorylation inhibits enzymes involved in glycogen and fatty acid metabolism. The simultaneous phosphorylation of multiple enzymes provides a coordinated response to the body's need to mobilize endogenous fuels during fasting or in response to stress.

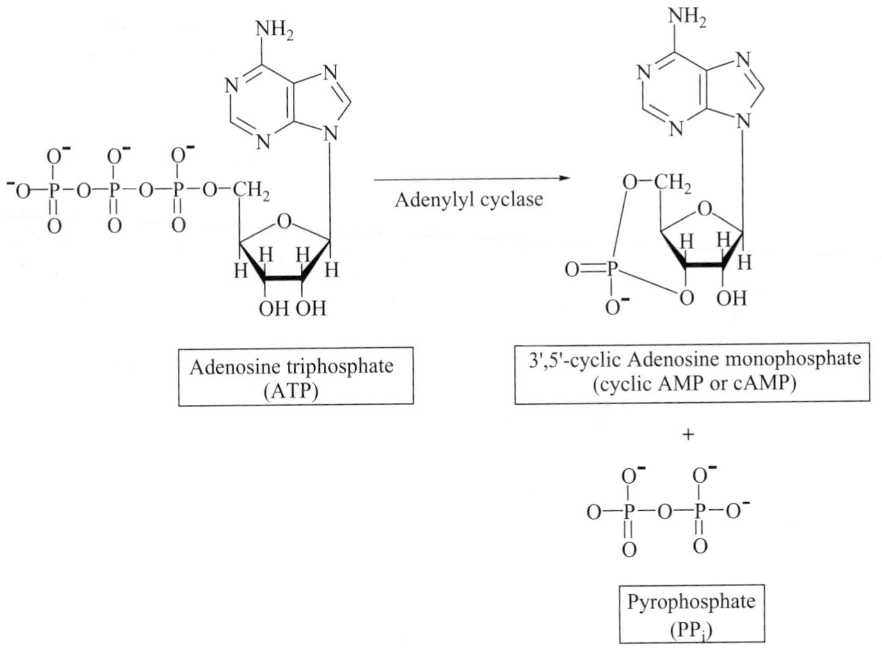

FIGURE 2-15 Synthesis of cyclic AMP by adenylyl cyclase.

Protein phosphorylation catalyzed by PKA can be reversed. Among the many effects of stimulation of cells by insulin is the activation of a signal-transduction cascade that results ultimately in the activation of protein phosphatase-1. Protein phosphatase-1 hydrolyzes the phosphate moieties from phosphoserine and phospho-threonine residues of many enzymes, thereby reversing the activation or inactivation that occurred when those enzymes were phosphorylated. Accordingly, insulin re-verses the metabolic effects of glucagon and epinephrine, and switches the direction of key metabolic processes to meet the body's needs in the fed state when there is active synthesis of triacylglycerol in adipocytes and hepatocytes as well as synthesis and storage of glycogen in muscle and liver.

2.5.4.2 Hydrolytic Cleavage of Inhibitory Peptides.
A number of enzymes, particularly digestive enzymes synthesized in the pancreas and the liver-synthesized proteases of the blood-clotting cascade, are secreted from their sites of synthesis in an inactive or zymogen form. Activation requires proteolytic hydrolysis of the pro-enzyme and release of a polypeptide, which then permits the remaining polypeptide fragment to alter its three-dimensional structure to one in which its active site and associated substrate binding pocket are configured correctly for catalysis. Thus, the digestive enzyme trypsin is secreted from the pancreas in the form of an inactive precursor, trypsinogen. Once in the lumen of the small intestine, a brush-border protease called enteropeptidase hydrolyzes one peptide bond within the trypsinogen

molecule, thereby releasing the inhibitory peptide and generating active trypsin. Secretion of trypsin in its zymogen form limits its activity to the digestive tract, thus protecting the pancreas and pancreatic duct from proteolytic damage.

2.5.5 Induction of Enzyme Synthesis

Hormonal regulation of enzymatic activity can also occur through stimulation or inhibition of transcription of genes that encode key metabolic enzymes. Hydrocortisone, a glucocorticoid hormone synthesized by the adrenal cortex, acts by entering the cell and binding to certain proteins in the cytosol that serve as glucocorticoid receptors. The hydrocortisone–glucocorticoid receptor complex then translocates to the nucleus, where it binds to specific hormone-response elements in DNA. The actions of hydrocortisone include induction of the synthesis of enzymes involved with gluconeogenesis, mobilization of adipose triacylglycerol, and degradation of muscle proteins. Hydrocortisone thus plays a major role in mediating long-term adaptations in the activities of metabolic pathways in response to starvation, sepsis, and stress.

2.6 DISEASE STATES ASSOCIATED WITH ABNORMAL ENZYME FUNCTIONING

2.6.1 Vitamin Deficiencies

Since vitamins are crucial components of many enzyme cofactors, an inadequate concentration of one or more of these essential dietary substances can result in impaired enzymatic activity. Vitamin deficiencies result from inadequate dietary intake or from impaired absorption or recycling of a vitamin. Impaired absorption of vitamins K and B_{12} is discussed in Chapters 3 and 22, respectively.

Unlike many of the other vitamin-based cofactors that attach to their respective enzymes through noncovalent bonds, biotin is covalently attached to lysyl residues of the enzymes with which it functions. Biotin deficiency can result from inadequate activity of the enzyme biotinidase, which normally hydrolyzes the biotinyl–lysyl bond and releases free biotin, thus permitting recycling of biotin when biotin-containing enzymes are degraded. A deficiency of biotin can also be induced by consumption of raw eggs, which contain avidin, a protein that binds biotin very tightly, thereby preventing absorption of biotin from the gut. Biotin deficiency reduces the activities of all four biotin-dependent enzymes: pyruvate carboxylase, acetyl-CoA carboxylase, propionyl-CoA carboxylase, and β-methylcrotonyl-CoA carboxylase.

Dietary deficiencies of particular vitamins usually result from restricted diets. In each case, there is impaired activity of all of the enzymes that utilize the particular vitamin-derived cofactor, and ultimately development of a specific vitamin-deficiency disease. Thus, deficiency of folic acid results in megaloblastic anemia and is associated with congenital neural tube defects, whereas the peripheral neuropathy and cardiac manifestations of beriberi are caused by a dietary deficiency of thiamine.

2.6.1.1 Scurvy. Scurvy is the result of a dietary deficiency of vitamin C (ascorbic acid), which is usually obtained from fresh fruits and vegetables, especially citrus fruits (e.g., oranges, grapefruit, limes), cabbage, mangoes, and tomatoes. Scurvy has been recognized since ancient times, and many indigenous cultures are known to have had remedies utilizing local plant sources, including teas brewed from pine needles. Scurvy was particularly rampant among European sailors on long ocean voyages.

Ascorbic acid is a cofactor in various hydroxylation reactions including hydroxylation of specific proline and lysine residues of procollagen (Fig. 2-10) and the hydroxylation of dopamine to form norepinephrine. Scurvy is primarily a disease of defective collagen synthesis, and is characterized by bleeding gums, hemorrhages, and impaired wound heeling. Ascorbic acid also plays an important non-cofactor role as an antioxidant; it regenerates the reduced forms of other antioxidants, such as vitamin E and glutathione, as well as inactivating potentially harmful reactive oxygen species and nitrogen radicals.

2.6.1.2 Pellagra. A deficiency of niacin results in pellagra, which is characterized by dermatitis (especially in areas of the skin exposed to sunlight), diarrhea, dementia, and ultimately–if untreated–death. Since niacin is the vitamin component of NAD^+ and $NADP^+$, which are cofactors in numerous oxidation–reduction reactions, it is involved with essentially all of the major metabolic pathways including glycolysis, β-oxidation of fatty acids, the TCA cycle, electron transport and oxidative phosphorylation, and the synthesis of fatty acids and cholesterol.

Pellagra was endemic in the American Southeast between the two world wars, primarily among poor mill workers, who consumed a limited diet consisting primarily of corn (maize) and lard. The niacin in maize has relatively low bioavailability unless the maize is treated with alkali. Europeans adopted corn as a crop but not the native tradition of grinding corn with lime (calcium oxide or calcium carbonate).

Niacin is unique among the B vitamins in that the dietary requirement for this vitamin can be partially satisfied by endogenous synthesis of niacin from the essential amino acid tryptophan; unfortunately, maize also happens to be a relatively poor source of tryptophan. Prevention of pellagra was accomplished through public health measures, particularly the fortification of cereal products (e.g., bread, biscuits, pasta) with niacin. Ironically, these measures were delayed for many years because public acceptance of pellagra as a disease of malnutrition was hampered by the eugenics movement, which stereotyped the victims of pellagra in the U.S. South as inherently inferior human beings.

2.6.2 Inborn Errors of Metabolism

Inborn errors of metabolism are genetic disorders resulting from partial loss of function or from null mutations (complete absence of activity) of genes coding for particular enzymes. Examples of inborn errors of metabolism include phenylketonuria (PKU, caused by a lack of phenylalanine hydroxylase), medium-chain acyl-CoA hydrogenase deficiency, and glucose 6-phosphate dehydrogenase deficiency. Other examples of inborn errors are the lysosomal storage diseases, which result from the loss of function of acid hydrolases required for lysosomal digestion of glycosamino-glycans, glycolipids, sphingomyelin, and glycogen.

2.6.3 Vitamin-Dependency Diseases

Inborn errors of metabolism often result from the synthesis of mutated enzymes, which have a decreased affinity (increased K_m) for their coenzyme or prosthetic group. In such cases the patient can often be treated successfully with exceptionally high intakes—or megadoses—of the vitamin precursor. For example, cystathionine synthase is a pyridoxal phosphate–dependent enzyme that synthesizes cystathionine from homocysteine and serine. Some forms of cystathionine β-synthase deficiency are responsive to treatment with pyridoxine (vitamin B_6). Similarly, some persons who are deficient in pyruvate dehydrogenase improve with thiamine therapy.

2.6.4 α_1-Antitrypsin Deficiency

α_1-Antitrypsin is a plasma glycoprotein that inhibits the activity of elastase and has inhibitory activity against a number of other serine proteases. Synthesized in and secreted by the liver, α_1-antitrypsin's major physiological function is to inhibit elastase released by neutrophils in the lung. In the absence of this bloodborne protease inhibitor, there is destruction of elastin in pulmonary alveoli, resulting in chronic obstructive pulmonary disease or emphysema. The disease occurs much earlier and is more severe in people who smoke, and some people with α_1-antitrypsin deficiency also develop cirrhosis of the liver.

2.6.5 Pancreatitis

Pancreatitis is an inflammation of the pancreas which can result from a number of conditions, including gallstones, chronic alcoholism, and the blockage of the pancreatic duct that can occur in cystic fibrosis. Damage to pancreatic cells results in premature activation of digestive proteases within the pancreas and resulting auto-digestion of the pancreas. Elevated serum levels of pancreatic enzymes, particularly pancreatic lipase and amylase, are a laboratory-based diagnostic criterion of acute pancreatitis.

2.6.6 Enzymes as Markers of Disease

Many tissues produce enzymes that are relatively cell-specific. Because these enzymes are released into the circulation as a result of tissue damage, assays of the levels of certain enzymes in blood can provide useful diagnostic information. Probably the most widely requested plasma enzyme assays are those for alanine aminotransferase (ALT) and aspartate aminotransferase (AST), both of which are present in high concentration in hepatocytes. When these cells are injured, for example by viral hepatitis or acetaminophen overdose, ALT and AST are released into the blood. As indicated above, elevated serum levels of pancreatic lipase and amylase are indicative of acute pancreatic disease. The level of the MB isozyme of creatine kinase (CPK), a marker for myocardial infarction, often rises rapidly in the plasma of a person who has experienced a heart attack. An increased level of the MM-isozyme of CPK is usually indicative of injury to skeletal muscle. Other examples of tissue-specific enzyme markers of cellular injury include bone-specific alkaline phosphatase (b-ALKP), which serves as a marker of bone turnover in patients with osteoporosis or Paget's disease, and acid phosphatase, which is elevated in plasma of patients with metastatic cancer of the prostate.

CHAPTER 3

DIGESTION AND ABSORPTION

3.1 FUNCTIONS OF DIGESTION AND ABSORPTION

The main function of digestion is to hydrolyze large dietary components such as starch, proteins, and fats into their absorbable component parts (e.g., glucose, amino acids, fatty acids). Relatively small dietary substances such as the disaccharides lactose and sucrose are also hydrolyzed into their component simple sugars. In addition, digestion serves to release some vitamins, such as biotin and vitamin B_{12}, from their protein-bound forms.

The well-orchestrated functioning of many organs of the gastrointestinal system, including the stomach, liver, gallbladder, pancreas, small intestine, and colon, is required for efficient digestion and absorption of the essential nutrients in foods. The stomach produces hydrochloric acid, which denatures proteins, rendering them more susceptible to proteolysis both by pepsin, produced by the stomach, and by pancreas-derived proteases. The liver produces and secretes bile salts, which are required for the digestion and absorption of triacylglycerols, which are comprised of long-chain fatty acids esterified to glycerol. In addition to secreting numerous digestive enzymes that act in the small intestine, the pancreas secretes large amounts of sodium bicarbonate, which neutralize stomach acid, thus providing the nearly neutral pH required for the activity of pancreatic enzymes in the lumen of the small intestine.

The enterocytes that line the lumen of the small intestine not only provide the surface to which disaccharidases and peptidases are attached, but are also the site

Medical Biochemistry: Human Metabolism in Health and Disease By Miriam D. Rosenthal and Robert H. Glew
Copyright © 2009 John Wiley & Sons, Inc.

where most of the small-molecular-mass products of digestion are absorbed. The ileum is an integral element of the enterohepatic circulation that accounts for the recycling of bile salts and the absorption of essential nutrients such as vitamin B_{12}. In addition, the small intestine is the body's largest endocrine organ, as it produces a variety of hormones that regulate digestion and energy balance. The colon is a major site of absorption of water and sodium and chloride ions. The colon is also the site of absorption of some of the metabolic by-products of colonic bacteria, particularly lactate, short-chain fatty acids such as propionate and butyrate, and ammonia, which is generated by hydrolysis of urea by bacterial urease.

3.2 DIGESTION AND ABSORPTION OF CARBOHYDRATES

3.2.1 Dietary Carbohydrates

The major dietary carbohydrates are starch, sucrose, and lactose. Starch, the polymeric form of glucose stored in plants, is a mixture of two macromolecular structures: amylose and amylopectin (Fig. 3-1). Amylose is a straight-chain polymer in which the glucose units are attached to one another through α-1,4 linkages. Amylopectin is a branched structure with branches formed by α-1,6 glycosidic linkages to the α-1,4 chains. Animal foods contain small quantities of glycogen, a glucose polymer that is similar to amylopectin but is more highly branched. Cellulose, the structural glucose polymer of plants, contains β-1,4 glycosidic bonds which are not hydrolyzed by human digestive enzymes. Cellulose is thus a dietary fiber rather than a bioavailable source of carbohydrate for the body.

Sucrose and lactose are disaccharides; that is, they are composed of two sugar units in glycosidic linkage (Fig. 3-2). Sucrose (table sugar), commonly extracted from sugarcane or sugar beets, consists of glucose (Glc) and fructose (Fru) and has the structure α-Glc($1 \rightarrow 2$)β-Fru. Lactose is the sugar found in milk and is comprised of β-galactose linked to C4 of glucose [β-Gal($1 \rightarrow 4$)Glc]. Fructose and glucose are also present as monosaccharides in honey and many fruits.

Most common monosaccharides and disaccharides are reducing sugars since they have a free aldehyde or ketone group. In an alkaline solution, a reducing sugar will reduce cupric ion (Cu^{2+}) to cuprous ion (Cu^+). By contrast, sucrose is not a reducing sugar.

3.2.2 Digestion of Starch

Salivary and pancreatic amylases are both endoglycosidases that randomly hydrolyze internal α-1,4 glycosidic bonds of amylose and amylopectin to form smaller polysaccharides called *dextrins*. Hydrolysis of the glucose polymers is initiated by salivary amylase (ptyalin), which hydrolyzes as much as 40% of dietary starch before the enzyme is inactivated by the low pH in the stomach. Pancreatic α-amylase contin-ues the starch digestion process in the small intestine, producing maltose [α-Glc

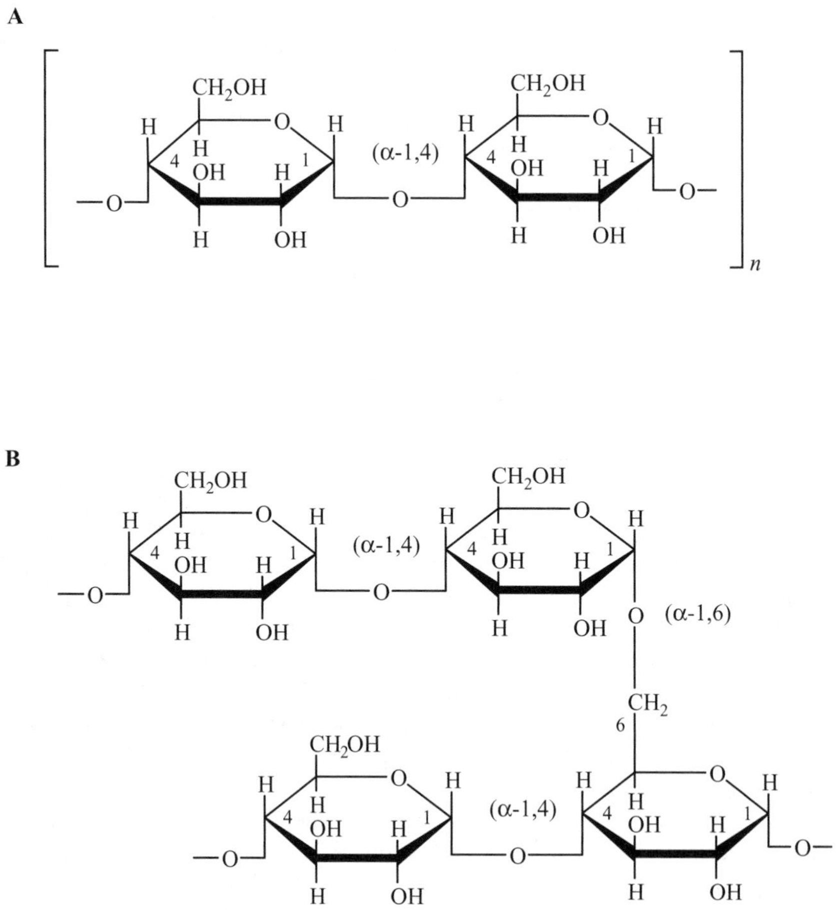

A

B

FIGURE 3-1 Structure of dietary starch: (A) the straight-chain structure of amylose (*n* indicates the number of repeating units, which may be in the thousands); (B) an α-1,6 glycosidic branch point in amylopectin.

(1 → 4)Glc], isomaltose [α-Glc(1 → 6)Glc], and limit dextrins, which are a mixture of oligosaccharides comprised of three to eight glucose units, including occasional α-1,6 branches.

3.2.3 Digestion of Oligosaccharides

The dietary disaccharides, sucrose and lactose, and the maltose, isomaltose, and oligosaccharides produced by partial digestion of dietary starch are hydrolyzed by enzymes that are localized on the surface of the brush border of the intestinal mucosa.

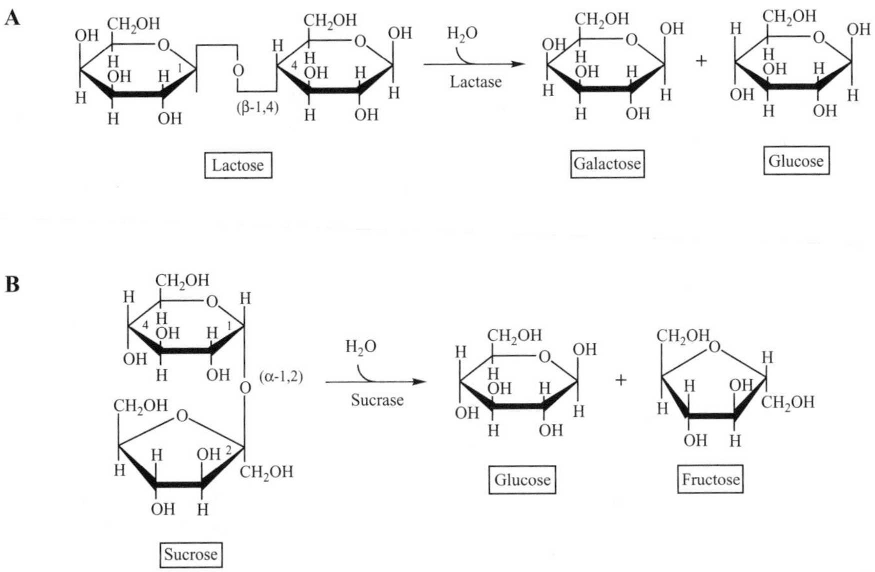

FIGURE 3-2   The disaccharides lactose (A) and sucrose (B) and their hydrolysis to simple sugars.

3.2.3.1 Maltase. Maltase is an α-glucosidase that hydrolyzes both maltose (Fig. 3-3A) and maltotriose:

$$\alpha\text{-Glc}(1 \rightarrow 4)\alpha\text{-Glc}(1 \rightarrow 4)\text{Glc [maltotriose]} + H_2O \rightarrow \text{maltose} + \text{glucose}$$

$$\alpha\text{-Glc}(1 \rightarrow 4)\text{Glc [maltose]} + H_2O \rightarrow 2 \text{ glucose}$$

3.2.3.2 Isomaltase. Isomaltase is an α-glycosidase that hydrolyzes the α-1,6 glycosidic bond of isomaltose and limit dextrans (Fig. 3-3B):

$$\alpha\text{-Glc}(1 \rightarrow 6)\text{Glc [isomaltose]} + H_2O \rightarrow 2 \text{ glucose}$$

3.2.3.3 Lactase. Lactase is a β-galactosidase that hydrolyzes lactose to glucose and galactose (Fig. 3-2A):

$$\beta\text{-Gal}(1 \rightarrow 4)\text{Glc [lactose]} + H_2O \rightarrow \text{glucose} + \text{galactose}$$

3.2.3.4 Sucrase. Sucrase is a disaccharidase that hydrolyzes sucrose (Fig. 3-2B):

$$\alpha\text{-Glc}(1 \rightarrow 2)\beta\text{-Fru[sucrose]} + H_2O \rightarrow \text{glucose} + \text{fructose}$$

It should be noted that the two polypeptides that have sucrase and isomaltase activity, respectively, are initially synthesized as a single polypeptide chain.

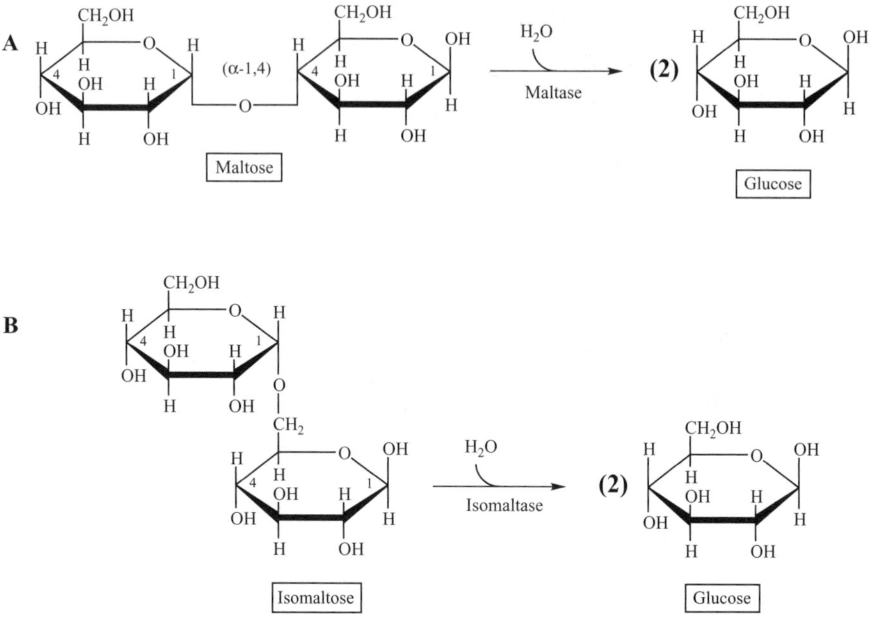

FIGURE 3-3 Hydrolysis of maltose (A) and isomaltose (B) generates glucose.

3.2.3.5 α-Dextrinase. This exoglycosidase hydrolyzes glucose α-1,4-glucose linkages starting at the nonreducing end of the oligosaccharide chain. Although α-dextrinase has greater activity for oligosaccharides with relatively longer chains, it also hydrolyzes maltose and maltotriose.

3.2.4 Absorption of Sugars

3.2.4.1 Glucose and Galactose. Glucose is absorbed into the cells of the intestinal mucosa in cotransport with Na$^+$ by GLUT1, the sodium glucose–dependent symporter. This process is driven by the active transport of Na$^+$ out of the cell through the basolateral membrane, which also serves to maintain a low concentration of intracellular Na$^+$. Galactose binds to the glucose-binding site of GLUT1 and is transported into the mucosa by the same cotransporter. There is also a facilitative transport mechanism for glucose absorption.

3.2.4.2 Fructose. Fructose is absorbed by facilitative diffusion, a process by which transport proteins facilitate the passage of a polar molecule across the plasma membrane. Fructose transport is driven by the concentration gradient of fructose across the membrane. All of the common dietary monosaccharides leave the enterocyte through the basolateral membrane by means of facilitated diffusion.

3.3 DIGESTION AND ABSORPTION OF DIETARY LIPIDS

3.3.1 Dietary Lipids

The major dietary lipids are triacylglycerols containing three long-chain fatty acids (usually C_{16}–C_{20}) esterified to glycerol (Fig. 3-4). Animal products also contain both free cholesterol and cholesteryl esters. Other dietary lipids include phospholipids, vitamins A, D, E, and K, and the carotenoids.

Since lipids are hydrophobic and poorly soluble in water, they have a strong tendency to aggregate into large lipid droplets. Efficient digestion of these droplets requires emulsification, the process by which large lipid droplets are dispersed into smaller ones, thus providing greater surface area for access by hydrolytic enzymes to their substrates. The process of emulsification involves both the physical effects of peristaltic churning of the food and the chemical dispersion of the droplets by the detergent action of bile salts.

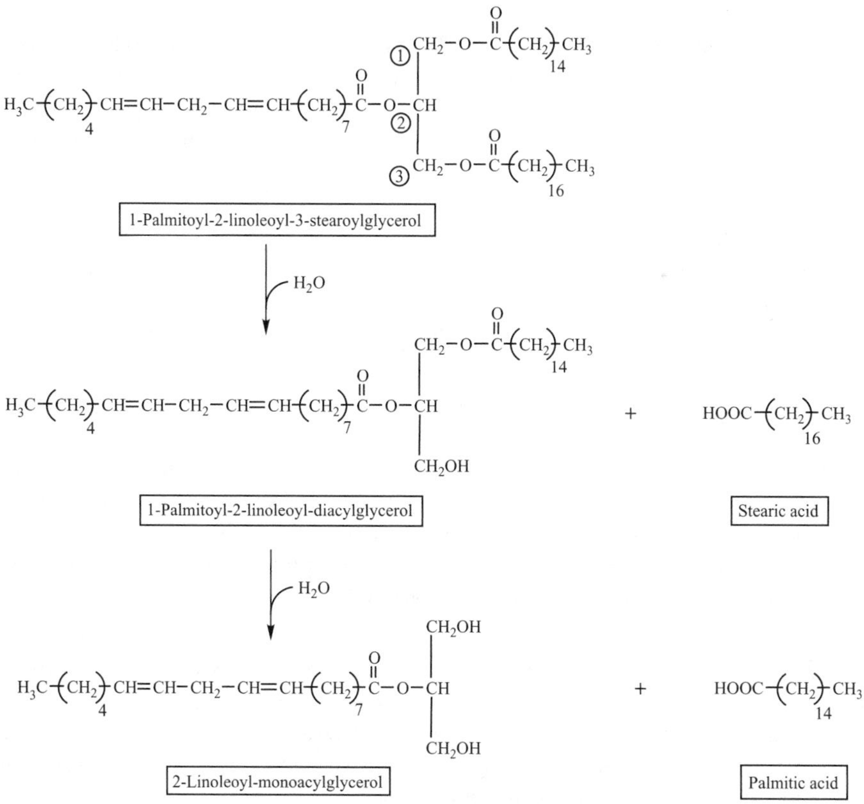

FIGURE 3-4 Structure of a typical dietary triacylglycerol and its hydrolysis in the intestine to a 2-monoacylglycerol plus two free fatty acids.

3.3.2   Bile Acid and Bile Salts

Effective digestion and absorption of dietary lipids requires both digestive enzymes and conjugated *bile acids* (a.k.a. *bile salts*). Bile acids are oxygenated derivatives of cholesterol that have several hydroxyl groups on the sterol rings and a shortened hydrocarbon tail ending in a carboxyl group (Fig. 3-5). Bile acids are weak acids with a pK_a value of about 6. The term *bile salts* refers to conjugated bile acids which contain either glycine or taurine linked via an amide bond to the carboxyl group of a bile acid (Fig. 3-5C). Conjugation decreases the pK_a of the bile salts; glycocholic acid has a pK_a of about 4, whereas the pK_a of taurocholic acid is about 2. The stronger hydrophilic domains of the bile salts renders them more amphipathic than bile acids and thus more effective emulsifiers.

3.3.2.1   *Bile Salts Emulsify Dietary Lipids.* The physical properties of the bile salts enable them to emulsify lipid droplets, thereby enhancing lipid digestion. Bile salts containing three hydroxyl groups (e.g., cholic acid) are better emulsifiers than those that have only two hydroxyl groups (e.g., deoxycholic acid).

3.3.2.2   *Bile Salts Stabilize Mixed Micelles.* A second role that bile salts play in the process of digestion and absorption of dietary lipids is the solubilization of the relatively hydrophobic products of lipid hydrolysis in the form of small aggregates called *mixed micelles*. As shown in Figure 3-6, the stereochemistry of the hydroxyl groups of the bile salts gives the planar ring structure a hydrophobic face and a hydrophilic face. Mixed micelles have structures similar to small disks cut out of a membrane bilayer, with the bile salts stabilizing the cut edges. Stabilization of mixed micelles by bile salts is required for the products of lipid hydrolysis to diffuse through the unstirred water layer near the surface of the intestine to the plasma membrane of the enterocyte brush border where they are absorbed.

3.3.3   Hydrolysis of Dietary Lipids

One major difference between the digestion of carbohydrates and lipids is that hydrolysis of lipids is only partial. Whereas carbohydrates are absorbed only as monosaccharides, lipids are absorbed as a mixture of monoacylglycerols, diacylglycerols, and lysophospholipids, as well as free fatty acids. These products are then reassembled within the enterocytes for transport within the body.

3.3.3.1   *Triacylglycerols.* The enzymes that hydrolyze triacylglycerols (triglycerides) are called *lipases*. The digestive lipases catalyze the partial hydrolysis of dietary fats containing long-chain fatty acids to a mixture consisting primarily of free fatty acids and 2-monoacylglycerols. Several lipases contribute to triacylglycerol digestion, the major one being pancreatic lipase:

$$\text{triacylglycerol} + 2H_2O \rightarrow \text{2-monoacylglycerol} + 2 \text{ free fatty acids}$$

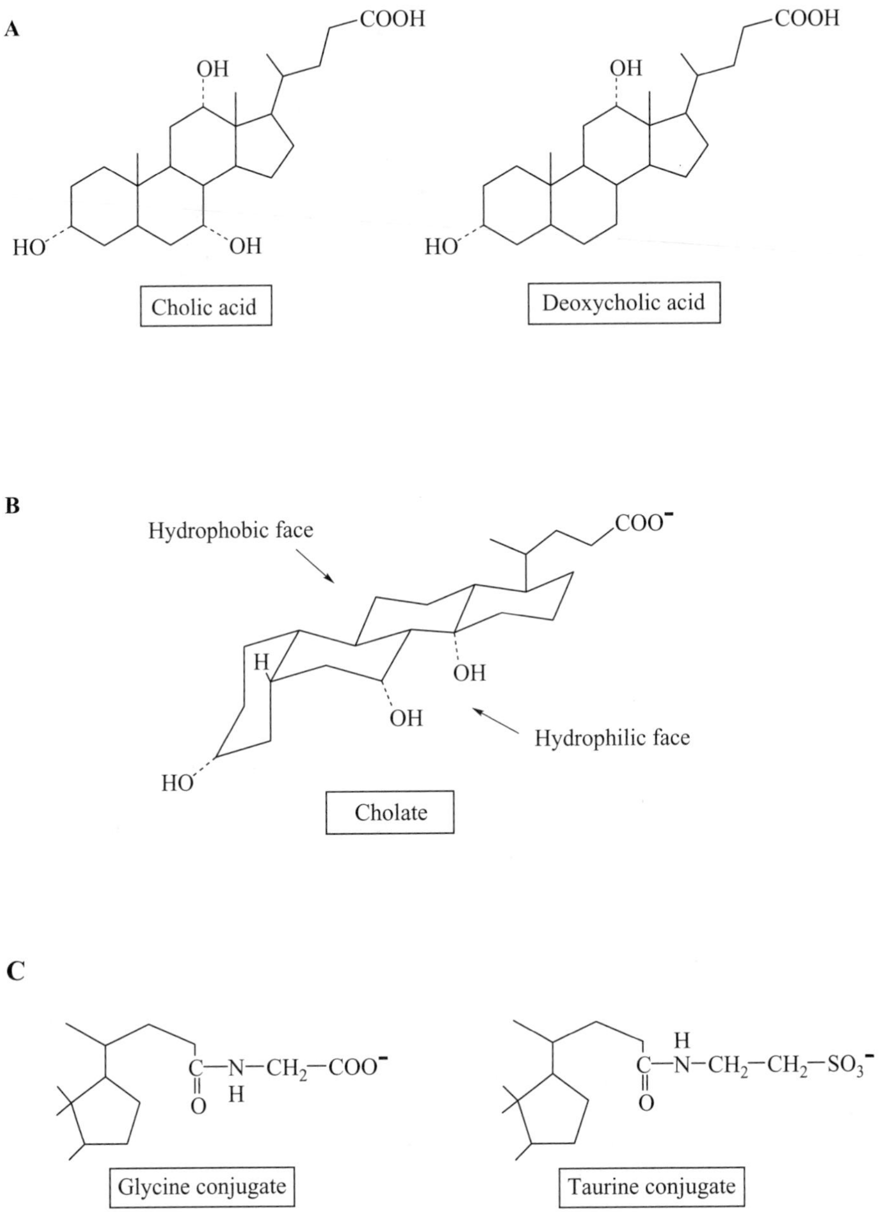

A

Cholic acid

Deoxycholic acid

B

Hydrophobic face

Hydrophilic face

Cholate

C

Glycine conjugate

Taurine conjugate

FIGURE 3-5 Structures of bile acids and bile salts: (A) the primary bile acids, cholic acid and deoxycholic acid; (B) the hydroxyl groups of bile salts such as taurocholic acid generate a polar face on an otherwise nonpolar steroid ring; (C) bile salts are formed by conjugation of bile acids with glycine or taurine.

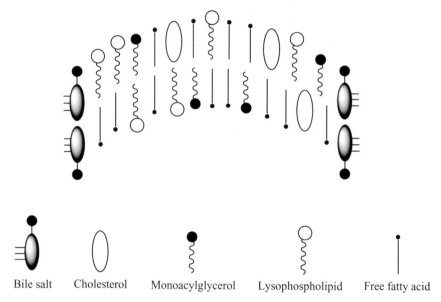

| Bile salt | Cholesterol | Monoacylglycerol | Lysophospholipid | Free fatty acid |

FIGURE 3-6   Cross-section of a discoidal mixed micelle.

Catalysis by pancreatic lipases requires the presence of a second pancreatic product called *colipase*. This 10-kDa nonenzyme protein reduces the surface tension at the lipid–aqueous interface, facilitating the interaction between the lipase and the lipid droplet.

The body also produces lingual and gastric lipases. Although gastric lipase is primarily active against substrates containing short-chain (C4–C6) and medium-chain (C8–C12) fatty acids, and is thus particularly important in the infant, it may also account for 10 to 30% of the hydrolysis of triacylglycerols, comprised of long-chain (C16–C20) fatty acids. The contribution of lingual lipase to fat digestion is normally quite low. Interestingly, however, lingual lipase is not inactivated by the acid pH of the stomach and, in the absence of pancreatic secretion of bicarbonate, remains active in the small intestine as well. Thus, particularly in the absence of pancreatic lipase activity, lingual lipase can contribute significantly to the digestion of dietary triacylglycerols. A neutral pH optimum, bile salt–stimulated lipase, present in human milk but not in cow's milk, contributes substantially to triacylglycerol hydrolysis in the intestine of breast-fed infants.

3.3.3.2 *Hydrolysis of Phospholipids.* Pancreatic phospholipase A_2 is secreted as a zymogen (inactive proenzyme), which is activated by trypsin-catalyzed hydrolysis. Phospholipase A_2 catalyzes the partial hydrolysis of both dietary phospholipids and the phosphatidylcholine secreted by the liver and contained in the bile. Pancreatic phospholipase A_2 is specific for fatty acids in the 2-position of phospholipids but has a broad specificity with respect to both the phospholipid polar head groups and the chain length of the target fatty acid:

$$\text{phosphatidylcholine} + H_2O \rightarrow \text{2-lysophosphatidylcholine} + \text{free fatty acid}$$

3.3.3.3 Hydrolysis of Cholesteryl Esters. Dietary cholesterol is a mixture of free cholesterol and cholesterol esterified with long-chain fatty acids. Pancreatic juice also contains a cholesterol esterase or cholesteryl ester hydrolase which catalyzes the following reaction:

$$\text{cholesteryl ester} + H_2O \rightarrow \text{cholesterol} + \text{free fatty acid}$$

3.3.4 Absorption of Dietary Lipids and Chylomicron Formation

The products of lipid digestion include a mixture of partially hydrolyzed lipids (primarily monoacylglycerols and lysophospholipids), free fatty acids, cholesterol, fat-soluble vitamins, and other lipophilic molecules (e.g., carotenoids). All the products of lipid digestion ultimately become solubilized by bile salts to form small mixed micelles that diffuse from the lumen of the intestine toward the apical surface of the epithelium of the duodenum and jejunum, where the dietary lipids are absorbed. Whereas absorbed glucose can readily be transported as such to the liver and other tissues in the bloodstream, this process is not suitable for free fatty acids, because of both their limited solubility and their detergent properties, which could disrupt cell membranes and inhibit enzymes. Although free fatty acids released from adipocytes are transported in plasma bound to serum albumin, the higher concentrations of free fatty acids present after a meal would overwhelm this transport system. Instead, the absorbed fatty acids are reesterified into less polar products for transport in the form of large lipoprotein aggregates called *chylomicrons* (Fig. 3-7).

The hydrophobic core of the chylomicrons consists primarily of triacylglycerol molecules. It also contains cholesteryl esters and other absorbed lipophilic molecules, such as fat-soluble vitamins. The chylomicron particle is surrounded by a surface layer of phospholipids, free cholesterol, and proteins, primarily apoprotein B48 (apo B48) and apo A1. After assembly, the chylomicrons are secreted from the enterocytes into the lymphatic circulation, from whence they eventually enter the blood via the thoracic duct. The subsequent hydrolysis of the triacylglycerols of circulating chylomicrons is discussed in Chapter 12.

3.3.4.1 Resynthesis of Triacylglycerol. In enterocytes, synthesis of triacylglycerol occurs through the sequential action of monoacylglycerol acyltransferase and diacylglycerol acyltransferase, for a net reaction

$$\text{2-monoacylglycerol} + \text{2 fatty acyl-CoA} \rightarrow \text{triacylglycerol} + \text{2CoASH}$$

This pathway is distinct from the triacylglycerol synthesis pathway in other cells, such as hepatocytes and adipocytes, which utilizes glycerol 3-phosphate as the acceptor

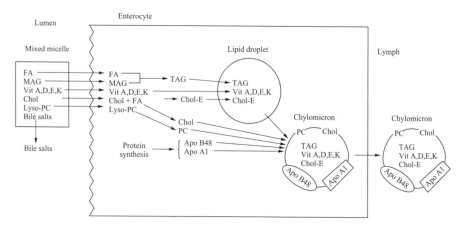

FIGURE 3-7 Absorption of dietary lipids into the enterocyte leads to chylomicron formation. apo A1, apoprotein A1; apo B48; apoprotein B48; Chol, cholesterol; Chol-E, cholesteryl ester; FA, free fatty acid; lyso-PC, 2-lysophosphatidylcholine; MAG, 2-monoacylglycerol; PC, phosphatidylcholine; TAG, triacylglycerol.

of acyl groups. The conversion of free fatty acids to fatty acyl-CoA in enterocytes utilizes the ubiquitous fatty acyl-CoA synthetase reaction

$$\text{fatty acid} + \text{ATP} + \text{CoASH} \rightarrow \text{acyl-CoA} + \text{AMP} + \text{PP}_i$$

3.3.4.2 *Reesterification of Other Absorbed Lipids.* As described above, the surface of chylomicrons contains cholesterol and phosphatidylcholine as well as apo B48. Secretion of cholesterol and phosphatidylcholine in bile and reabsorption of cholesterol and lysophosphatidylcholine in the small intestine provides enterocytes with a supply of the components involved in chylomicron assembly. Lysophosphatidylcholine acyltransferase catalyzes the reassembly of the phospholipid:

$$\text{2-lysophosphatidylcholine} + \text{fatty acyl-CoA} \rightarrow \text{phosphatidylcholine} + \text{CoASH}$$

Any free (nonesterified) cholesterol in excess of that which can be accommodated on the surface of the chylomicron particle is esterified by acyl-CoA:cholesterol acyltransferase (ACAT) and transported in the core of the chylomicron:

$$\text{cholesterol} + \text{fatty acyl-CoA} \rightarrow \text{cholesteryl ester} + \text{CoASH}$$

3.3.4.3 *Absorption of Bile Acids.* Bile salts are not absorbed together with the products of hydrolysis of dietary triglycerides, phospholipids, and cholesteryl esters, and they are not incorporated into chylomicrons. Instead, bile salts remain in the intestinal lumen until they reach the distal ileum (Fig. 3-8), where most are absorbed by an active transport mechanism that utilizes a Na^+-bile salt cotransport

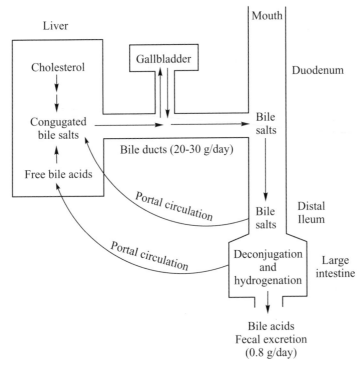

FIGURE 3-8   Enterohepatic circulation of bile salts.

system. The bile salts are transported through the portal vein to the liver, where they are extracted from the circulation by hepatocytes and then secreted back into bile. Specific transporters on both ileal and hepatic cells are required for this process. This enterohepatic circulation results in the secretion and reabsorption of the same pool of bile salts some 4 to 10 times a day, thus enabling the bile salts to be efficient promoters of fat digestion and absorption.

Those bile salts that are not reabsorbed in the ileum pass to the large intestine, where they are deconjugated through the hydrolytic removal of glycine or taurine. Bacterial metabolism also produces secondary bile acids, which have one less hydroxyl group than that of their respective primary bile acids. Some of these secondary bile acids are reabsorbed from the large intestine and returned to the liver, where they are reconjugated and reutilized. The remainder of the bile salts, approximately 0.8 g/day, is excreted in the feces.

3.3.4.4 Digestion and Absorption of Triacylglycerols Containing Medium-Chain Fatty Acids.
The digestion and absorption of triacylglycerols containing short- and medium-chain fatty acids differs in several ways from that of the more common triacylglycerols that contain long-chain fatty acids. First, additional lipases are available for the hydrolysis of the shorter-chain fatty acids. Gastric lipase

preferentially hydrolyzes triacylglycerols in breast milk and some tropical oils (e.g., coconut) that contain large amounts of medium-chain fatty acids. Nonhydrolyzed triacylglycerols containing medium-chain (C6–C12) fatty acids are also absorbed intact into cells of the intestinal mucosa, where they are hydrolyzed by a mucosal lipase. Collectively, gastric- and mucosal-catalyzed lipolysis permits utilization of medium-chain triglycerides as a dietary lipid in persons who produce insufficient amounts of pancreatic lipase (e.g., patients with cystic fibrosis). Medium-chain triacylglycerols, such as those in coconut oil and human milk, can also be digested and absorbed in the absence of bile salts, although the presence of bile salts does enhance their absorption.

Since short- and medium-chain fatty acids are more soluble than C16 and C18 fatty acids, they can be transported as free fatty acids in the portal blood and are not reesterified and incorporated into chylomicrons. Thus, people with a reduced capacity for hydrolyzing triacylglycerols in circulating chylomicrons are sometimes prescribed diets in which triacylglycerols that contain medium-chain fatty acids are used in place of common dietary fats.

3.4 DIGESTION AND ABSORPTION OF PROTEINS

3.4.1 Substrates for Protein Digestion

The proteases of the digestive tract hydrolyze both exogenous or dietary proteins and endogenous proteins. Endogenous proteins include the proteases themselves as well as the proteins derived from the lining of the gastrointestinal tract. In fact, the amino acids absorbed by an average person are derived almost equally from endogenous protein (70 g/day) and dietary protein (60 to 90 g/day).

3.4.2 Enzymes That Contribute to Protein Digestion

3.4.2.1 Proteases. Proteases hydrolyze internal peptide bonds of polypeptides, producing smaller peptides and polypeptides (Fig. 3-9). The proteases involved in digestion are relatively specific for the amino acid side chain, designated R′ in Figure 3-9.

Pepsin, which is secreted by the stomach and active at acidic pH, is a relatively nonspecific protease that recognizes the R group of many different amino acids, including those that are dicarboxylic (Asp, Glu), aromatic (Phe, Tyr), or contain large, bulky side groups (Leu, Met). Pepsin can digest as much as 10 to 20% of the protein in a meal. Hydrolysis of dietary collagen by pepsin also facilitates the subsequent access of pancreatic proteases to proteins in ingested meats.

The pancreas secretes several proteases, each with its own particular substrate specificity. Trypsin cleaves peptide bonds on the C-terminal side of the basic amino acids Arg and Lys, whereas chymotrypsin cleaves peptide bonds on the C-terminal side of Leu, Met, Asn, and the aromatic amino acids Phe and Tyr. Elastase cleaves on the C-terminal side of amino acids that have a small side chain, such as Ala, Gly, and Ser.

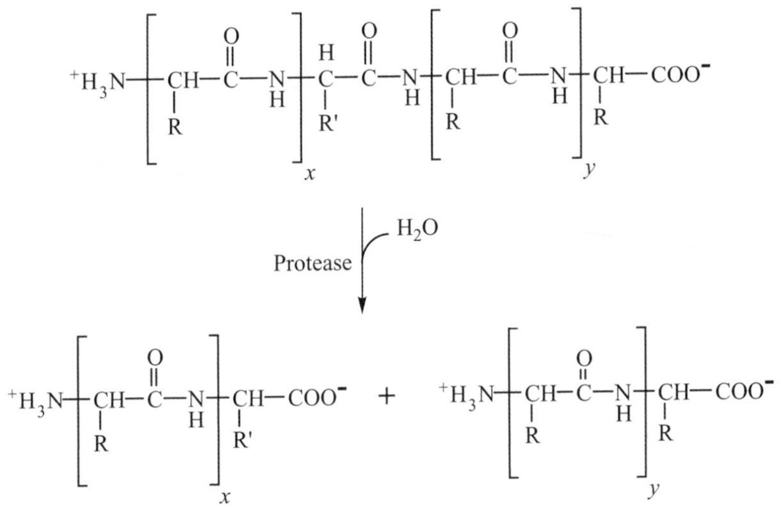

FIGURE 3-9 Hydrolysis of dietary proteins. R′ denotes the side chain of an amino acid for which the particular protease is selective.

Activation of Proteases. Pepsin is secreted in its zymogen form, pepsinogen, and is then converted to the active protease by HCl. Once activated, pepsin can hydrolyze other molecules of pepsinogen to generate additional molecules of pepsin.

Pancreatic trypsinogen is activated by the hydrolytic action of enteropeptidase, a protease that is synthesized by the brush-border cells of the small intestine. Once activated, trypsin can activate additional molecules of trypsinogen as well as other pancreatic enzymes:

$$\text{trypsinogen} \rightarrow \text{trypsin}$$

$$\text{chymotrypsinogen} \rightarrow \text{chymotrypsin}$$

$$\text{proelastase} \rightarrow \text{elastase}$$

In all cases, activation of the proenzyme involves hydrolysis of one or more peptide bonds, which results in the release of a segment of the polypeptide chain and permits the enzyme to assume a three-dimensional conformation that has a correctly configured active site.

Pancreatic Trypsin Inhibitor. The pancreas also secretes a small (6-kDa) protein called *pancreatic trypsin inhibitor* that binds very tightly to the active site of trypsin. Pancreatic trypsin inhibitor blocks the activity of any trypsin that may have resulted from premature conversion of trypsinogen to trypsin. This inhibitor thus acts to prevent a few active trypsin molecules from activating the full range of pancreatic digestive enzymes, which would otherwise damage the pancreas or pancreatic ducts.

3.4.2.2 *Carboxypeptidases.* Pancreatic juice also contains carboxypeptidases A and B, which are zinc-dependent exopeptidases that cleave peptide bonds and release amino acids one at a time from the C-terminal end of peptides. Both enzymes are secreted as zymogens and activated by trypsin. Carboxypeptidase A is specific for amino acids with hydrophobic side chains (e.g., valine, phenylalanine), whereas carboxypeptidase B is specific for basic amino acids (e.g., lysine, arginine).

3.4.2.3 *Aminopeptidases.* Cells of the intestinal mucosa produce a number of intra- and extracellular aminopeptidases which release amino acids one at a time from the N-terminal end of peptide chains.

3.4.3 Absorption of Components of Dietary Protein

Enterocytes of the small intestine absorb both amino acids and oligopeptides, particularly dipeptides and tripeptides. Indeed, oligopeptides may account for as much as two-thirds of the absorbed amino acids. There are numerous transport systems on the apical surface of the enterocyte for amino acids and peptides. Many but not all of these transport systems require cotransport of sodium. Once inside the enterocytes, the peptides are hydrolyzed to free amino acids by intracellular aminopeptidases. Free amino acids are then transported across the basolateral membrane and enter the blood.

3.5 DIGESTION AND ABSORPTION OF MICRONUTRIENTS

The body also absorbs a variety of vitamins and minerals that are termed *micronutrients* because they are required in relatively small quantities. Many of these micronutrients require specific mechanisms for absorption. In some cases, digestive processes are also required to release a bound cofactor such as vitamin B_{12} from the protein to which it is bound. The bioavailability and absorption of folate and vitamin B_{12} are discussed in Chapter 22, and the regulation of iron absorption is discussed in Chapter 23. Processes of digestion and absorption of other selected micronutrients are outlined briefly below.

3.5.1 Fat-Soluble Vitamins

Vitamins A (retinol), D (cholecalciferol), E (α-tocopherol), and K (phylloquinone and menaquinone) are lipids with limited solubility in water. In the gastrointestinal tract they are solubilized by bile salts, incorporated into mixed micelles along with the products of lipid digestion, and internalized by the intestinal mucosa. Once inside the enterocytes, the fat-soluble vitamins are incorporated into chylomicrons for transport through the lymph into the blood and eventually to the liver. Thus, conditions that impair the digestion and absorption of dietary lipids, particularly the absence of bile salts, will also compromise the absorption of fat-soluble vitamins to an extent that could lead to deficiencies of these vitamins.

β-Carotene and related retinoids are also lipids and require bile salts and mixed micelle formation for absorption. Once inside the enterocyte, β-carotene is cleaved

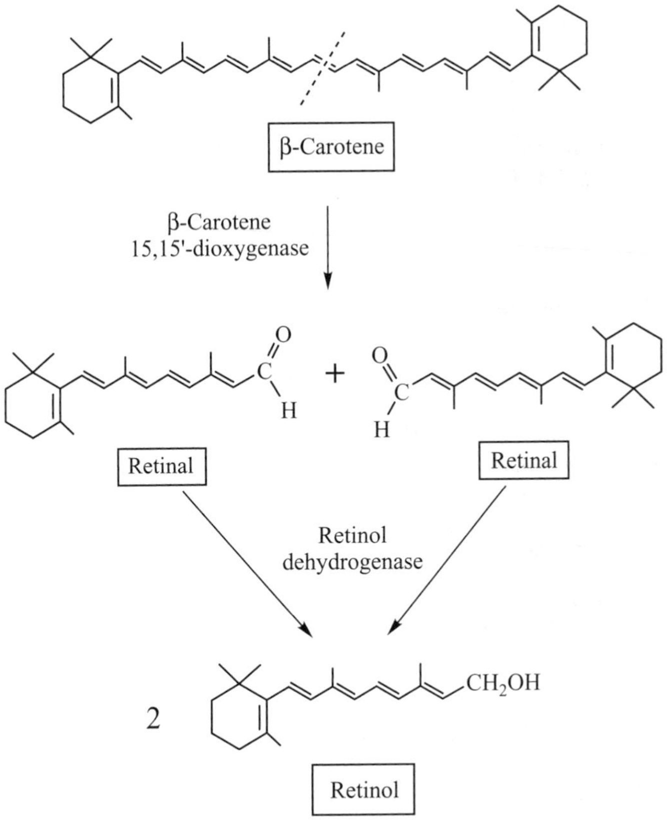

FIGURE 3-10 Hydrolysis of dietary β-carotene and the reduction of the resultant retinal to retinol.

by 15,15′-carotene dioxygenase to two molecules of *all-trans*-retinal, which are then reduced by NADPH-dependent retinol dehydrogenase to *all-trans*-retinol and incorporated into chylomicrons for transport throughout the body (Fig. 3-10).

3.5.2 Absorption of Zinc and Copper Ions

Levels of Zn^{2+} and Cu^{2+} in the body are regulated primarily by the extent of their absorption from the gut. Digestion of proteins is required to release both of these divalent cations from protein-bound dietary sources.

Once inside the enterocytes, Zn^{2+} is initially bound to cysteine-rich intestinal proteins (CRIPs), which serve as intracellular binding proteins for divalent cations. Increased plasma concentrations of zinc lead to increased synthesis of thionein, a low-molecular weight, cysteine-rich protein that binds zinc and other divalent cations. The resulting Zn^{2+}–thionein complex (metallothionein) sequesters Zn^{2+} within the enterocyte and limits its transport across the basolateral membrane into the plasma.

At the end of their lifespan, enterocytes are sloughed, returning the Zn^{2+} to the lumen of the intestine, where it is eventually excreted in the feces. This process serves to prevent absorption of excess zinc by the body.

Thionein also binds absorbed Cu^{2+} ions and prevents excess absorption of copper. Since zinc ions induce synthesis of thionein, excess dietary or pharmaceutical intakes of zinc increase the sequestration of copper ions within the enterocytes, which can lead to copper deficiency.

3.6   REGULATION OF DIGESTION

The gastrointestinal (GI) tract is a major endocrine organ. The overall function of the hormones secreted by the gut is to optimize digestion and absorption of nutrients from the gut by regulating GI motility and secretory processes. Following is a brief description of the role that some of these hormones play in digestion and absorption.

3.6.1   Gut Hormones

3.6.1.1   Gastrin. Gastrin regulates HCl secretion by the stomach and has a growth-promoting effect on the gastric mucosa. Histamine and acetylcholine also promote HCl secretion by ligand receptor–dependent mechanisms.

3.6.1.2   Cholecystokinin (CCK). Cholecystokinin stimulates secretion of pancreatic enzymes as well as contraction of the gallbladder, which enhances bile flow. It is secreted by endocrine cells located mainly in the duodenum.

3.6.1.3   Secretin. Secretin is a small polypeptide secreted by endocrine cells in the small intestine in response to a low pH (<5). It stimulates secretion of pancreatic juice containing digestive enzymes and sodium bicarbonate, which neutralizes gastric acid.

3.7   ABNORMAL FUNCTIONING OF DIGESTION AND ABSORPTION

3.7.1   Lactase Deficiency

Lack of the enzyme lactase leads to lactose intolerance (i.e., development of diarrhea and gaseous abdominal distension following ingestion of lactose or milk sugar). Congenital lactase deficiency is a rare condition characterized by a total lack of lactase activity. More commonly, the inability of adults to tolerate lactose occurs due to lactase nonpersistence (a.k.a. lactose intolerance), in which a person is born producing sufficient lactase to digest milk sugar, but within the first decade of life gradually loses the ability to produce the enzyme. Lactose nonpersistence is actually the normal condition in humans and other mammals. Multiple occurrences of genetic mutations in the promoter region of the lactase gene enabled some members of cattle-raising populations in north-central Europe and sub-Saharan Africa to consume milk as well as meat. The mutations conveyed a powerful survival advantage and were thus subject to positive genetic selection. Loss of lactase expression may also occur secondary to disorders that damage the normal structure and function of the intestinal mucosa, such as acute diarrheal disease (e.g., enteritis), gastrointestinal parasites (e.g., giardiasis), enteropathies (e.g., celiac disease), and chronic inflammatory bowel disease (e.g., Crohn's disease).

3.7.2 Celiac Disease

Celiac disease, also called *celiac sprue*, *nontropical sprue*, or *gluten-sensitive enteropathy*, is an autoimmune enteropathy characterized by intestinal inflammation and malabsorption following ingestion of gliadin, a component of a family of wheat proteins called *glutens*. In celiac disease there is villous atrophy ("flattening"), crypt hyperplasia, and accumulation of lymphocytes in the connective tissue immediately under the intestinal epithelium. Patients with celiac disease produce antibodies not only to gliadin but also to other proteins present in connective tissue surrounding smooth muscle cells in the intestinal wall. Loss of the intestinal villus and the enzymes associated with it deprives the gastrointestinal tract of important digestive enzymes (e.g., lactase). Impaired functioning of enterocytes also results in malabsorption of the products of digestion, especially amino acids, fatty acids, and fat-soluble vitamins, as well as minerals (e.g., copper, calcium). The disease can be treated with a gluten-free diet.

3.7.3 Gallstones

Gallstones are solids that form when crystals of cholesterol or bile pigments precipitate out of the liquid stored in the gallbladder. Stones that remain in the gallbladder and do not cause blockage are said to be *silent*. However, when the stones lodge in the ducts that carry bile from the liver to the small intestine, they can lead to inflammation of the ducts, the gallbladder, the pancreas, or the liver. Gallstone attacks usually occur after high-fat meals but may also occur in the middle of the night. Acute inflammation can be extremely painful and chronic obstruction can lead to life-threatening pancreatic or liver disease.

Approximately 80% of gallstones are composed of cholesterol, which forms large yellow-green crystals or multiple tiny sandlike particles. They are most likely to form when bile contains too much cholesterol or not enough bile salts, or when the gallbladder does not empty as rapidly as it should. The other 20% of gallstone cases result from formation of solid precipitates in which the major component is bilirubin, the breakdown product of hemoglobin, which gives the stool its dark color.

The most common treatment for both types of gallstones involves surgical removal of the gallbladder. Digestion of a fatty meal proceeds relatively normally even in the absence of a gallbladder, since bile flows out of the liver directly into the small intestine. Oral treatment with bile salts can dissolve small cholesterol stones, thereby obviating the need for surgery. Since bile salt therapy requires months of treatment and is often followed by reoccurrence of stones, it is reserved for persons for whom surgery is not an appropriate option.

3.7.4 Steatorrhea

Impaired functioning of any of the components of lipid digestion and absorption can result in steatorrhea or excretion of fat in foul-smelling, bulky stools, and poor absorption of fat-soluble vitamins.

3.7.4.1 Impaired Hydrolysis of Triacylglycerol. Many conditions impair hydrolysis of dietary triacylglycerols. One of the most common of these conditions is chronic pancreatitis, which can result in decreased secretion of pancreatic lipase. Decreased intestinal activity of pancreatic lipase is also observed in patients with gastrinomas or other conditions that result in excess production of gastric HCl. Steatorrhea is also a common side effect of the diet drug Xenical (orlistat), which inhibits pancreatic lipase activity.

3.7.4.2 Insufficient Secretion of Bile. When blockage of the bile duct reduces secretion of bile into the small intestine, the stools appear gray rather than reddish brown, reflecting the lack of excreted bile pigments. Lack of bile salts results in steatorrhea with excretion of free fatty acids rather than triacylglycerol in the stool. This occurs because bile salts are required primarily for absorption of dietary fat and because significant hydrolysis of triacylglycerol by pancreatic lipase is possible even in the absence of bile salts. Bile salt insufficiency also results in impaired absorption of other dietary lipids, including the fat-soluble vitamins.

3.7.4.3 Impaired Absorption by the Intestinal Mucosa. As indicated above, celiac disease or glutin-sensitive enteropathy results in malabsorption of dietary fat as well as other nutrients. Malabsorption of lipids may also occur as a result of mucosal inflammation, cystic fibrosis, bacterial overgrowth syndrome, and surgical resection of the small intestine.

3.7.5 Hypergastrinemia

Excessive secretion of gastrin (hypergastrinemia) is the cause of Zollinger–Ellison syndrome. The hallmark of this syndrome is gastric and duodenal ulceration due to excessive and unregulated secretion of gastric acid. Hypergastrinemia can also be caused by gastrin-secreting tumors that develop in the pancreas or duodenum or by infection with *Helicobacter pylori*, which induces mucosal inflammation.

3.7.6 Parenteral Feeding

There are many clinical conditions in which patients benefit from nutritional support which is parenteral, in that it completely bypasses the digestive system. They include various severe malabsorptive syndromes, such as necrotizing colitis in infants, severe short bowel syndrome, and mechanical obstruction not immediately remediable by surgery. Since nutrition is provided intravenously rather than via the digestive tract, it is necessary to provide these patients with glucose rather than starch, and with amino acids rather than protein. By contrast, parenteral nutrition may include lipid emulsions (e.g., Intralipid, Liposyn) containing triacylglycerols stabilized by a surface layer of phospholipids. The lipid particles in these lipid emulsions are similar in size to chylomicrons, and like chylomicrons, are hydrolyzed in the blood, releasing free fatty acids and glycerol. Parenteral nutrition solutions must also supply essential minerals and vitamins.

CHAPTER 4

GLYCOLYSIS

4.1 FUNCTIONS OF GLYCOLYSIS

Glycolysis is a metabolic pathway that cleaves glucose into two molecules of pyruvate or lactate. During glycolysis, some of the energy in the glucose molecule is converted into ATP. Although glycolysis to pyruvate is an oxidative process, it is not dependent on molecular oxygen. By contrast, the subsequent fate of pyruvate depends on the presence of both mitochondria and sufficient oxygen. In the presence of oxygen, the product of glycolysis is pyruvate, which is further oxidized by the pyruvate dehydrogenase enzyme complex and the dehydrogenases of the tricarboxylic acid (TCA) cycle. In the absence of sufficient oxygen, or in red blood cells, where mitochondria are absent, lactate is the final product of glycolysis and there is no net oxidation. Glycolysis is not only the major pathway for oxidizing glucose, but also the main pathway for metabolizing other dietary sugars, such as galactose and fructose.

Once glucose has been trapped inside a cell in the form of glucose 6-phosphate, there may be as many as three metabolic options available for the glucose moiety (Fig. 4-1). In a hepatocyte, for example, glucose can be oxidized via glycolysis for the primary purpose of ATP production, stored as glycogen, or oxidized in the pentose phosphate pathway to generate NADPH and ribose for nucleic acid synthesis. Red blood cells, on the other hand, cannot synthesize glycogen; they can, however, metabolize glucose through the pentose phosphate pathway or through glycolysis.

Medical Biochemistry: Human Metabolism in Health and Disease   By Miriam D. Rosenthal and Robert H. Glew
Copyright © 2009 John Wiley & Sons, Inc.

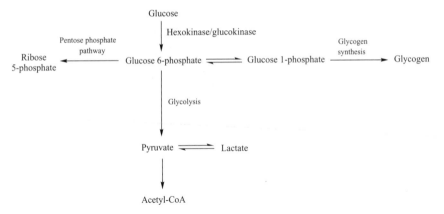

FIGURE 4-1   Three possible metabolic fates of glucose: glycolysis, the pentose phosphate pathway, and glycogen synthesis.

4.1.1   Glycolysis Provides Energy

The main function of glycolysis is energy (ATP) production. The conversion of one molecule of glucose to pyruvate or lactate is associated directly with the net production of two ATP molecules. Since the maximum number of ATP molecules that can be realized from the complete oxidation of one molecule of glucose to CO_2 and water is 30 to 32, the ATP yield from glycolysis is clearly relatively small.

Generation of energy via glycolysis is oxygen independent. In the presence of oxygen in tissues such as muscle and liver that contain mitochondria, for example, the end product of glycolysis is pyruvate. In contrast, in red blood cells lacking mitochondria or in mitochondria-containing tissues that are insufficiently oxygenated, lactate is the end product of glycolysis.

4.1.2   Glycolysis Provides Substrate for Further Oxidation

Pyruvate, the end product of aerobic glycolysis, can be further oxidized in mitochondria. The pathway (described in Chapter 5) involves the pyruvate dehydrogenase–catalyzed oxidation of pyruvate to acetyl-CoA and CO_2, followed by subsequent oxidation of acetyl-CoA to CO_2 and water in the TCA cycle.

4.1.3   Intermediates and Products of Glycolysis Can Provide Substrates for Other Pathways

Tissues such as adipose and liver that have a high capacity for triacylglycerol synthesis contain glycerol 3-phosphate dehydrogenase, which converts dihydroxyacetone phosphate (DHAP) into glycerol 3-phosphate, which is a critical substrate in the pathways of triacylglycerol and glycerophospholipid (i.e., phospholipid) synthesis:

$$\text{dihydroxyacetonephosphate} + \text{NADH} + \text{H}^+ \rightleftharpoons \text{glycerol 3-phosphate} + \text{NAD}^+$$

Red blood cells use 1,3-bisphosphoglycerate, another intermediate in glycolysis, to generate 2,3-bisphosphoglycerate, which is an allosteric regulator of the interaction of oxygen with hemoglobin. The interconversion of these two bisphosphoglycerates is catalyzed by bisphosphoglycerate mutase:

$$1,3\text{-bisphophoglycerate} \rightleftharpoons 2,3\text{-bisphosphoglycerate}$$

Pyruvate, the end product of aerobic glycolysis, can acquire an amino group by transamination, thus producing the amino acid alanine. In addition, acetyl-CoA, produced by the mitochondrial oxidation of pyruvate, is a substrate for the synthesis of both fatty acids and cholesterol.

4.2   LOCALIZATION OF GLYCOLYSIS

Glucose is the universal fuel in humans in the sense that literally every type of cell in the body possesses the glycolytic pathway in its cytosol and can therefore metabolize glucose at least to the level of pyruvate or lactate. Although most cells can also utilize fatty acids as an energy source, some cells, such as erythrocytes and those in both the lens and cornea of the eye, contain few or no mitochondria and rely on glycolysis for essentially all of their ATP production. Even though brain cells do contain mitochondria, the impermeability of the blood–brain barrier to most long-chain fatty acids prevents fatty acids from being an important fuel source for the brain. Thus, although the brain is a highly aerobic organ that derives much of its ATP from the oxidation of acetyl-CoA in the tricarboxylic acid cycle, under all circumstances the brain metabolizes large quantities of glucose via glycolysis to generate pyruvate, which can then be oxidized to acetyl-CoA.

4.3   PHYSIOLOGICAL AND PATHOPHYSIOLOGICAL CONDITIONS IN WHICH GLYCOLYSIS IS ESPECIALLY ACTIVE

4.3.1   Fed State

Glycolysis is especially active in the fed state when the body is actively digesting, absorbing, and processing nutrients. Insulin, which is secreted by the β-cells of the pancreas in response to elevated postprandial blood glucose levels, stimulates glucose metabolism in muscle, liver, and fat cells, but not in the brain.

4.3.2 Exercising Muscle

Although muscle cells at rest derive most of their energy from the oxidation of fatty acids, exercising muscle oxidizes glucose as well as fatty acids. As the intensity of exercise increases, progressively more of the energy in muscle will be derived from the oxidation of glucose.

There are two sources of glucose for muscle: localized glycogen stores within the muscle and glucose extracted from the blood. In vigorously exercising muscle, the demand for oxygen can become so great as to outstrip the oxygen supply. Under these conditions, the muscle relies extensively on glycolysis to satisfy its ATP needs, and lactate production increases.

4.3.3 Cancer Cells

Cancer cells consume glucose at a much higher rate and produce much more lactic acid than their normal counterparts, even under aerobic conditions. This phenomenon is known as the *Warburg effect*. A large fraction of the increased ATP produced by glycolysis in cancer cells is used for fatty acid, protein, and DNA synthesis, all three of which are increased in cancer cells. The Warburg effect also provides tumors with the large amounts of lactate and pyruvate that are precursors to the acetyl-CoA substrate that fatty acid synthesis requires.

4.4 PATHWAY OF GLYCOLYSIS AND RELATED REACTIONS

4.4.1 Uptake of Glucose into Cells Is Facilitated by Tissue-Specific Glucose Transporters

A family of glucose transporters (designated GLUT1 through GLUT5) facilitates movement of glucose across the plasma membrane. Glucose enters cardiac and skeletal muscle cells as well as adipocytes via the insulin-stimulated glucose-4 transporter designated GLUT4. In the absence of a strong insulin signal (i.e., a low insulin/glucagon ratio), GLUT4 is bound to intracellular vesicles. Following insulin stimulation, GLUT4-containing vesicles in muscle and adipocytes translocate to and fuse with the plasma membrane, thus providing the mechanism by which insulin stimulates uptake of glucose from the blood (Fig. 4-2). Glucose transport via GLUT4 is a major regulatory step in glucose metabolism in muscle and adipocytes.

GLUT2, which is constitutively present on the plasma membrane of liver cells, has a lower affinity for glucose then does GLUT4, but is expressed in such abundance that intracellular glucose equilibrates essentially instantaneously with glucose in the blood. As a result, the hepatic flux of glucose through glycolysis is proportional to the glucose concentration in the circulation.

Red blood cells, brain, and kidney contain GLUT1 and GLUT3 transporters, which, like GLUT2, are insulin-independent. Unlike GLUT2, GLUT1 and GLUT3 have a high affinity for glucose, thus promoting glucose uptake in the fasted state.

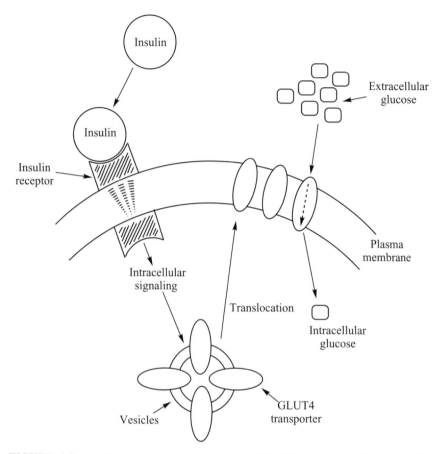

FIGURE 4-2   Insulin promotes the translocation of GLUT4 transporters from intracellular vesicles to the plasma membrane of adipocytes and muscle cells.

GLUT5 is expressed at high levels in the small intestine, where it functions primarily as a transporter of fructose rather than of glucose.

4.4.2 Trapping Glucose Intracellularly

Once glucose enters a cell it must be trapped; otherwise, it will diffuse back into the blood. All cells, including hepatocytes, contain hexokinase, which phosphorylates glucose, thereby trapping it in the cytosol (Fig. 4-3):

$$\text{glucose} + \text{ATP} \rightarrow \text{glucose 6-phosphate} + \text{ADP}$$

Phosphorylation of glucose also serves to activate the sugar for metabolism. As more and more glucose is trapped in hepatocytes, however, the concentration

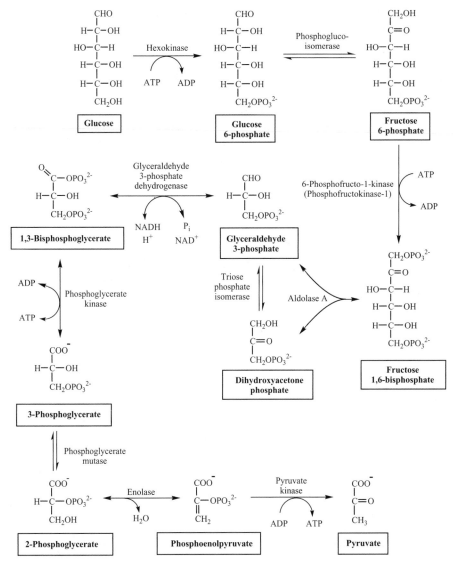

FIGURE 4-3 Glycolytic pathway. Although the structures of the sugars are illustrated in their open-chain form, in solution they exist in their respective ring forms.

of glucose 6-phosphate increases to the point where it inhibits hexokinase. This phenomenon is an example of product inhibition. If hepatocytes had no other glucose-trapping enzyme than hexokinase, the liver would soon cease extracting glucose from the blood and the body would experience a period of prolonged hyperglycemia. The problem is solved by the induction of the enzyme glucokinase by insulin.

As described in Chapter 2, glucokinase and hexokinase constitute an isoenzyme pair. Unlike hexokinase, glucokinase is not inhibited by glucose 6-phosphate. The function of glucokinase is to trap glucose when the blood glucose concentration rises after a meal.

The reaction catalyzed by both hexokinase and glucokinase is physiologically irreversible. Therefore, hepatocytes cannot run the hexokinase reaction in reverse to release the glucose moiety of glucose 6-phosphate generated when glycogen is broken down (glycogenolysis) or during gluconeogenesis. A separate enzyme called glucose 6-phosphatase dephosphorylates glucose 6-phosphate, thereby releasing free glucose that can leave the cell and enter the blood:

$$\text{glucose 6-phosphate} + H_2O \rightarrow \text{glucose} + P_i$$

4.4.3 The Individual Steps of Glycolysis

The first step following phosphorylation and trapping of glucose is the isomerization of glucose 6-phosphate to fructose 6-phosphate (Fig. 4-3). This reversible reaction is catalyzed by glucose 6-phosphate isomerase:

$$\text{glucose 6-phosphate} \rightleftharpoons \text{fructose 6-phosphate}$$

The next step in the pathway, which is catalyzed by phosphofructokinase-1 (PFK-1), is irreversible and commits fructose 6-phosphate to glycolysis. PFK-1 also represents the major regulated step in the glycolytic pathway:

$$\text{fructose 6-phosphate} + \text{ATP} \rightarrow \text{fructose 1,6-bisphosphate} + \text{ADP}$$

Next, aldolase A cleaves fructose-1,6-bisphosphate into two three-carbon fragments (called *triose phosphates*), glyceraldehyde 3-phosphate and dihydroxyacetone phosphate (DHAP):

$$\text{fructose-1,6-bisphosphate} \rightleftharpoons \text{glyceraldehyde 3-phosphate}$$
$$+ \text{dihydroxyacetone phosphate}$$

Since aldolase-type reactions are reversible, aldolase A can also participate in the pathway that is essentially the reverse of glycolysis: namely, *gluconeogenesis*. The liver contains a second aldolase, designated aldolase B, which participates in the metabolism of fructose. Aldolase A and aldolase B are not isozymes.

Instead of utilizing two separate pathways for converting each of the trioses from the aldolase A reaction into pyruvate, nature has evolved a more economical strategy that involves the conversion of one of the trioses, dihydroxyacetone phosphate, into the other, glyceraldehyde 3-phosphate. The enzyme that accomplishes this freely reversible interconversion of the two triose phosphates is triosephosphate isomerase:

dihydroxyacetone phosphate $\rightleftharpoons$ glyceraldehyde 3-phosphate

The next step is the only oxidation–reduction reaction of glycolysis: namely, the NAD^+-dependent oxidation of glyceraldehyde 3-phosphate to 1,3-bisphosphoglycerate. This reversible reaction is catalyzed by glyceraldehyde 3-phosphate dehydrogenase:

glyceraldehyde 3-phosphate $+ NAD^+ + P_i \rightleftharpoons$ 1,3-bisphosphoglycerate

$+ NADH + H^+$

The glyceraldehyde 3-phosphate dehydrogenase reaction couples an oxidation–reduction reaction with a reaction that incorporates inorganic phosphate (P_i) into an organic compound. The overall reaction generates a high-energy phosphate ester linkage whose bond energy is greater than that of the terminal (γ) phosphate of ATP. The energy in the phosphoanhydride bond in 1,3-bisphosphoglycerate can then be transferred to ADP in a reversible reaction catalyzed by phosphoglycerate kinase:

1,3-bisphosphoglycerate $+ ADP \rightleftharpoons$ 3-phosphoglycerate $+ ATP$

Next, phosphoglyceromutase catalyzes the transfer of the phosphate group from the 3-position to the 2-position of phosphoglycerate:

3-phosphoglycerate $\rightleftharpoons$ 2-phosphoglycerate

Enolase then dehydrates 2-phosphoglycerate to generate phosphoenolpyruvate (PEP):

2-phosphoglycerate $\rightleftharpoons$ phosphoenolpyruvate $+ H_2O$

The phosphoenol configuration of atoms in phosphoenolpyruvate causes the carbon–oxygen–phosphorus linkage to be high energy. In fact, the free-energy change associated with the hydrolysis of the phosphate group of PEP is twofold more negative than that associated with the hydrolysis of the terminal (γ) phosphate group of ATP.

The next step in glycolysis involves the transfer of the phosphate group of PEP to ADP in a reaction catalyzed by pyruvate kinase:

phosphoenolpyruvate $+ ADP \rightarrow$ pyruvate $+ ATP$

The pyruvate kinase reaction is irreversible and therefore cannot play a role in the pathway of gluconeogenesis. As we will see in Chapter 9, surmounting the thermodynamic barrier constituted by the direct conversion of pyruvate to phosphoenolpyruvate requires two enzymatic steps, both of which utilize the energy of nucleotide triphosphates.

4.4.4 Energy Yield from Glycolysis

The direct energy yield of glycolysis starting from free glucose and ending with either pyruvate or lactate is 2 ATP. ATP is generated in two reactions in the pathway, the glycerate kinase and pyruvate kinase reactions. Since one molecule of glucose gives rise to two molecules of glyceraldehyde 3-phosphate in glycolysis, a total of four molecules of ATP are produced; however, two molecules of ATP are consumed in the conversion of free glucose into fructose 1,6-bisphosphate.

4.4.5 Glycolysis Requires NAD$^+$ and Inorganic Phosphate

No matter how available glucose is, glycolysis will not function if the concentrations of two other critical substrates, NAD$^+$ and inorganic phosphate (orthophosphate or P$_i$), are suboptimal. Since inorganic phosphate is a substrate in the glyceraldehyde 3-phosphate dehydrogenase reaction, glycolysis will cease to function if sufficient inorganic phosphate is not available in the cytosol. In certain metabolic diseases, such as hereditary fructose intolerance, where cellular phosphate accumulates in a sugar phosphate form, the concentration of inorganic phosphate becomes too low to support the glyceraldehyde 3-phosphate dehydrogenase reaction, thereby reducing the rate of glycolysis.

Similarly, if cytosolic NADH is not reoxidized to NAD$^+$, glycolysis will not be possible. There are two ways that a cell can regenerate NAD$^+$: One depends on the mitochondrial electron transport system, the other on lactate dehydrogenase.

In the case of well-oxygenated tissues that contain mitochondria, the reducing equivalents in NADH are ultimately transported into mitochondria. However, since the mitochondrial inner membrane is impermeable to NADH, there is a need for alternate mechanisms for transporting the two electrons from NADH from the cytosol into the mitochondrial matrix. The two systems that transport glycolysis-derived electrons from the cytosolic NADPH into the mitochondria, the glycerol 3-phosphate shuttle and the malate–aspartate shuttle, are discussed in Chapter 6.

Alternatively, NADH can be reoxidized to NAD$^+$ by lactate dehydrogenase catalyzes the following reversible reaction (Fig. 4-4):

$$\text{pyruvate} + \text{NADH} + \text{H}^+ \rightleftharpoons \text{lactate} + \text{NAD}^+$$

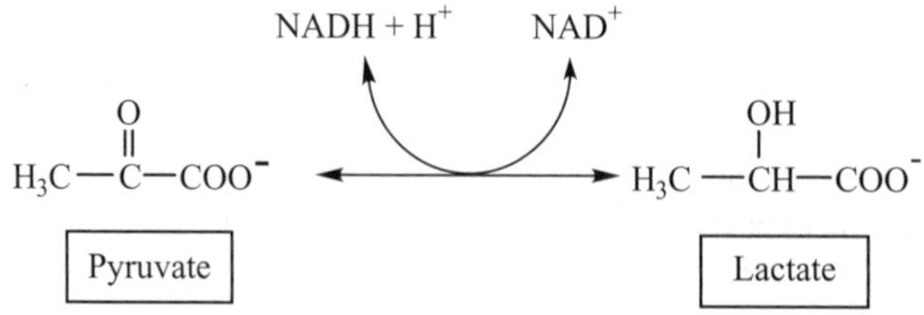

FIGURE 4-4 Lactate dehydrogenase reaction.

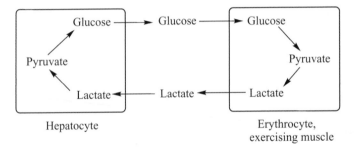

FIGURE 4-5 Pyruvate/lactate (Cori) cycle.

When glycolysis produces lactic acid, the process is referred to as *anaerobic glycolysis*. Lactate is produced both by red blood cells, which lack mitochondria, and by vigorously exercising muscle when the oxygen demand outstrips the oxygen supply. In both cases, lactate is transported in the blood to the liver, where it can provide substrate for gluconeogenesis. The newly synthesized glucose molecules can then be secreted by the liver into the blood, where they can be taken up and oxidized by red cells and muscle. This process is known as the *Cori cycle* (Fig. 4-5).

4.4.6 Metabolism of Fructose

Fructose and glucose are the monosaccharide components of sucrose, which is hydrolyzed by the enzyme sucrase that is localized on the outer surface of the brush border of the small intestine. Fructose is also present in honey and many fruits. Seminal vesicles secrete fructose into seminal fluid, where it functions as the major fuel for sperm cells.

Following its absorption into the blood, dietary fructose is extracted by the liver, where it is metabolized by glycolysis. Under normal circumstances, fructose is phosphorylated and trapped inside hepatocytes by fructokinase (Fig. 4-6):

$$fructose + ATP \rightarrow fructose\ 1\text{-phosphate} + ADP$$

Fructose can also be phosphorylated by hexokinase, but since the K_m of hexokinase for fructose is very high, hexokinase acts on fructose only when the fructose concentration is elevated abnormally. In contrast to fructokinase, which produces fructose 1-phosphate, hexokinase acting on fructose produces fructose 6-phosphate.

Fructose 1-phosphate is split by aldolase B into a triose (glyceraldehyde) and a triose phosphate (dihydroxyacetone phosphate):

$$fructose\ 1\text{-phosphate} \rightleftharpoons glyceraldehyde + dihydroxyacetone\ phosphate$$

Unlike aldolase A, which is present in all cells of the body, aldolase B is found only in liver, kidney, and the small intestine.

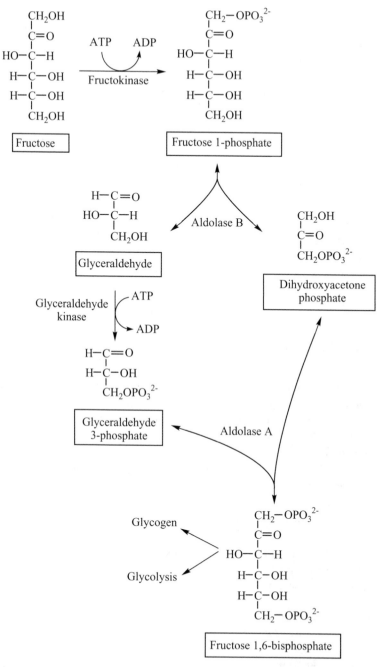

FIGURE 4-6   Metabolism of fructose to fructose 1,6-bisphosphate.

Glyceraldehyde arising from the aldolase B reaction is phosphorylated and trapped inside the hepatocyte by the action of glyceraldehyde kinase:

$$\text{glyceraldehyde} + \text{ATP} \rightarrow \text{glyceraldehyde 3-phosphate} + \text{ADP}$$

At this point, fructose has been converted into glyceraldehyde 3-phosphate and dihydroxyacetone phosphate, two triose phosphates that are the intermediates in glycolysis.

Alternatively, in hepatocytes, glyceraldehyde may be reduced by alcohol dehydrogenase:

$$\text{glyceraldehyde} + \text{NADH} + \text{H}^+ \rightarrow \text{glycerol} + \text{NAD}^+$$

The glycerol can then be phosphorylated by glycerol kinase to generate glycerol 3-phosphate, which is then available for the synthesis of triacylglycerol and glycerophospholipids.

4.4.7 The Metabolism of Galactose

Lactose (galactosyl-β-1,4-glucose) in the milk of mammals (including humans) is the major dietary source of galactose. As described in Chapter 3, lactose is hydrolyzed in the intestine by lactase. Galactose produced by hydrolysis of dietary lactose is mostly in the form of the β-isomer. β-Galactose is first converted to α-galactose by galactose mutarotase and then phosphorylated and trapped in hepatocytes by a galactose-specific kinase called *galactokinase* (Fig. 4-7):

$$\text{galactose} + \text{ATP} \rightarrow \text{galactose 1-phosphate} + \text{ADP}$$

Unlike hexokinase and glucokinase, both of which phosphorylate the C6 hydroxyl group of glucose, galactokinase and fructokinase phosphorylate the C1 hydroxyl group of their respective sugar substrates.

Isomerization of galactose 1-phosphate to glucose 1-phosphate requires two reactions. First, galactose must be incorporated into a uridine-based sugar nucleotide in a reaction catalyzed by uridyltransferase (formal name, UDP-glucose:galactose 1-phosphate uridyltransferase):

(1) galactose 1-phosphate + UDP-1-glucose $\rightleftharpoons$ UDP-1-galactose

+ glucose 1-phosphate

The second step in the generation of glucose 1-phosphate is the epimerization of the galactose moiety. UDP-glucose 4-epimerase inverts the hydroxyl group on C4 of the galactose residue in UDP-1-galactose, thereby converting UDP-1-galactose into UDP-1-glucose (Fig. 4-7). NAD$^+$ is a cofactor in this epimerase reaction:

(2) UDP-1-galactose $\rightleftharpoons$ UDP-1-glucose

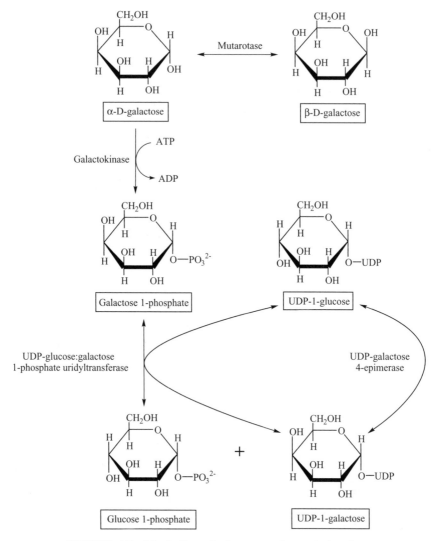

FIGURE 4-7   Metabolism of galactose to glucose 1-phosphate.

The net effect of reactions (1) and (2) is

$$\text{galactose 1-phosphate} \rightleftharpoons \text{glucose 1-phosphate}$$

Thus, UDP-1-glucose functions essentially as a catalyst in the pathway of galactose metabolism. It is consumed in the uridyltransferase reaction and regenerated in the UDP-galactose 4-epimerase reaction.

Before glucose 1-phosphate can enter glycolysis, it must first be converted into glucose 6-phosphate in the reversible reaction catalyzed by phosphoglucomutase:

$$\text{glucose 1-phosphate} \rightleftharpoons \text{glucose 6-phosphate}$$

UDP-1-glucose is also an intermediate in the pathways that synthesize glycogen and many different glycoconjugates. Indeed, it is for these reasons that lactose is an ideal dietary sugar for infants, who relative to adults, are more dependent on hepatic glycogen stores. Since UDP-1-glucose is in the direct pathway for the synthesis of glycogen, dietary galactose is converted directly into hepatic glycogen. In contrast, fructose derived from dietary sucrose must first be metabolized to triose phosphates, which are glycolytic intermediates. In the fed state, when dietary glucose stimulates insulin secretion, hepatic gluconeogenesis is suppressed and glycolytic intermediates are oxidized to pyruvate as opposed to being used to synthesize glycogen.

Humans can also synthesize UDP-1-galactose from glucose when the diet does not contain lactose or any other source of galactose. Since both the uridyltransferase and epimerase reactions are reversible, the pathway shown in Figure 4-7 can run in either direction. Synthesis of UDP-1-galactose is particularly important in the lactating mammary gland, which uses galactosyl transferase to synthesize lactose:

$$\text{UDP-1-galactose} + \text{glucose} \rightarrow \text{lactose} + \text{UDP}$$

4.5 REGULATION OF GLYCOLYSIS

The main factor that regulates glycolysis is the energy charge of the cell. When the intracellular ATP/ADP ratio is high, glycolysis is inhibited; conversely, when the ATP concentration is low and the concentrations of ADP and AMP are high, the flux of glucose through glycolysis is increased. Similarly, when the ATP/ADP ratio is high, the TCA-cycle enzyme isocitrate dehydrogenase is inhibited, resulting ultimately in an increase in the concentration of citrate in the cytosol of the cell (see Chapter 9).

Three enzymes are involved in regulating glycolysis (Fig. 4-8): phosphofructokinase-1, hexokinase, and pyruvate kinase, each of which catalyzes an irreversible reaction. In addition, translocation of GLUT4 transporters from intracellular vesicles to the plasma membrane regulates glucose metabolism in muscle and adipocytes.

4.5.1 Phosphofructokinase-1

Phosphofructokinase-1 (PFK-1) is the major regulatory enzyme in glycolysis. Allosteric regulation of PFK-1 permits the enzyme to respond to the energy needs of the cell and to hormonal signaling by insulin and glucagon. PFK-1 is inhibited by ATP and citrate, and activated by AMP. Citrate is an intermediate in the mitochondrial TCA cycle. When the energy charge of the cell is high, ATP inhibits mitochondrial

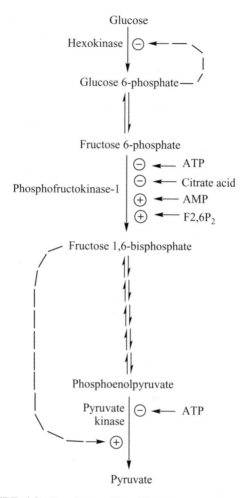

FIGURE 4-8   Regulation of hepatic glycolysis by metabolites.

isocitrate dehydrogenase, the key regulatory enzyme of the TCA cycle, resulting in a backup of the TCA cycle and accumulation of citrate.

Fructose 2,6-bisphosphate ($F2,6P_2$) is an allosteric activator of PFK-1. The intracellular concentration of $F2,6P_2$ is regulated by, and correlated directly with, the insulin/glucagon ratio. The two enzymes that determine the $F2,6P_2$ level directly are phosphofructokinase-2 (PFK-2) and fructose 2,6-bisphosphatase (FBPase-2). The activities of these two enzymes are regulated such that when PFK-2 is active, FBP-2 is inactive, and vice versa (Fig. 4-9). PFK-2 and FBPase-2 are encoded by one gene and the two catalytic activities are contained in a single polypeptide chain, hence the term *double-headed* enzyme. The activities of this double-headed enzyme system are regulated differently in different tissues.

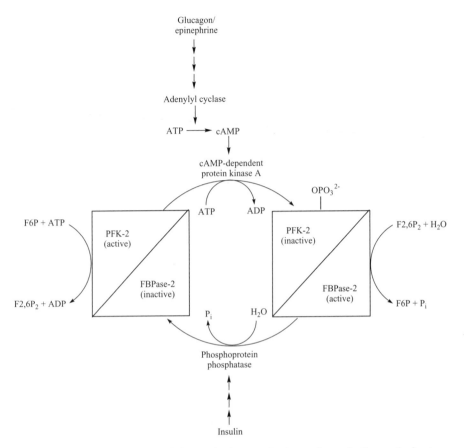

FIGURE 4-9 Hormonal regulation of the synthesis and catabolism of fructose 2,6-bisphosphate (F2,6P$_2$) in liver. PFK-2, phosphofructokinase-2; FBPase-2, fructose 2,6-bisphosphatase.

The hepatic double-headed enzyme is regulated by protein phosphorylation. As shown in Figure 4-9, when the hepatic PFK-2/FBPase-2 protein is phosphorylated by cAMP-activated protein kinase, PFK-2 is inactive and FBPase-2 is active, causing the F2,6P$_2$ level in the cell to decline. Glucagon, which activates adenylyl cyclase and thus raises the cAMP level of the cell, therefore has the effect of decreasing the F2,6P2 level, decreasing phosphofructokinase-1 activity, and ultimately slowing the flux of glucose through glycolysis. In contrast, binding of insulin to its receptor results in the activation of a phosphoprotein phosphatase that dephosphorylates the doubled-headed enzyme, thereby causing the F2,6P$_2$ concentration to rise and increasing the flux of glucose through glycolysis.

In contrast to the case in liver, the PFK-2/FBPase-2 protein in skeletal muscle has an alanine residue in place of the serine residue that is the substrate for cAMP-activated protein kinase A. As a result, the protein constitutively synthesizes F2,6P$_2$ and glycolysis is not inhibited by epinephrine-induced intracellular signaling.

A third isozyme of PFK-2/FBP-2, present in heart muscle, has multiple phosphorylation sites including one phosphorylated by AMP-activated kinase (AMPK). Energy depletion in heart muscle results in a high AMP/ATP ratio which promotes activation of AMPK. Activated AMPK in turn phosphorylates PFK-2, which increases the intracellular concentration of F-2,6-P2, thereby stimulating glycolysis and energy production.

4.5.2   Hexokinase

As discussed above, hexokinase is regulated by direct feedback inhibition by one of the products of the reaction it catalyzes: glucose 6-phosphate.

4.5.3   Pyruvate Kinase

This enzyme, which catalyzes the last step of glycolysis, is inhibited by ATP and activated by fructose-1,6-bisphosphate (Fig. 4-8). Fructose-1,6-bisphosphate activation of pyruvate kinase is an example of feed forward activation. The activity of the liver isozyme of pyruvate kinase is also regulated by phosphorylation and dephosphorylation. Glucagon stimulates cAMP synthesis in hepatocytes, causing cAMP-activated protein kinase A in turn to phosphorylate pyruvate kinase. The phosphorylated form of protein kinase is inactive.

4.6   DISEASES INVOLVING GLYCOLYTIC ENZYMES

4.6.1   Genetic Deficiencies Directly Affecting Glycolysis

An absolute deficiency of either of the two ATP-producing enzymes of glycolysis would be lethal since virtually every cell in the body depends on glycolysis for energy production. A partial lack of pyruvate kinase or phosphoglycerate kinase activity in red blood cells depletes them of ATP, thereby compromising the cells' ability to export electrolytes and maintain proper osmotic balance between the cytosol and extracellular compartment. The end result is swelling and premature destruction of red cells, which manifests as hemolytic anemia.

4.6.2   Mercury Poisoning

Many enzymes in the body contain critical sulfhydryl groups that can react with mercury ions. The sulfhydryl group of glyceraldehyde 3-phosphate dehydrogenase that is essential for catalysis has an unusually high affinity for mercury. Binding of mercury to the enzyme's active-site thiol group inactivates this critical dehydrogenase and impedes glycolysis. Mercury can enter the body through consumption of ocean fish such as swordfish and tuna, water sources that are contaminated by industrial waste, or by consumption of grains that had been treated with a mercury-containing rodenticide.

4.6.3 Arsenic Poisoning

In the body, arsenate is reduced to arsenite, which is the more toxic form of the metal. One deleterious effect of arsenite results from its ability to substitute for inorganic phosphate in the glyceraldehyde 3-phosphate dehydrogenase reaction. This results in the synthesis of an unstable 1-arseno-3-phosphoglycerate molecule, which hydrolyzes rapidly and spontaneously. The net effect of arsenic on glycolysis is to bypass the substrate-level phosphorylation reaction catalyzed by phosphoglycerate kinase, reducing the net energy yield from glycolysis to zero.

4.6.4 Essential Fructosuria

A genetic deficiency of fructokinase is a benign condition. A person who is fructo-kinase deficient will experience only a transient fructosemia (fructose in the blood) and fructosuria (fructose in the urine) following consumption of sucrose, high-fructose corn syrup, or invert sugar (an equimolar mixture of glucose and fructose). Unlike glucose, fructose is readily excreted by the kidney. In addition, hexokinase has broad substrate specificity and is capable of phosphorylating fructose as well as glucose, albeit at a much lower rate:

$$\text{fructose} + \text{ATP} \rightarrow \text{fructose 6-phosphate} + \text{ADP}$$

The fructose 6-phosphate generated by this reaction is then metabolized by the glycolytic pathway.

4.6.5 Hereditary Fructose Intolerance

In contrast to essential fructosuria, hereditary fructose intolerance is a life-threatening metabolic disorder. It is caused by a genetically based deficiency of aldolase B. Lack of aldolase B causes any ingested fructose to accumulate inside hepatocytes as fructose 1-phosphate, which has the effect of tying up intracellular phosphate in the form of a sugar phosphate. This, in turn, depletes the cell off inorganic phosphate, which is required for glycolysis in the cytosol and oxidative phosphorylation in mitochondria. Thus, a lack of aldolase B activity prevents the liver cell from making sufficient ATP from either glycolysis or mitochondrial respiration to maintain energy-requiring cellular functions.

4.6.6 Galactosemia

A genetic deficiency of any one of the enzymes of the pathway of galactose metabolism—galactokinase, uridyltransferase and galactose epimerase—will impede galactose metabolism and will result in galactosemia if the affected infant ingests lactose. However, if a deficiency of one of these enzymes is identified in the first few days of life through a genetic-disease screening program, the potentially devastating effects of the mutation can be prevented by eliminating milk and other lactose-containing foods from the diet.

The classic and most severe form of galactosemia is caused by a deficiency of galactose 1-phosphate uridyltransferase. The resulting accumulation of galactose 1-phosphate in tissues leads to multiple-organ-system pathology, including neurological damage, cataracts, coma, and eventually death. By contrast, deficiency of galactokinase causes cataract formation but is otherwise less severe than classic galactosemia. The enzyme aldose reductase plays a role in the cataract formation associated with galactosemia by reducing galactose to galactitol and trapping this polyol inside cells of the lens. High intracellular concentrations of galactitol draw water into the tissue, thereby promoting osmotic damage to the lens.

Deficiencies of UDP-1-galactose 4-epimerase prevent both utilization of exogenous galactose and endogenous galactose synthesis. For this reason, patients are usually placed on a galactose-restricted rather than a galactose-free diet. Patients may also benefit from supplementation with *N*-acetylgalactosamine, which is needed for the synthesis of glycolipids and other glycoconjugates.

4.6.7 Diabetes Mellitus

The hallmarks of diabetes mellitus are hyperglycemia and glucosuria. The transport of glucose into muscle and adipose tissue and the stimulation of glycolysis in many tissues of a person in the fed state are dependent on normal signaling by insulin. Insulin also regulates and coordinates lipid and amino acid metabolism and glycogen storage, as well as gluconeogenesis. Type I diabetes results from inadequate synthesis of insulin by the β-cells of the pancreas, while type II diabetes is characterized primarily by insulin resistance, a condition in which peripheral cells do not respond normally to insulin.

Many of the pathological consequences of diabetes are the result of the formation of covalent bonds between glucose and proteins in various tissues (e.g., proteins in the lens of the eye). Glucose is an aldose sugar; that is, it contains an aldehyde group. Aldehydes are notoriously reactive, especially with the amino groups of proteins. Nearly all plasma and tissue proteins have an unblocked amino terminus as well as multiple lysine side chains. Even at the normal blood glucose concentration of 100 mg/dL, glucose reacts nonenzymatically with these amino groups, forming irreversible covalent sugar–protein adducts, referred to generically as *glycated proteins*. Hemoglobin A1c (HbA1c) is one such glycated protein. About 5.5% of the hemoglobin in a healthy person is glycated. In a patient with diabetes, there is a positive correlation between the percentage of HbA1c and his or her average blood glucose level over time; measurement of HbA1c therefore provides a measure of the efficacy of treatments to control the diabetes.

CHAPTER 5

PYRUVATE DEHYDROGENASE AND THE TRICARBOXYLIC ACID CYCLE

5.1 FUNCTIONS OF PYRUVATE DEHYDROGENASE AND THE TRICARBOXYLIC ACID CYCLE

5.1.1 Functions of the Pyruvate Dehydrogenase Reaction

Collectively, the pyruvate dehydrogenase (PDH) reaction and the tricarboxylic acid (TCA) cycle account for the complete combustion of pyruvate to CO_2 and water. The main functions of pyruvate dehydrogenase are to produce acetyl-CoA, which can be oxidized completely to CO_2 and water for energy in the tricarboxylic acid cycle, and to generate NADH, which can be oxidized by the mitochondrial electron transport system to support the production of ATP.

5.1.1.1 Pyruvate Dehydrogenase Generates Acetyl-CoA. Of the 30 to 32 ATP molecules that can be obtained by oxidizing one molecule of glucose to CO_2 and water, only 2 ATP are generated directly during glycolysis. An additional 3 to 5 ATP can be generated by mitochondrial oxidation of the two NADH molecules generated per molecule of glucose metabolized by glycolysis to two molecules of pyruvate. The remaining 25 ATP are produced when the two molecules of pyruvate enter the mitochondrion and are oxidized to acetyl-CoA and, in turn, the acetyl-CoA is oxidized to CO_2 and water.

Conceptually, PDH is the bridge between glycolysis and the TCA cycle. However, this bridge is unidirectional because while humans can oxidize pyruvate to

Medical Biochemistry: Human Metabolism in Health and Disease   By Miriam D. Rosenthal and Robert H. Glew
Copyright © 2009 John Wiley & Sons, Inc.

acetyl-CoA (and ultimately to CO_2), they cannot carry out the opposite reaction of converting acetyl-CoA into pyruvate. It is the irreversibility of the PDH reaction that explains why the liver cannot use acetyl-CoA as a substrate in gluconeogenesis.

5.1.1.2 PDH Is an Important Site of Regulation. By regulating the flow of pyruvate into the TCA cycle, pyruvate dehydrogenase serves as an important site for regulating energy metabolism. Inhibition of PDH thus preserves glucose and gluconeogenic precursors such as alanine when other fuels, such as acetyl-CoA generated by the oxidation of fatty acids, are available for utilization.

5.1.2 Functions of the Tricarboxylic Acid Cycle

Each turn of the TCA cycle involves the entry into the pathway of one molecule of acetate from acetyl-CoA. The two-carbon acetate moiety combines with the four-carbon acid, oxaloacetate, to form citrate, which has six carbons. Subsequent reactions of the cycle result in the release of two carbon atoms as CO_2 and the regeneration of oxaloacetate. It is the regeneration and reutilization of oxaloacetate, which confers on the TCA cycle its cyclical character. Important functions of the TCA cycle are described below.

5.1.2.1 The TCA Cycle Generates Energy. The TCA cycle is the main source of energy in humans. It is responsible for the total oxidation of acetyl-CoA molecules that arise from the pyruvate dehydrogenase reaction, fatty acid β-oxidation, the oxidation of amino acids, ketone body catabolism, and the oxidation of ethanol. For each turn of the TCA cycle only a single molecule of high-energy nucleotide triphosphate (GTP) is produced. Most of the energy in acetyl-CoA that is released by the oxidative reactions of the TCA cycle is captured in the form of reduced electron carriers, specifically NADH and $FADH_2$. It is only when these two reduced electron carriers give up their electrons to the electron transport chain that their energy materializes in the form of ATP through the process of oxidative phosphorylation.

5.1.2.2 The TCA Cycle Provides Intermediates for Other Pathways. Some of the intermediates in the TCA cycle can be withdrawn and used in the synthesis of other cellular substances. For example, succinyl-CoA is a substrate in the pathway of heme synthesis. Similarly, the α-ketoglutarate generated when certain amino acids are broken down can enter the TCA cycle and be metabolized to malate which in turn can be exported from mitochondria into the cytosol. After cytosolic malate dehydrogenase oxidizes malate to oxaloacetate, the latter can be used to synthesize glucose by means of the gluconeogenesis pathway.

5.1.2.3 Enzymes of the TCA Cycle Make Acetyl-CoA Available for Fatty Acid and Cholesterol Synthesis. The two types of cells with the greatest capacity for fatty acid synthesis are the hepatocyte and the adipocyte. Almost all of the acetyl-CoA substrate utilized for fatty acid synthesis is generated inside mitochondria by the pyruvate dehydrogenase reaction. The acetyl moieties are transported out of the mitochondrion in the form of the TCA-cycle intermediate citrate. Regeneration of acetyl-CoA from citrate in the cytosol then provides substrate for the synthesis of fatty acids and cholesterol.

5.1.2.4 The TCA Cycle Generates a Metabolite That Regulates Other Metabolic Pathways.
The citrate concentration in a cell is strongly dependent on the energy charge of the cell. When the energy charge of the cell is high, ATP depresses TCA-cycle activity, and the citrate concentration increases. When the mitochondrial ATP concentration is high, citrate moves from the mitochondrial matrix into the cytosol, where it regulates glycolysis, gluconeogenesis, and fatty acid synthesis, all three of which are localized to the cytosol. Cytosolic citrate inhibits glycolysis by inhibiting phosphofructokinase-1 (PFK-1), stimulates gluconeogenesis by stimulating fructose 1,6-bisphosphatase (see Chapter 9), and promotes de novo fatty acid synthesis by activating acetyl-CoA carboxylase (see Chapter 11).

5.1.2.5 The TCA-Cycle Enzymes Participate in Pathways That Shuttle Reducing Equivalents into the Mitochondrion.
Glycolysis generates NADH in the cytosol, but NADH cannot enter the mitochondrion. The shuttle systems that transport NADH-reducing equivalents from the cytosol into the mitochondrion for electron transport and oxidative phosphorylation utilize enzymes of the TCA cycle.

5.2 LOCATION OF PYRUVATE DEHYDROGENASE AND THE TCA-CYCLE ENZYMES

All mitochondria-containing cells possess PDH and the TCA-cycle enzymes. The pyruvate dehydrogenase complex is associated with the matrix-facing surface of the inner mitochondrial membrane. The TCA-cycle enzyme succinate dehydrogenase is an integral protein of the inner mitochondrial membrane; all the other enzymes of the TCA cycle are located in the mitochondrial matrix. In most cells, pyruvate arises mainly from glycolysis of glucose and is therefore generated in the cytosol. Pyruvate enters mitochondria through an organic-acid carrier that facilitates its transport across the inner mitochondrial membrane.

5.3 PHYSIOLOGICAL STATES WHEN PDH AND THE TCA CYCLE ARE ESPECIALLY ACTIVE OR INACTIVE

The TCA cycle is the main oxidative pathway for generating reducing equivalents in the form of NADH and $FADH_2$ that can be used to synthesize ATP. As such, flux through the TCA cycle increases in skeletal muscle and heart cells during aerobic exercise. By contrast, the TCA cycle in the liver is relatively inactive during fasting.

During a fast, intermediates of the TCA cycle are converted to malate, which is then transported out of the mitochondrion to provide substrate for gluconeogenesis. Under these conditions, the acetyl-CoA generated by β-oxidation of fatty acids in the liver is utilized to synthesize ketones, which are exported from hepatocytes into the blood. The ketones are metabolized to CO_2 and water by other tissues, primarily muscle.

PDH activity is increased in the fed state when many different types of cells are using mainly glucose as their fuel source, as opposed to fasted state when muscle, liver, and many other organs rely primarily on fatty acid oxidation to generate ATP. In addition, PDH activity in muscle increases with increased aerobic exercise, resulting in greater reliance on glucose as a fuel source.

5.4   PDH AND THE REACTIONS OF THE TCA CYCLE

5.4.1   Pyruvate Dehydrogenase

The PDH complex catalyzes a series of reactions (Fig. 5-1), the net result of which is

$$\text{pyruvate} + \text{NAD}^+ + \text{CoASH} \rightarrow \text{acetyl-CoA} + \text{NADH} + \text{H}^+ + \text{CO}_2$$

The reaction is complex in two respects. First, the pyruvate dehydrogenase enzyme is an extremely large, multisubunit complex comprised of multiple copies of

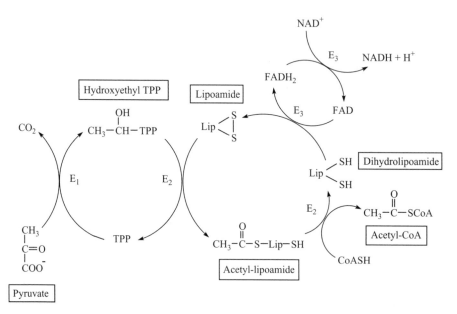

FIGURE 5-1 Sequence of reactions catalyzed by the pyruvate dehydrogenase complex. TPP, thiamine pyrophosphate.

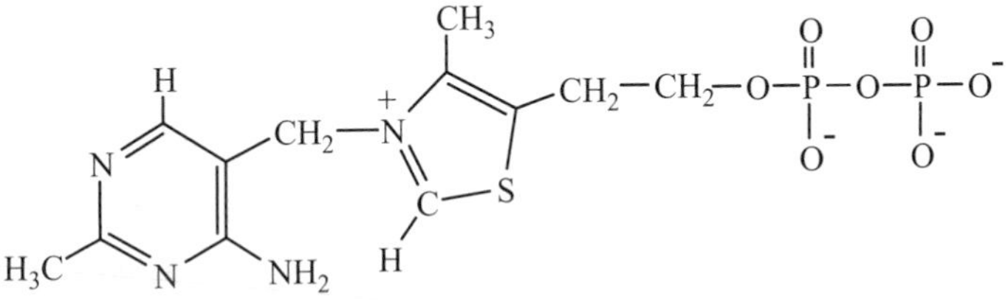

FIGURE 5-2   Structure of thiamine pyrophosphate.

each of three catalytic enzymes: pyruvate dehydrogenase, dihydrolipoamide acetyl-transferase, and dihydrolipoamide dehydrogenase and two regulatory enzymes, PDH kinase and PDH phosphatase. Second, the PDH reaction requires five cofactors: coenzyme A (CoASH), NAD^+, FAD, lipoic acid, and thiamine pyrophosphate.

The pyruvate dehydrogenase subunit, designated E_1, catalyzes the oxidative decarboxylation of pyruvate and the transfer of the resultant acetyl unit to the cofactor thiamine pyrophosphate (TPP) (Fig. 5-2), which is tightly bound to E_1. The dihydrolipoamide acetyltransferase subunit, designated E_2, catalyzes the transfer of the acetyl group from the thiamine pyrophosphate of E_1 to CoASH. The prosthetic group of E_2 is lipoic acid, which is covalently attached to the ε-amino group of a lysine residue of dihydrolipoamide acetyltransferase, thereby forming lipoamide. In the process of this reaction, (oxidized) lipoamide is reduced to dihydrolipoamide (Fig. 5-3). The E_3 subunit, dihydrolipamide dehydrogenase, contains FAD as its prosthetic group and catalyzes the regeneration of lipoamide from dihydrolipoamide, generating NADH in the process.

5.4.2 The Tricarboxylic Acid Cycle

The overall process by which acetyl-CoA is oxidized to CO_2 in the TCA cycle is irreversible. The pathway itself produces just one equivalent of high-energy nucleotide triphosphate (GTP) per molecule of acetyl-CoA oxidized by the cycle (Fig. 5-4). However, for each acetyl-CoA molecule, four energy-rich reduced electron carriers (three NADH and one $FADH_2$) are generated per turn of the cycle:

$$\text{acetyl-CoA} + 3NAD^+ + P_i + GDP + FAD + 2H_2O \rightarrow$$

$$2CO_2 + CoASH + 3NADH + 3H^+ + GTP + FADH_2$$

The three NADH and one $FADH_2$ generated in one complete cycle support the production of 9 ATP by means of oxidative phosphorylation. Thus, the TCA cycle is capable of producing 10 equivalents of ATP (9 ATP and 1 GTP) from one molecule of acetyl-CoA.

The TCA cycle is initiated by the condensation of the carbon chain of acetyl-CoA with oxaloacetic acid in an irreversible reaction catalyzed by citrate synthase:

$$\text{acetyl-CoA} + \text{oxaloacetate} \rightarrow \text{citrate} + \text{CoASH}$$

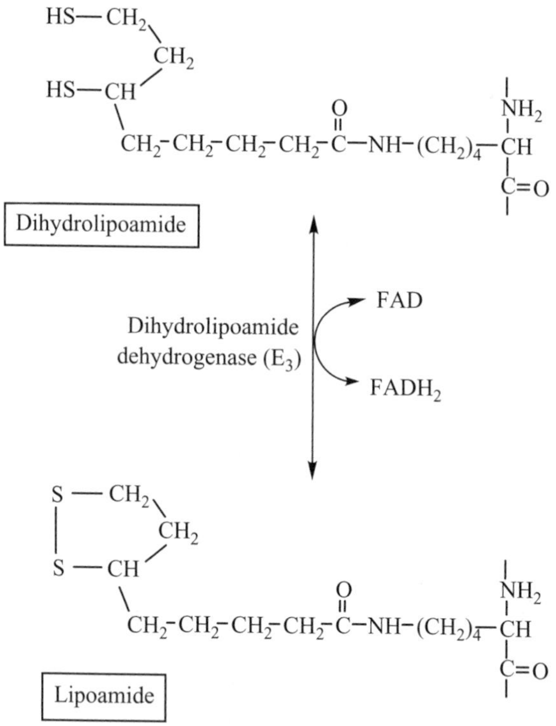

FIGURE 5-3 Reaction catalyzed by dihydrolipoamide dehydrogenase (E3) of the pyruvate dehydrogenase complex. Both oxidized lipoic acid (lipoamide) and reduced lipoic acid (dihydrolipoamide) are linked to the E2 subunit of PDH through an amide linkage.

Oxaloacetate functions in a catalytic fashion in the TCA cycle: It is the substrate that admits acetyl units into the cycle, and it is regenerated in the last step of the cycle.

The citrate formed in the first step of the TCA cycle is isomerized by aconitase to isocitrate:

$$citrate \rightleftharpoons isocitrate$$

The aconitase reaction is reversible and the equilibrium favors citrate over isocitrate by a factor of 20 to 1. Thus, when subsequent metabolism of isocitrate is inhibited, citrate levels (instead of isocitrate) increase in mitochondria. Aconitase has another function in addition to its role in the TCA cycle. The cytosolic form of the enzyme is involved in iron transport into cells and the regulation of the iron levels in the body.

The first redox reaction of the TCA cycle is catalyzed by the NAD^+-linked enzyme, isocitrate dehydrogenase:

$$isocitrate + NAD^+ \rightarrow \alpha\text{-ketoglutarate} + CO_2 + NADH + H^+$$

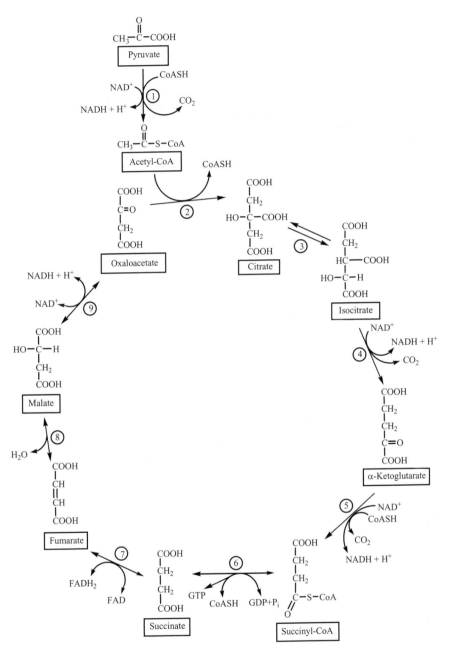

FIGURE 5-4 Reactions of the tricarboxylic acid cycle: ①, pyruvate dehydrogenase complex; ②, citrate synthase; ③, aconitase; ④, isocitrate dehydrogenase; ⑤, α-ketoglutarate dehydrogenase; ⑥, succinate thiokinase; ⑦, succinate dehydrogenase; ⑧, fumarase; ⑨, malate dehydrogenase.

When the electrons from NADH pass through the electron transport chain that is coupled to the phosphorylation of ADP, the energy that was transferred from isocitrate to NAD^+ in the isocitrate dehydrogenase reaction is ultimately used to synthesize ATP.

Next, α-ketoglutarate is oxidatively decarboxylated to produce succinyl-CoA in a reaction catalyzed by α-ketoglutarate dehydrogenase:

$$\alpha\text{-ketoglutarate} + NAD^+ + CoASH \rightarrow \text{succinyl-CoA} + CO_2 + NADH + H^+$$

The oxidative decarboxylation reaction catalyzed by α-ketoglutarate dehydrogenase is similar to the pyruvate dehydrogenase reaction, and FAD, thiamine pyrophosphate, and lipoic acid are cofactors in this reaction as well as that catalyzed by pyruvate dehydrogenase.

Succinyl-CoA contains a high-energy thioester bond which can be used to drive the synthesis of GTP. This is the only reaction in the TCA cycle that produces a high-energy nucleotide triphosphate directly, without the involvement of the machinery of oxidative phosphorylation; it is catalyzed by succinate thiokinase:

$$\text{succinyl-CoA} + GDP + P_i \rightarrow \text{succinate} + GTP + CoASH$$

The high-energy γ-phosphate bond of GTP can be transferred to ADP in a reversible reaction catalyzed by nucleotide kinase:

$$GTP + ADP \rightleftharpoons GDP + ATP$$

Proceeding through the TCA cycle, succinate is oxidized to fumarate in a reaction in which FAD is the electron acceptor:

$$\text{succinate} + FAD \rightleftharpoons \text{fumarate} + FADH_2$$

The enzyme that catalyzes this reaction, succinate dehydrogenase, is an integral membrane-bound protein localized to the inner mitochondrial membrane, thereby allowing for efficient energy transfer of the two electrons from $FADH_2$ to the co-localized electron transport chain.

Fumarate is then hydrated by fumarase:

$$\text{fumarate} + H_2O \rightleftharpoons \text{malate}$$

Fumarase exists in cytosolic and mitochondrial isoforms; however, both are encoded by the same gene. Cytosolic fumarase is active in liver, where it contributes to regenerating aspartate from the fumarate produced during the synthesis of urea.

Malate is then oxidized to oxaloacetate by malate dehydrogenase, completing the TCA cycle and bringing it full circle:

$$\text{malate} + NAD^+ \rightleftharpoons \text{oxaloacetate} + NADH + H^+$$

5.4.3 Synthesis of Catalytic Intermediates for the TCA Cycle

Since oxaloacetate functions as a catalyst in the TCA cycle, continued operation of the TCA cycle is critically dependent on maintaining the intramitochondrial concentration of oxaloacetate. Reactions that produce oxaloacetate directly or other compounds which can be metabolized to oxaloacetate are said to be anaplerotic in the sense that they can replenish oxaloacetate.

5.4.3.1 Synthesis of Oxaloacetate. The main reaction that replenishes oxaloacetate in the mitochondrion is catalyzed by the biotin-containing enzyme pyruvate carboxylase:

$$\text{pyruvate} + CO_2 + ATP \rightarrow \text{oxaloacetate} + ADP + P_i$$

Oxaloacetate is also synthesized by aspartate aminotransferase (AST), which transfers the amino group of aspartate to α-ketoglutarate:

$$\text{aspartate} + \alpha\text{-ketoglutarate} \rightleftharpoons \text{oxaloacetate} + \text{glutamate}$$

5.4.3.2 Synthesis of Other TCA-Cycle Intermediates. Reactions that generate TCA-cycle intermediates such as α-ketoglutarate and succinyl-CoA are also anaplerotic, since these TCA-cycle intermediates can readily be metabolized to oxaloacetate. The major precursors of α-ketoglutarate and other TCA-cycle intermediates are the glucogenic amino acids, whose catabolism gives rise to substrates for gluconeogenesis. Another source of succinyl-CoA is propionyl-CoA, which arises from the catabolism of valine and isoleucine, the oxidation of methyl-branched fatty acids (e.g., phytanic acid) and the β-oxidation of relatively rare odd-carbon fatty acids.

5.4.4 Reactions That Deplete the TCA Cycle of Intermediates

As noted above, mitochondrial malate can be transported into the cytosol and utilized for gluconeogenesis. Intermediates in the TCA cycle can also serve as precursors in the synthesis of other cellular molecules. For example, succinyl-CoA is a substrate in the first step of heme synthesis, the δ-aminolevulinic acid synthase reaction:

$$\text{succinyl-CoA} + \text{glycine} \rightarrow \delta\text{-aminolevulinic acid} + \text{CoASH}$$

Other such examples are the synthesis of glutamate and aspartate from α-ketoglutarate and oxaloacetate, respectively:

$$\alpha\text{-ketoglutarate} + \text{alanine} \rightleftharpoons \text{glutamate} + \text{pyruvate}$$
$$\text{oxaloacetate} + \text{glutamate} \rightleftharpoons \text{aspartate} + \alpha\text{-ketoglutarate}$$

Glutamate, in turn, is the precursor of glutamine, arginine, proline, and the neuro-transmitter γ-aminobutyric acid (GABA). Aspartate can be converted to asparagine.

5.5   REGULATION OF PDH AND THE TCA CYCLE

5.5.1   Pyruvate Dehydrogenase

The enzyme is regulated in two ways: by feedback inhibition involving metabolites and by phosphorylation/dephosphorylation.

5.5.1.1   Regulation of PDH by Means of Phosphorylation/Dephos-phorylation.   Pyruvate dehydrogenase is active in the dephosphorylated state and inactive when phosphorylated (Fig. 5-5). The enzyme is phosphorylated by a specific kinase named *pyruvate dehydrogenase kinase* (PDH kinase). PDH kinase is regulated not by cAMP but by molecules that signal changes in the energy charge of the cell. When the mitochondrial concentration of NADH, ATP, or acetyl-CoA is elevated, PDH kinase activity is stimulated, and pyruvate dehydrogenase becomes phosphorylated and inhibited. In contrast, high concentrations of pyruvate bind to PDH kinase and prevent the kinase from phosphorylating and inactivating pyruvate dehydrogenase. A PDH-specific calcium-activated phosphoprotein phosphatase (designated PDH phosphatase) removes a phosphate group from phosphorylated pyruvate dehydrogenase, thereby activating the enzyme.

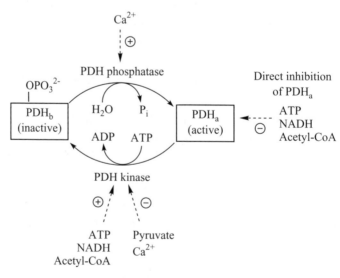

FIGURE 5-5   Regulation of pyruvate dehydrogenase (PDH).

5.5.1.2 *Direct Regulation of Pyruvate Dehydrogenase by Metabolites.*

The activity of the active, dephosphorylated form of pyruvate dehydrogenase is regulated directly by NADH and acetyl-CoA, which competitively inhibit the enzyme. ATP also stimulates PDH kinase, thereby inactivating PDH. PDH kinase is inhibited by a high intramitochondrial concentration of free calcium; this provides the mechanism by which epinephrine activates PDH in heart muscle.

5.5.2 Regulation of the TCA Cycle

The major locus of regulation of the TCA cycle is the isocitrate dehydrogenase reaction, and the key regulatory substance is ATP. A high-energy charge is associated with a high ATP concentration in the mitochondrion, which in turn inhibits the activity of isocitrate dehydrogenase. The TCA cycle is also inhibited by a high NADH/NAD$^+$ ratio, with NADH acting as a product inhibitor of isocitrate dehydrogenase, α-ketoglutarate dehydrogenase, and malate dehydrogenase.

5.6 ABNORMAL FUNCTION OF PDH AND THE TCA CYCLE

5.6.1 Genetic Deficiency of Pyruvate Dehydrogenase

Impaired activity of PDH can result from mutations in the genes encoding any one of the three subunits of the enzyme. Most of the mutations that affect the pyruvate dehydrogenase complex adversely occur in the gene encoding the α subunit of E_1, which is located on the X chromosome; both hemizygous males and heterozygous females are affected.

Regardless of the molecular basis, a deficiency of pyruvate dehydrogenase activity is associated with severe metabolic acidosis due to the resulting high circulating levels of pyruvate and lactate, and with fatigue and hypotonia due to insufficient ATP production. Patients with mutations affecting the E_3 component of PDH also have reduced levels of activity of both α-ketoglutarate dehydrogenase and the branched-chain α-ketoacid dehydrogenase that is involved in the catabolism of leucine, isoleucine, and valine since dihydrolipoamide dehydrogenase is a common component of these enzymes as well as PDH.

Energy production is impaired in patients with PDH deficiency, especially in neural cells, which are normally dependent on a constant supply of glucose as a fuel source. Therapy in PDH-deficient persons involves a ketogenic diet (80% fat, 20% protein + carbohydrates), which creates conditions under which the brain increases its utilization of ketone bodies as an alternative fuel source. Some patients with PDH deficiency carry mutations that result in decreased affinity of their PDH for thiamine pyrophosphate cofactor; these people benefit from very high doses of thiamine. A drug called *dichloroacetate* (DCA) has also been used in some instances in an effort to stimulate residual PDH activity. DCA inhibits PDH kinase, thereby maintaining PDH in the dephosphorylated, active form and increasing the intracellular ATP level.

5.6.2 Genetic Diseases of the TCA Cycle

There are several distinct but very rare diseases that result from defective enzymes in the TCA cycle. Fumarase deficiency results in elevated plasma, urine, and tissue concentrations of fumarate, pyruvate, and lactate, muscular hypotonia, and severe neurologic impairment (encephalopathy with developmental delay, seizures, microcephaly, enlarged cerebral ventricles). Inherited deficiencies of succinate dehydrogenase and α-ketoglutarate dehydrogenase are also associated with lactic acidosis and major neurological problems.

5.6.3 Thiamine Deficiency

The classical presentation of thiamine deficiency is beriberi. A lack of dietary thiamine results in low levels of thiamine pyrophosphate (TPP) and impaired activity of pyruvate dehydrogenase and α-ketoglutarate dehydrogenase, as well as TPP-dependent transketolase, which is a component of the pentose phosphate pathway. Thiamine deficiency is especially damaging to the heart and brain, which have large energy requirements. Some of the neurological damage associated with thiamine deficiency may also be due to the impairment of direct actions of thiamine triphosphate on neural membrane proteins.

CHAPTER 6

ELECTRON TRANSPORT AND OXIDATIVE PHOSPHORYLATION

6.1 FUNCTION OF ELECTRON TRANSPORT AND OXIDATIVE PHOSPHORYLATION

The pathway of mitochondrial electron transport represents the final stage in the oxidation of carbohydrates, fats, and amino acids. This pathway transfers the reducing equivalents in NADH and $FADH_2$ to molecular oxygen. The oxidation of NADH and $FADH_2$ is accompanied by a substantial decrease in free energy, much of which is captured by the concurrent formation of ATP from ADP and P_i in a process termed *oxidative phosphorylation*. Although there is also some substrate-level phosphorylation which generates high-energy phosphate bonds during both glycolysis and the TCA cycle, oxidative phosphorylation provides most of the ATP that humans generate.

6.2 LOCALIZATION OF ELECTRON TRANSPORT AND OXIDATIVE PHOSPHORYLATION

Electron transport and oxidative phosphorylation occur within all cells except red blood cells and the cornea and lens of the eye, which lack mitochondria. The electron transport chain consists of four macromolecular complexes, three of which (complexes I, III, and IV) span the inner mitochondrial membrane, and one (complex II) which is embedded in but does not span the inner mitochondrial membrane. The

Medical Biochemistry: Human Metabolism in Health and Disease   By Miriam D. Rosenthal and Robert H. Glew
Copyright © 2009 John Wiley & Sons, Inc.

energy released as electrons flow through the electron-transport chain is used to pump protons across the inner mitochondrial membrane to the intermembrane space, which lies between the inner and outer mitochondrial membranes, producing a proton gradient across the inner membrane.

The macromolecular complex called *ATP synthase*—the enzyme that produces ATP—also spans the inner mitochondrial membrane, forming a channel between the matrix and the cytosol. ATP synthase utilizes the electrochemical potential of the proton gradient to provide the energy for the synthesis of ATP. Thus, ATP synthesis is coupled to the collapse of the proton gradient and the return of protons through the channel of the ATP synthase complex to the matrix of the mitochondrion.

With the exception of brown fat, the rates of electron transport and oxidative phosphorylation are generally greatest in those tissues that contain large numbers of mitochondria. Mitochondria-rich tissues include the brain and central nervous tissue, red muscle fibers (relative to white muscle fibers), and the renal cortex (relative to the renal medulla). Brown fat is also rich in mitochondria (relative to white fat) and has a high rate of electron transport. Unlike other tissues, however, electron transport in brown fat is not tightly coupled to oxidative phosphorylation and functions primarily to generate heat rather than ATP.

6.3 PHYSIOLOGICAL CONDITIONS IN WHICH ELECTRON TRANSPORT AND OXIDATIVE PHOSPHORYLATION ARE ESPECIALLY ACTIVE

Electron transport and oxidative phosphorylation are most active when there is an increased need for ATP. The utilization of oxygen is thus dramatically increased in exercising muscle relative to the relaxed state. Transport of electrons and oxidative phosphorylation are also increased in hypermetabolic states such as sepsis and trauma.

6.4 REACTIONS OF ELECTRON TRANSPORT AND OXIDATIVE PHOSPHORYLATION

6.4.1 Electron-Transport Chain

The electron-transport chain consists of four macromolecular respiratory complexes, each of which contains multiple prosthetic groups that serve as electron carriers. The electron carriers of the various respiratory complexes include iron–sulfur (Fe–S) proteins, heme-containing cytochromes, and flavin mononucleotides (FMN). Electrons are transferred between the respiratory complexes by two smaller electron carriers, ubiquinone (also called *coenzyme Q*) and the heme protein cytochrome *c* (Fig. 6-1).

6.4.1.1 Entry of Electrons from NADH into the Electron-Transport Chain. The pathway that the electrons of NADH take after they enter the electron-transport chain is shown in Figure 6-1. The two electrons from NADH are transferred

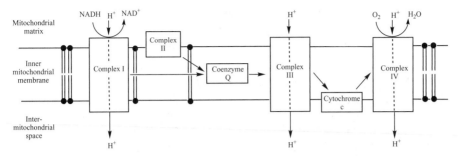

FIGURE 6-1 Components of the electron-transport chain, illustrating the pathway of electron flow from NADH to O_2. Shown are the sites where hydrogen ions are pumped out from the mitochondrial matrix into the intermitochondrial space.

to *respiratory complex I*, with the concomitant oxidation of NADH to NAD^+; after passing through the FMN and Fe–S components of the complex, these electrons are transferred to ubiquinone (CoQ):

$$NADH + H^+ + CoQ \rightarrow NAD^+ + CoQH_2$$

Reduced coenzyme Q ($CoQH_2$) in turn transfers the electrons to *respiratory complex III*, and in the process is oxidized back to the quinone state. The electrons then pass through the multiple prosthetic groups of respiratory complex III and into cytochrome *c*:

$$CoQH_2 + \text{cytochrome } c_{ox} \rightarrow CoQ + \text{cytochrome } c_{red}$$

Next, the reduced form of cytochrome *c* transfers electrons to *respiratory complex IV*, also called *cytochrome c oxidase*. Respiratory complex IV, which contains two hemeproteins (cytochrome *a* and cytochrome a_3), utilizes the electrons to reduce molecular oxygen to water:

$$\text{cytochrome } c_{red} + O_2 \rightarrow \text{cytochrome } c_{ox} + H_2O$$

The net effect of the flow of electrons from NADH through the various respiratory complexes is the oxidation of NADH by molecular oxygen:

$$NADH + H^+ + O_2 \rightarrow NAD^+ + H_2O$$

Movement of electrons through respiratory complexes I, III, and IV is accompanied by the pumping of protons from the mitochondrial matrix across the inner mitochondrial membrane into the intermembrane space. The accumulation of protons on the cytosolic side of the membrane establishes a proton gradient across the membrane and results in a cytosolic pH which is approximately one pH unit lower than the pH of the mitochondrial matrix. This pH gradient represents potential energy or the potential to accomplish chemical work (e.g., ATP synthesis).

6.4.1.2 Entry of Electrons from FADH₂ into the Electron-Transport Chain.

There are several ways that electrons enter the electron-transport chain from $FADH_2$ (Fig. 6-2). The major source of $FADH_2$ is that which is generated in mitochondria during β-oxidation of fatty acids. $FADH_2$ is also generated within mitochondria by the TCA cycle enzyme succinate dehydrogenase:

$$\text{succinate} + \text{FAD} \rightarrow \text{fumarate} + FADH_2$$

The succinate dehydrogenase complex of the respiratory chain contains both $FADH_2$ and Fe–S prosthetic groups and is designated *respiratory complex II*. However, the succinate dehydrogenase complex differs from respiratory complexes I, III, and IV in two important ways. First, although the succinate dehydrogenase complex is embedded in the inner mitochondrial membrane, it does not span the membrane. Second, the movement of electrons through the succinate dehydrogenase complex is not accompanied by the pumping of protons from the mitochondrial matrix across the inner mitochondrial membrane into the cytosol.

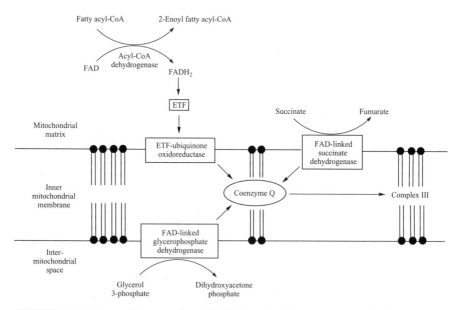

FIGURE 6-2   Entry of electrons from $FADH_2$ into the electron-transport chain at the level of coenzyme Q. ETF, electron-transfer flavoprotein.

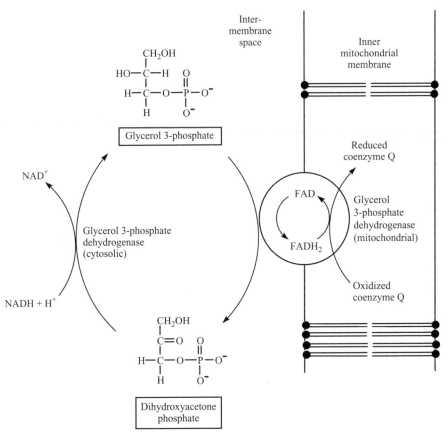

FIGURE 6-3 Entry of electrons from glycerol 3-phosphate dehydrogenase into the electron-transport chain.

$FADH_2$ is also generated in the cytosol by the action of glycerol 3-phosphate dehydrogenase (Fig. 6-3):

glycerol 3-phosphate + FAD → dihydroxyacetone phosphate + $FADH_2$

The $FADH_2$ is then oxidized by the mitochondrial isozyme of glycerol 3-phosphate dehydrogenase:

dihydroxyacetone phosphate + $FADH_2$ → glycerol 3-phosphate + FAD

This enzyme, also known as *flavoprotein dehydrogenase*, is embedded in the inner mitochondrial membrane, with its active site facing the cytosol, and transfers electrons through $FADH_2$ directly to coenzyme Q in the electron-transport chain.

As shown in Figure 6-2, electrons from $FADH_2$ generated by acyl-CoA dehydrogenases in the β-oxidation pathway, the succinate dehydrogenase complex, and mitochondrial glycerophosphate dehydrogenase are transferred to coenzyme Q:

$$FADH_2 + CoQ \rightarrow FAD + CoQH_2$$

Electrons are then transferred from $CoQH_2$ to respiratory complexes III and IV and, as described above, are ultimately used to reduce molecular oxygen:

$$FADH_2 + \tfrac{1}{2}O_2 \rightarrow FAD + H_2O$$

6.4.2 Synthesis of ATP

ATP synthase (a.k.a. *respiratory complex V*) catalyzes the synthesis of ATP in mitochondria (Fig. 6-4):

$$ADP + P_i \rightarrow ATP$$

ATP synthase, which was originally called mitochondrial ATPase because it also catalyzes the reverse reaction (i.e., the hydrolysis of ATP) is a large multiprotein complex comprised of two major subunits (F_0 and F_1). The F_0 *subunit* spans the inner mitochondrial membrane and contains a proton channel. Movement of protons through

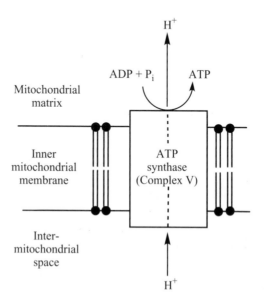

FIGURE 6-4   Generation of ATP by respiratory complex V (ATP synthase) is accompanied by the pumping of electrons back into the mitochondrial matrix.

the channel back into the mitochondrial matrix occurs down the proton gradient and provides the energy for the F_1 *subunit*, which catalyzes the synthesis of ATP.

6.4.2.1 How Much ATP Can Be Generated from the Oxidation of Mitochondrial NADH?

The actual value of the P/O ratio, the ratio of the number of ATP molecules synthesized to the number of oxygen atoms consumed, has long been a subject of considerable controversy. Older studies suggested that the value was 3.0 for electrons entering the electron-transport chain from NADH. This round number was aesthetically pleasing since it suggested synthesis of one ATP from the protons pumped out of the mitochondrion at each of respiratory complexes I, III, and IV. However, more recent estimates indicate that for each NADH molecule that donates its electrons to the respiratory chain, four protons are pumped out of the mitochondrion by respiratory complex I, four protons by respiratory complex III, and two protons by respiratory complex IV. Since approximately four protons are needed to provide the energy required to synthesize one ATP, the actual P/O ratio is closer to 2.5 than to 3.0.

6.4.2.2 Why do Electrons from FADH₂ Generate Less ATP Than Those from NADH?

As discussed above, electrons from $FADH_2$, including those arising from the succinate dehydrogenase complex, enter the electron-transport chain at coenzyme Q, bypassing respiratory complex I, which is the initial site at which protons are pumped out of the mitochondrial matrix. As a result, the passage of electrons from $FADH_2$ to molecular oxygen results in a lower net transfer to protons from the mitochondrial matrix to the cytosol, and a P/O ratio of 1.5 rather than 2.5.

6.4.3 Transport of Reducing Equivalents into the Mitochondrion

Much of the oxidative machinery of the cell is located within mitochondria, including the TCA cycle and enzymes that catalyze the β-oxidation of fatty acids and the oxidation of the carbon skeletons of most amino acids. Since the NADH and $FADH_2$ generated by these oxidative pathways are produced within the mitochondrion, the reducing equivalents in these cofactors can readily be transferred into the electron-transport chain. By contrast, the glycolytic pathway that oxidizes glucose generates NADH in the cytosol. Since the inner mitochondrial membrane is impermeable to NADH, cells need a mechanism for transporting the reducing equivalents in NADH from the cytosol into the mitochondrial matrix. Two such mechanisms exist: the malate–aspartate shuttle and the glycerol 3-phosphate shuttle.

6.4.3.1 Malate–Aspartate Shuttle.

The malate–aspartate shuttle involves the reduction of a cytosolic substrate by NADH and transport of the reduced product from the cytosol into the mitochondrion, where it is reoxidized in an NAD^+-dependent reaction, thus consuming cytosolic NADH and generating mitochondrial NADH (Fig. 6-5). NADH is utilized by cytosolic malate dehydrogenase to reduce oxaloacetate to malate:

$$\text{oxaloacetate} + \text{NADH} + H^+ \rightleftharpoons \text{malate} + NAD^+$$

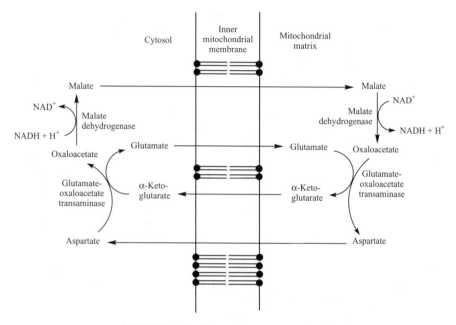

FIGURE 6-5   The malate–aspartate shuttle.

Malate is then transported from the cytosol into the mitochondrion, where it is oxidized to oxaloacetate by mitochondrial malate dehydrogenase isozyme, which also catalyzes the last step of the TCA cycle.

The actual pathway is complicated by the specific nature of the transporters that facilitate the movement of organic compounds in and out of the mitochondrion. Thus, the oxaloacetate produced by mitochondrial malate dehydrogenase is not transported directly out of the mitochondrion. Instead, the oxaloacetate molecule in the mitochondrial matrix first acquires an amino group from glutamate (through a process known as *transamination*):

$$oxaloacetate + glutamate \rightleftharpoons aspartate + \alpha\text{-ketoglutarate}$$

Aspartate is then transported out of the mitochondrion into the cytosol, where the transamination reaction occurs in the reverse direction, regenerating oxaloacetate. This process utilizes two mitochondrial transporters, one of which exchanges cytosolic malate for mitochondrial α-ketoglutarate, and a second, known as aralar 1/citrin, which exchanges mitochondrial aspartate for cytosolic glutamate.

The net effect of the malate–aspartate pathway is the transport of the reducing equivalents of cytosolic NADH into the mitochondrion:

$$NADH_{cytosolic} + NAD^{+}_{mitochondrial} \rightarrow NAD^{+}_{cytolsolic} + NADH_{mitochondrial}$$

6.4.3.2 Glycerol 3-Phosphate Shuttle. In the second electron-shuttle mechanism, the NADH generated during glycolysis is used by cytosolic glycerophosphate dehydrogenase to reduce dihydroxyacetone phosphate to glycerol 3-phosphate (Fig. 6-3). Glycerol 3-phosphate is then oxidized back to dihydroxyacetone phosphate by mitochondrial FAD-linked glycerol 3-phosphate dehydrogenase. Since the electrons from $FADH_2$ are transferred directly to ubiquinone, thus bypassing respiratory complex I, the glycerol 3-phosphate shuttle yields only 1.5 ATP per molecule of cytosolic NADH rather than the 2.5 ATP per NADH produced by means of the malate–aspartate shuttle.

6.4.3.3 Why Are There Two Different Shuttles for Transferring Reducing Equivalents from the Cytosol into the Mitochondrion? The two shuttle mechanisms are adapted to meet the needs of different metabolic conditions. Although the malate–aspartate shuttle is clearly the more efficient of the two shuttles in terms of the net yield of ATP, the pathway is readily reversible and transfers reducing equivalents into the mitochondrion only when the $NADH/NAD^+$ ratio is higher in the cytosol than in the mitochondrion. By contrast, the glycerol 3-phosphate shuttle is essentially irreversible and transfers reducing equivalents into the mitochondrion even when the $NADH/NAD^+$ ratio in the cytosol is lower than it is in the mitochondrion.

6.4.4 Export of ATP from Mitochondria

Transport of ATP out of the mitochondrion is accomplished by an adenine nucleotide translocase which moves one ADP into the mitochondrion for each ATP exported into the cytosol. Since the net charge of ATP is -4, whereas ADP has a charge of -3, the exchange is accompanied by the import of one proton per ATP into the mitochondrion and is driven by the pH gradient across the inner mitochondrial membrane.

Export of ATP from the mitochondrion in exchange for ADP also results in a net export of phosphate. Sustained ATP export thus requires a phosphate transporter, which imports cytosolic phosphate (P_i) in exchange for a hydroxyl ion. The exchange of P_i for a hydroxyl ion also results in a net import of one proton into the mitochondrion. Thus, the protons that are pumped out of the mitochondrion during electron transport provide some of the energy needed to drive the two transporters involved in ATP export from the mitochondrion.

6.4.5 Exchange of High-Energy Phosphate Bonds Between Nucleotides

Oxidative phosphorylation provides ATP for muscle work, membrane transporters, and a variety of biosynthetic reactions, such as protein synthesis. Other nucleotide triphosphates also play major roles in metabolism. For example, UTP is used to activate sugars for synthesis of glycogen and glycoconjugates, CTP in the synthesis of phospholipids, and GTP for both protein synthesis and the activation of the G proteins involved in signal transduction. Nucleoside triphosphates and deoxyribonucleoside triphosphates are also the substrates for the synthesis of DNA and RNA, respectively.

6.4.5.1 *Nucleotide Kinases.* The terminal high-energy bond of ATP can be transferred to various nucleoside diphosphates (e.g., UDP, CDP, GDP) by nucleotide kinases. For example, UDP generated from UTP in the pathway that incorporates glucose 1-phosphate into glycogen is regenerated using ATP as the high-energy phosphate donor:

$$\text{UDP} + \text{ATP} \rightleftharpoons \text{UTP} + \text{ADP}$$

6.4.5.2 *Myokinase.* Myokinase or adenylate kinase is a nucleotide kinase that generates ATP from two molecules of ADP:

$$\text{ADP} + \text{ADP} \rightleftharpoons \text{ATP} + \text{AMP}$$

High levels of myokinase are found in muscle cells, where it serves to regenerate ATP during periods of high energy demand. The AMP generated in the myokinase reaction is an allosteric activator of key regulatory enzymes of glycolysis, thus stimulating ATP synthesis.

6.4.6 Storage of High-Energy Phosphate Bonds

Creatine phosphate is used by muscle and other cells to store energy from ATP in a readily available form. *Creatine kinase* (CK), also called *creatine phosphokinase* (CPK), is an energy-transfer enzyme that catalyzes the synthesis of creatine phosphate from creatine and ATP (Fig. 6-6):

$$\text{creatine} + \text{ATP} \rightleftharpoons \text{creatine phosphate} + \text{ADP}$$

CK is present in many different types of cells (brain, colon), but the highest CK level in the body is found in the cytosol and mitochondria of skeletal and heart muscle. Mitochondrial CK is located on the outer surface of the inner mitochondrial membrane, where it uses the ATP generated by oxidative phosphorylation to catalyze the production of creatine phosphate. The high-energy creatine phosphate molecule diffuses into the cytosol, where it is stored until needed for muscle work. Cytosolic CK is closely associated with myofibrils and catalyzes the formation of ATP from creatine phosphate and ADP, thus providing ATP for muscle contraction.

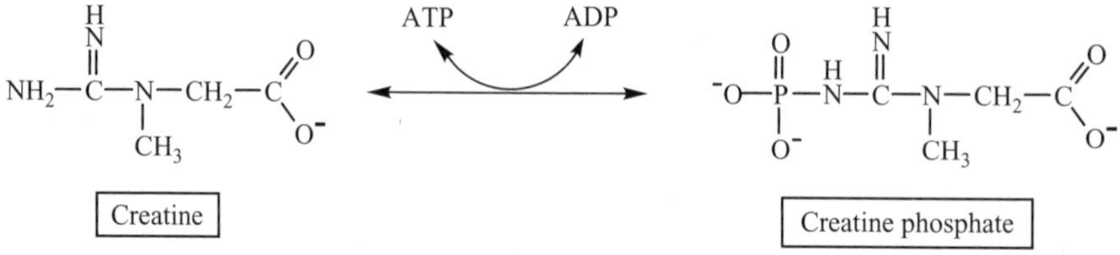

FIGURE 6-6   Synthesis of creatine phosphate.

6.4.6.1 *Creatine Kinase as a Marker of Heart and Skeletal Muscle Injury.*

For decades measurement of serum levels of CK served as a marker for heart or skeletal muscle injury. CK is a dimeric protein made up of two distinct polypeptide subunits, designated M (muscle) and B (brain). Three CK isoenzymes were found in human tissues: CK-MM in skeletal muscle, CK-BB in brain, and the heterodimer CK-MB, considered the myocardial isozyme. More recent studies have demonstrated, however, that cardiac muscle normally contains CK-MM, and that the CK-MB isozyme is present in cardiac muscle only after the muscle has sustained prior injury. The diagnostic value of plasma CK-MB measurements thus reflects the fact that many people who have myocardial infarctions have a prior history of subclinical cardiac damage. CK-MB is also present in skeletal muscle that has experienced prior injury, including muscle of highly trained long-distance runners. Although an increased plasma level of CK-MB is often still suggestive of myocardial injury, quantification of plasma levels of cardiac isoforms of the structural proteins called *troponins* has superseded CK as the definitive marker of myocardial injury.

6.4.7 Oxidative Phosphorylation as a Source of Reactive Oxygen Species

Mitochondria are the major source of oxygen free radicals in tissues. In the process of consuming large amounts of molecular oxygen while producing ATP coupled to the oxidation of respiratory substrates, some of the oxygen is only partially reduced, generating superoxide ($O_2^{\bullet-}$) and other reactive oxygen species (ROS). Superoxide anion is produced in the mitochondrial matrix by complex I (NADH–ubiquinone oxidoreductase) and complex III (ubiquinone–cytochrome c oxidoreductase). In contrast, cytochrome c oxidase (complex IV) has such a high oxidative capacity that it reduces oxygen completely to H_2O and does not produce ROS.

Fortunately, there are many cellular defenses against ROS. For example, superoxide anion is converted to hydrogen peroxide (H_2O_2) by mitochondrial and cytosolic superoxide dismutases:

$$2O_2^{\bullet-} + 2H^+ \rightarrow O_2 + H_2O_2$$

H_2O_2 can, in turn, be neutralized by several cellular scavenger systems, including glutathione peroxidase, catalase, thioredoxin, and glutaredoxin.

6.5 REGULATION OF OXIDATIVE PHOSPHORYLATION

Electron transport and oxidative phosphorylation are normally tightly coupled; that is, electron transport depends on concurrent ATP synthesis. The process by which electron transport pumps protons out of the mitochondrion ceases unless there is a mechanism for returning protons to the matrix of the mitochondrion. Under normal conditions, this return of protons from the intermembrane space into the mitochondrion is accomplished through the action of ATP synthase. Thus, both electron transport and oxidative phosphorylation occur only when ADP is available as a substrate for oxidative phosphorylation. Both pathways are less active when the energy needs of the cell are low and the ATP/ADP ratio is high. Inhibition of electron transport, in turn, increases the intramitochondrial $NADH/NAD^+$ ratio, which results in inhibition of both the TCA cycle and fatty acid oxidation.

6.5.1 Brown Fat

One tissue in which electron transport is normally dissociated from oxidative phosphorylation is brown fat, which derives its color and name from the fact that it has a much greater mitochondrial content than that of white adipose tissue and more extensive vascularization. Although virtually no brown fat is present in adults, brown fat may account for as much as 5% of the body weight of neonates. The main function of brown fat is thermogenesis, which provides heat when the infant is exposed to a cold environment. The molecular basis for this nonshivering thermogenesis is the presence of thermogenin or mitochondrial uncoupling protein 1. Thermogenin is a transmembrane protein in the inner mitochondrial membrane that contains a proton channel which allows protons to reenter the mitochondrial matrix independent of the activity of ATP synthase. Thermogenin thus acts to uncouple electron transport from oxidative phosphorylation, which results in the dissipation of the energy of the proton gradient as heat.

6.6   ABNORMAL FUNCTIONING OF ELECTRON TRANSPORT AND OXIDATIVE PHOSPHORYLATION

6.6.1 Genetic Diseases

Many degenerative diseases result from mutations in genes that code for the components of electron transport and the oxidative phosphorylation apparatus. Whereas some of these mutations alter nuclear DNA and follow classical Mendelian inheritance patterns, others result from mutations in mitochondrial DNA. Mitochondrial DNA (mtDNA) encodes 13 of the approximately 1000 proteins that comprise the mitochondrion, including seven subunits of respiratory complex I and one or more of the subunits of respiratory complexes III and IV and ATP synthase. In addition, mtDNA encodes ribosomal and transfer RNA molecules required for the intramitochondrial synthesis of these proteins. Point mutations in or deletions of mitochondrial genes can impair electron transport and oxidative phosphorylation.

The syndromes resulting from disorders of mtDNA-encoding protein components of electron transport and oxidative phosphorylation include LHON (Leber hereditary optic neuropathy) and some forms of Leigh disease. Disease resulting from point mutations in tRNA include MERRF (myoclonic epilepsy associated with ragged-red fibers) and MELAS (mitochondrial myopathy, encephalomyopathy, lactic acidosis, and strokelike episodes).

Mitochondrial disorders resulting from mutations in mtDNA are usually sporadic or follow matrilineal inheritance. Since most cells contain numerous mitochondria, each with multiple DNA molecules, individual cells are usually heteroplasmic in that they have a mixture of mitochondrial genomes. Furthermore, cellular replication may divide the mitochondria unevenly between the daughter cells, resulting in cells with an increased mutant load and correspondingly increased impairment of oxidative phosphorylation. The earliest manifestations of mitochondrial mutations are neurological and muscular abnormalities and are a consequence of the greater dependence of brain, heart, and skeletal muscle on mitochondrial ATP synthesis to maintain cellular functions.

6.6.2 Inhibitors of Electron Transport

Substances that inhibit electron transport and thus ATP generation are highly toxic. Among these are carbon monoxide, azide (N_3^-), and cyanide, all of which inhibit the cytochrome c oxidase activity of respiratory complex IV. Other inhibitors of electron transport include the rat poison rotenone, which inhibits respiratory complex I, and antimycin A, which inhibits respiratory complex III. Electron transport is also inhibited by the antifungal agent oligomycin, which inhibits ATP synthase.

6.6.3 Uncoupling Agents

Lipophilic weak acids readily penetrate the inner mitochondrial membrane and can transport protons back into the mitochondrial matrix. These acids thereby uncouple electron transport from oxidative phosphorylation, thus dissipating the energy derived from electron transport as heat. As such, they are the toxicologic equivalent of thermogenin and prevent ATP synthesis. One well-known uncoupling agent is 2,4-dinitrophenol, which was used briefly in the 1930s as a weight-loss pharmaceutical. Since 2,4-dinitrophenol causes cataracts and fatal fevers, the drug was quickly removed from the market.

6.6.4 Thyroid Disease

One of the major functions of thyroid hormones (e.g., thyroxine) is to maintain energy homeostasis. Thyroxine stimulates the synthesis of many of the proteins of the electron transport system and oxidative phosphorylation, including cytochrome oxidase, ATP synthase, and the adenine nucleotide transporter, resulting in an increased capacity for mitochondrial ATP synthesis. At the same time, hyperthyroidism results in a decreased ratio of ATP generated to O_2 utilized, with an increased dissipation of the energy of electron transport as heat. By contrast, hypothyroid individuals have both an increase in the efficiency of ATP generation and a decrease in the maximal rate of ATP synthesis.

CHAPTER 7

THE PENTOSE PHOSPHATE PATHWAY

7.1 FUNCTIONS OF THE PENTOSE PHOSPHATE PATHWAY

We learned in Chapter 4 that the main function of glycolysis is to oxidize glucose. The *pentose phosphate pathway* is a second pathway for the oxidation of glucose that synthesizes NADPH and pentose phosphates, which are five-carbon sugar phosphates. Because the pathway for the synthesis of the pentose phosphate ribulose 5-phosphate branches from glycolysis after the formation of glucose 6-phosphate, this alternate pathway is sometimes referred to as the *hexose monophosphate shunt*. All the enzymes involved in the pentose phosphate pathway are localized to the cytosol.

Glycolysis is a catabolic pathway. Under aerobic conditions, its products are pyruvate, ATP, and reducing equivalents in the form of NADH. By contrast, the pentose phosphate pathway serves two distinct anabolic roles. First, as its name implies, the pathway generates pentose 5-phosphates, which are substrates for nucleic acid synthesis. Second, the pathway generates reducing equivalents in the form of NADPH. In contrast to glycolysis, the pentose phosphate pathway does not generate ATP.

7.1.1 Functions of NADPH

NADPH is formed from $NADP^+$ (or nicotinamide-adenine dinucleotide phosphate) by the addition of two electrons and a hydrogen ion (proton). The $NADP^+$/NADPH couple is thus very similar to the NAD^+/NADH couple. However, whereas NADH is

Medical Biochemistry: Human Metabolism in Health and Disease   By Miriam D. Rosenthal and Robert H. Glew
Copyright © 2009 John Wiley & Sons, Inc.

primarily an electron carrier involved in fuel oxidation and ATP generation, NADPH serves to provide reducing equivalents for biosynthetic reactions such as fatty acid synthesis and cholesterol synthesis. Other roles for NADPH include recycling of oxidized glutathione to its reduced form and providing the reductant in the cytochrome P450–catalyzed hydroxylation of sterols.

NADH and NADPH provide the cell with two distinct pools of reducing equivalents, which can coexist at different redox levels. Since the electrons of cytosolic NADH are usually shuttled efficiently into the mitochondria, the cytosolic $NADH/NAD^+$ ratio is normally relatively low (usually < 0.01). By contrast, a high ratio of $NADPH/NADP^+$ is required for biosynthetic pathways. Since electrons are not transferred directly from cytosolic NADPH to NAD^+ or from NADH to $NADP^+$, the reducing equivalents associated with NADH and NADPH, respectively, can be effectively targeted for different needs of the cell.

7.2   TISSUES IN WHICH THE PENTOSE PHOSPHATE PATHWAY IS ACTIVE

Although the enzymes of the pentose phosphate pathway are present in all cells, their level of expression varies greatly from tissue to tissue. Relatively high levels of the pentose phosphate pathway enzymes are found in the liver where large quantities of fatty acids and cholesterol are synthesized, and in endocrine glands such as the ovaries, testes, and adrenal cortex, which synthesize cholesterol and steroid hormones. High levels of the pentose phosphate pathway enzymes are also found in cells of the early embryo and in other rapidly dividing cells, such as enterocytes, all of which require substantial amounts of ribose 5-phosphate for nucleic acid synthesis. By contrast, only low levels of hexose monophosphate shunt enzymes are present in skeletal muscle.

7.2.1   Phagocytic Cells

The highest levels of glucose 6-phosphate dehydrogenase, the first enzyme in the pentose phosphate pathway, are found in neutrophils and macrophages. In these phagocytic cells, NADPH is used to generate superoxide radicals from molecular oxygen in a reaction catalyzed by NADPH oxidase:

$$2O_2 + NADPH \rightarrow 2O_2^{\bullet} + NADP^+ + H^+$$

Superoxide anion, in turn, can serve to generate other reactive oxygen species (ROS), such as hydrogen peroxide (H_2O_2), hydrochlorous acid (HOCl), and hydroxyl radical (OH·), that kill phagocytized microorganisms (Fig. 7-1). Superoxide anion can also react with nitric oxide (NO) to generate peroxynitrite ($ONOO^-$), which in turn can lead to the formation of other nitrogen-containing radicals. The marked increase in

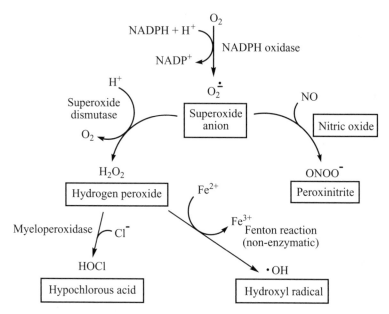

FIGURE 7-1   Role of NADPH oxidase in the generation of reactive oxygen species.

the rate of O_2 consumption by phagocytic cells following exposure to bacteria and other stimuli is often referred to as the *oxygen burst*.

7.2.2 Erythrocytes

Reactive oxygen species can attack proteins, membrane lipids, and DNA. Red blood cells rely on NADPH for protection against hemolysis caused by exposure to ROS such as hydrogen peroxide and lipid peroxides. The major intracellular antioxidant in red blood cells, as in most other cells, is glutathione, a tripeptide (γ-glutamylcysteinylglycine), which in the reduced state contains a sulfhydryl group, hence the abbreviation GSH. Vitamin E (Fig. 7-2) also plays a role in eliminating free

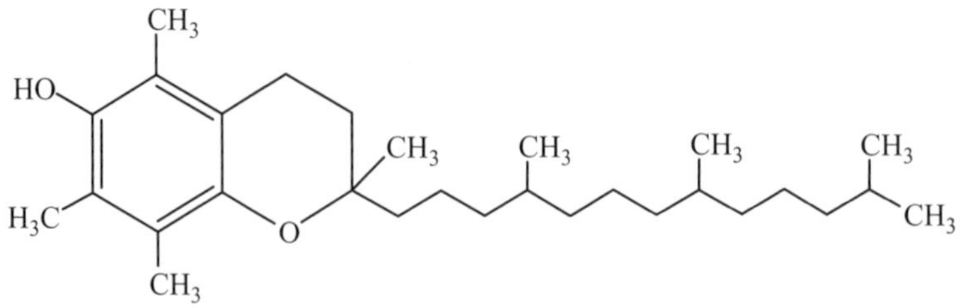

FIGURE 7-2 α-Tocopherol (vitamin E).

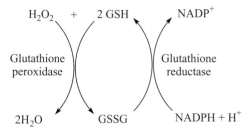

FIGURE 7-3 Role of NADPH in the generation of reduced glutathione. GSH, reduced glutathione; GSSG, oxidized glutathione.

radicals. Since it is fat soluble, vitamin E tends to partition into cellular membranes, which is where most of its antioxidant function is exerted.

The destruction of hydrogen peroxide is catalyzed by glutathione peroxidase, a ubiquitous, cytosolic, selenium-containing enzyme. The enzymatic destruction of hydrogen peroxide generates oxidized glutathione dimer (GSSG) (Fig. 7-3):

$$H_2O_2 + 2GSH \rightarrow 2H_2O + GSSG$$

NADPH is then used by glutathione reductase to restore oxidized glutathione to its original, reduced state:

$$GSSG + NADPH + H^+ \rightarrow 2GSH + NADP^+$$

Not surprisingly, glucose flux through the hexose monophosphate shunt increases when there is infection or exposure to certain drugs (e.g., primaquine, paraquat, naphthalene) that increase oxidant stress. In the absence of sufficient NADPH, the glutathione defense system is compromised and the risk of hemolysis is increased.

7.3 PHYSIOLOGICAL STATES AND CONDITIONS DURING WHICH THE PENTOSE PHOSPHATE PATHWAY IS PARTICULARLY ACTIVE

The pentose phosphate pathway is most active under conditions that require increased production of NADPH. Its activity is enhanced in granulocytes (e.g., neutrophils) during phagocytosis and in erythrocytes in response to oxidant stress.

NADPH provides reducing equivalents for biosynthetic processes such as fatty acid and cholesterol synthesis. In the liver, the pentose phosphate pathway is active in the fed state, when excess dietary carbohydrates are being converted into fatty acids and then into triacylglycerols. Indeed, the two NADPH-generating enzymes of the pentose phosphate pathway, glucose 6-phosphate dehydrogenase and 6-phosphogluconate dehydrogenase, are both induced by high levels of blood glucose and a high insulin/glucagon ratio (as is the case after a meal).

7.4   REACTIONS OF THE PENTOSE PHOSPHATE PATHWAY

Conceptually, the pentose phosphate pathway can be viewed as consisting of two phases. The first phase, which starts with glucose 6-phosphate, is the oxidative component of the pathway and the one that generates NADPH. In the process, 6-carbon glucose 6-phosphate is converted to five-carbon ribulose 5-phosphate and CO_2. The oxidative phase of the pentose phosphate pathway is irreversible. The second phase of the pentose phosphate pathway allows for the nonoxidative interconversion of sugar phosphates and functions to recycle excess pentose phosphates back into intermediates of the glycolytic pathway at the level of fructose 6-phosphate and glyceraldehyde 3-phosphate. Alternatively, the second phase of the pentose phosphate pathway can be utilized to generate pentose phosphates for nucleic acid synthesis when the synthesis of NADPH is not required and glucose flux through the oxidative phase does not occur.

7.4.1   Oxidative Phase

The three reactions that comprise the oxidative phase of the pentose phosphate pathway are shown in Fig. 7-4. Glucose 6-phosphate dehydrogenase catalyzes the initial oxidation of glucose 6-phosphate to 6-phosphoglucono-δ-lactone. This is the first of the two steps that generate NADPH and, as discussed below, is the regulated step of the pathway:

$$glucose\ 6\text{-phosphate} + NADP^+ \rightarrow 6\text{-phosphoglucono-}\delta\text{-lactone}$$

$$+NADPH + H^+$$

Gluconolactonase then catalyzes the hydrolysis of the lactone bond of the cyclical 6-phosphoglucono-δ-lactone to produce a linear structure:

$$6\text{-phosphoglucono-}\delta\text{-lactone} + H_2O \rightarrow 6\text{-phosphogluconate} + H^+$$

6-Phosphogluconate dehydrogenase catalyzes the $NADP^+$-dependent oxidative decarboxylation of 6-phosphoglucononate to produce ribulose 5-phosphate and CO_2, thereby generating the second molecule of NADPH in the process:

$$6\text{-phosphogluconate} + NADP^+ \rightarrow ribulose\ 5\text{-phosphate} + CO_2 + NADPH + H^+$$

7.4.2   Nonoxidative Phase

The nonoxidative phase of the pentose phosphate pathway (Fig. 7-5) involves reshuffling the carbon skeletons of various sugar phosphates to reversibly interconvert pentose phosphates and hexose phosphates. It is best visualized as having three rearrangement steps:

$$C_5 + C_5 \rightleftharpoons C_3 + C_7 \qquad \text{(transketolase)}$$

$$C_3 + C_7 \rightleftharpoons C_6 + C_4 \qquad \text{(transketolase)}$$

$$C_4 + C_5 \rightleftharpoons C_6 + C_3 \qquad \text{(transketolase)}$$

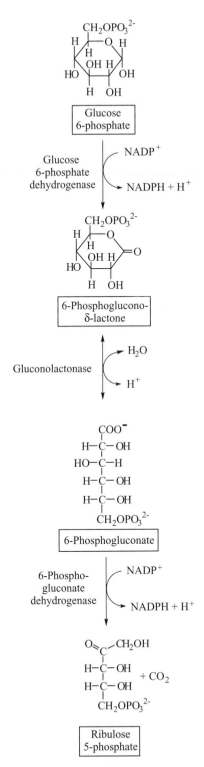

FIGURE 7-4 Oxidative phase of the pentose phosphate pathway.

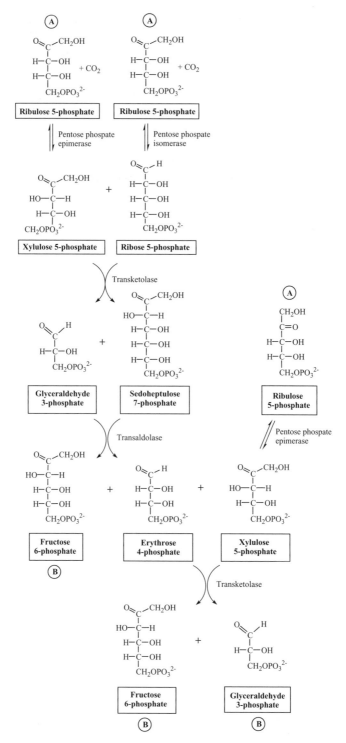

FIGURE 7-5 Nonoxidative phase of the pentose phosphate pathway. Ⓐ, substrates; Ⓑ, products.

The net effect of these reactions is

$$3C_5 \rightleftharpoons 2C_6 + C_3$$

where C_6 in fructose 6-phosphate and C_3 in glyceraldehyde 3-phosphate.

The nonoxidative phase of the pentose phosphate pathway starts with two pentose phosphates, ribose 5-phosphate and xylulose 5-phosphate, both of which are synthesized from ribulose 5-phosphate, which is the product of the oxidative phase of the pathway (Fig. 7-4). Ribose 5-phosphate is generated from ribulose 5-phosphate by phosphopentose isomerase, a step that is also required to generate ribose 5-phosphate for nucleotide synthesis. The reversible isomerization of the aldose sugar phosphate and the ketosugar phosphate is analogous to the conversion of glucose 6-phosphate to fructose 6-phosphate during glycolysis. Phosphopentose epimerase converts ribulose 5-phosphate to a second ketosugar phosphate, xylulose 5-phosphate.

The nonoxidative phase of the pentose phosphate pathway consists of three successive transfers of two- or three-carbon fragments between sugar-phosphates. First, transketolase transfers a two-carbon unit from xylulose 5-phosphate to ribose 5-phosphate, producing glyceraldehyde 3-phosphate plus the seven-carbon sedoheptulose 7-phosphate. Transketolase requires thiamine pyrophosphate as a cofactor. In the second step, transaldolase transfers a three-carbon unit from sedoheptulose 7-phosphate to glyceraldehyde 3-phosphate, forming four-carbon erythrose-4-phosphate plus fructose 6-phosphate. Transketolase then catalyzes a second two-carbon transfer reaction from the donor molecule, xylulose 5-phosphate. This time the acceptor molecule is erythrose 4-phosphate, thereby forming glyceraldehyde 3-phosphate plus a second molecule of fructose 6-phosphate.

7.5   REGULATION OF THE PENTOSE PHOSPHATE PATHWAY

The rate-limiting step of the oxidative phase is the first enzyme, glucose 6-phosphate dehydrogenase. The activity of this enzyme is regulated by the availability of the electron acceptor, $NADP^+$. Competition by the product, NADPH, with $NADP^+$ for binding to glucose 6-phosphate dehydrogenase acts to inhibit the enzyme. Under normal physiological conditions, most of the $NADP^+$ is in the reduced form (NADPH), and the activity of glucose 6-phosphate dehydrogenase is low. Activation of biosynthetic pathways such as fatty acid synthesis and cholesterogenesis results in consumption of NADPH and increases the flux of glucose 6-phosphate through glucose 6-phosphate dehydrogenase.

The oxidative phase is also regulated at the level of gene expression. In the liver, both glucose 6-phosphate dehydrogenase and 6-phosphogluconate dehydrogenase are induced upon refeeding after starvation when the availability of exogenous glucose promotes fatty acid synthesis and increases the need for NADPH.

As noted above, the oxidative and nonoxidative phases of the pentose phosphate pathway can function independently of each other. Flux through the nonoxidative phase of the pathway is regulated by the supply and demand for ribose 5-phosphate. The nonoxidative phase is inactive when all the pentose phosphates produced by the oxidative phase are converted into ribose 5-phosphate for RNA and DNA synthesis. On the other hand, when the oxidative phase is producing pentose phosphates in excess of cellular requirements, the nonoxidative phase becomes active and converts the excess pentose phosphates into fructose 6-phosphate plus glyceraldehyde 3-phosphate. Since the pentose phosphate pathway usually functions during the fed state, the fructose 6-phosphate and glyceraldehyde 3-phosphate generated by the nonoxidative phase are utilized primarily for glycolysis, producing ATP for biosynthetic processes.

When the demand for ribose 5-phosphate exceeds that of the NADPH which is concurrently produced by the oxidative phase of the pathway, the nonoxidative phase operates in the reverse direction; that is, it converts fructose 6-phosphate plus glyceraldehyde 3-phosphate into ribose-5-phosphate. Since all of the reactions of the nonoxidative phase are reversible, a decline in concentrations of ribose 5-phosphate will stimulate pentose phosphate synthesis without a concomitant increase in the flux of glucose through the oxidative phase of the pathway.

7.6 ABNORMAL FUNCTIONING OF THE PENTOSE PHOSPHATE PATHWAY

7.6.1 Glucose 6-Phosphate Dehydrogenase Deficiency

Globally, deficiency of glucose 6-phosphate dehydrogenase (G6PD) is the most common inborn error of metabolism. The major clinical manifestation of G6PD deficiency is acute hemolytic anemia following exposure to oxidative stress. Glucose 6-phosphate dehydrogenase deficiency may also present as neonatal jaundice during the first few days of life. The G6PD gene is X-linked and affects hemizygous males and homozygous females; heterozygous females are usually relatively unaffected unless X chromosome inactivation is skewed unfavorably. Many persons with G6PD deficiency are normally asymptomatic. However, an episode of hemolytic anemia usually occurs within hours of exposure to an oxidative stress which markedly increases the demand for NADPH to restore oxidized glutathione to its reduced form. Viral and bacterial infections are the most common triggers of acute hemolytic episodes. Hemolysis may also result from the ingestion of specific drugs, such as antimalarials or sulfonamide antibiotics, or certain foods, such as unripe fava beans, that contain the pyrimidine β-glycosides, vicine, and convicine that react nonenzymatically with O_2 to produce reactive oxygen species. The anemic crisis is usually self-limiting and ends when older erythrocytes, those most deficient in G6PD, have been hemolyzed.

7.6.1.1 *Why Are Erythrocytes Particularly Vulnerable to Deficiencies of Glucose 6-Phosphate Dehydrogenase?* Red blood cells rely on NADPH to maintain glutathione in its reduced form, thereby protecting the cells from oxidative

stress. Other cells, such as neutrophils and hepatocytes, also require high levels of NADPH; however, they have alternative pathways for generating NADPH. The major alternative source of NADPH, particularly for fatty acid and cholesterol synthesis, is the malic enzyme, which catalyzes the reaction

$$malate + NADP^+ \rightarrow pyruvate + CO_2 + NADPH + H^+$$

The malate utilized in this reaction is a component of the metabolic pathway that shuttles acetyl-CoA out of the mitochondrion in the form of citrate (see Chapter 11). Since erythrocytes do not contain mitochondria, they lack this alternative source of NADPH.

A second problem for red blood cells is that the most common genetically based deficiency in glucose 6-phosphate dehydrogenase, G6PD A$^-$, results in a relatively unstable G6PD enzyme that has a shortened half-life. Whereas cells that have a nucleus can synthesize new molecules of the G6PD enzyme, erythrocytes lack this capacity.

7.6.1.2 What Happens with More Severe G6PD Deficiencies? The rare patient who has an exceeding low level of G6PD activity in his or her cells is likely to have chronic hemolytic anemia (even without added oxidative stress) and an increased susceptibility to infection, because the NADPH supply of neutrophils is inadequate to support sufficient generation of H_2O_2 during phagocytosis.

7.6.1.3 Why Is There Such a High Gene Frequency for Defective Alleles of G6PD? G6PD deficiency has a prevalence of 5 to 25% in areas where malaria is endemic. Male hemizygotes and female heterozygotes both have significant protection against severe malaria. Erythrocytes of persons with G6PD deficiency are more sensitive than normal to hydrogen peroxide generated by the malarial parasite. Free-radical damage to erythrocyte membrane lipids causes hemolysis and death of the intracellular parasite before the parasite can reach maturity.

7.6.2 Chronic Granulatomous Disease

Chronic granulatomous disease (CGD) is a relatively rare genetic disorder affecting phagocytic cells normally involved in killing pathogenic microorganisms. Mutations in any one of the proteins that comprise the phagocyte NADPH oxidase system can prevent the enzyme from generating superoxide anion and downstream reactive oxygen species (e.g., H_2O_2, hydroxyl free radical, hypochlorite) involved in killing infectious organisms. People with CGD are at increased risk for pneumonia, osteomyelitis, and abscesses.

CHAPTER 8

GLYCOGEN

8.1 FUNCTION OF GLYCOGEN

Glycogen is a highly branched homopolymer of glucose that serves as the main carbohydrate-based energy store in the body. Although crucial to both homeostasis of blood glucose and to muscle work, glycogen actually represents less than 1% of the body's caloric stores, with triacylglycerol and mobilizable proteins accounting for the rest.

Glycogen granules are located in the cytosol. They consist of linear chains of glucose in α-1,4 glycosidic linkage, with α-1,6 glycosidic linkages forming branches after approximately every 8 to 10 glucose residues (Fig. 8-1). Starch, the glucose homopolymer in plants, consists of two types of molecules: amylose, which is a linear structure with glucose units in α-1,4 glycosidic linkages, and amylopectin, which contains α-1,6 glycosidic branches off the linear α-1,4 glycosidic chain. Glycogen is thus similar to, but more highly branched than, amylopectin. Because the anomeric carbons of the outermost glucose moieties of glycogen are all in glycosidic linkages with adjacent glucose moieties and thus not free to open up into the aldehyde form, the outer ends of the glycogen branches are all nonreducing. The glucose moiety in the core of glycogen is also nonreactive because it is covalently bound to glycogenin, the protein which, as described below, serves as the primer that initiates glycogen synthesis.

Glycogenolysis (glycogen breakdown) provides a readily available source of glucose when it is needed. Glycogen synthesis functions to replenish the glycogen stores in liver and muscle when dietary carbohydrates are available.

Medical Biochemistry: Human Metabolism in Health and Disease   By Miriam D. Rosenthal and Robert H. Glew
Copyright © 2009 John Wiley & Sons, Inc.

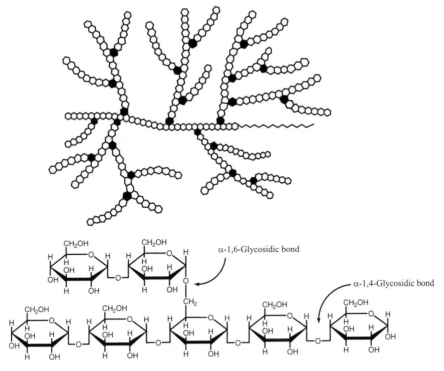

FIGURE 8-1   Structure of glycogen. The solid hexagons represent glucose units at α1,6 branch points. The zig-zag line represents glycogenin.

8.2   TISSUES THAT STORE GLYCOGEN

Although nearly all cells of the body contain trace amounts of glycogen, the major glycogen stores are found in liver and skeletal muscle. The liver contains the greatest concentration of glycogen of any organ, accounting for as much as 10% of tissue wet weight. On a mass basis, however, the largest portion of the body's total glycogen stores is in muscle. In a healthy adult in the fed state, the liver contains about 400 g of glycogen, whereas muscle contains approximately 1200 g of glycogen. Liver and muscle glycogen stores serve two distinctly different purposes. When mobilized, the glucose derived from liver glycogen is secreted by hepatocytes and used to maintain the concentration of glucose in the blood. By contrast, glucose arising from glycogen breakdown in skeletal and heart muscle remains in the muscle cells and is used to provide energy for muscle work.

The brain also contains a small but significant amount of glycogen, which is localized primarily in astrocytes. Brain glycogen accumulates during sleep and is mobilized upon waking, suggesting a functional role for brain glycogen in the conscious brain. The glycogen reserves in the central nervous system also provide at least a moderate degree of protection against hypoglycemia.

Another organ with a specialized role for glycogen stores is the fetal lung. Type II pulmonary cells begin to accumulate glycogen at about 26 weeks of gestation. Late in gestation, these cells shift their metabolism toward the synthesis of pulmonary surfactant, with the intracellular glycogen serving as a major substrate for the synthesis of surfactant lipids, of which dipalmitoylphosphatidylcholine is the major component.

8.3 PHYSIOLOGICAL STATES AND CONDITIONS DURING WHICH GLYCOGEN METABOLISM IS ESPECIALLY ACTIVE

Glycogenolysis in liver occurs in the fasted state and is stimulated primarily by the hormone glucagon. Resynthesis of glycogen occurs during the fed state and is stimulated both by insulin and the increased availability of glucose.

Glycogenolysis in muscle occurs during muscle contraction and affects only those muscles that are actually engaged in physical activity. Resynthesis of glycogen in muscle following exercise is more rapid when blood glucose levels are high and insulin is available to stimulate the activity of the GLUT4 transporters that facilitate the entry of glucose into muscle cells.

8.4 REACTIONS OF GLYCOGEN SYNTHESIS AND GLYCOGENOLYSIS

8.4.1 Glycogenolysis

Glycogenolysis (glycogen mobilization) is the process by which glucose units are removed, one at a time, from the numerous nonreducing ends of glycogen. In contrast to the cleavage of the glycosidic bonds of starch by α-amylase in the intestinal tract, cleavage of α-1,4-glycosidic bonds during glycogenolysis in muscle and liver is a phosphorolytic rather than a hydrolytic process and utilizes orthophosphate (P_i) rather than water.

8.4.1.1 Glycogen Phosphorylase. The first step in glycogenolysis is catalyzed by glycogen phosphorylase, commonly called *phosphorylase*. This enzyme cleaves the α-1,4 glycosidic bonds of glycogen (Fig. 8-2):

$$\text{glycogen}_{(n \text{ glucose residues})} + HPO_4^{2-} \rightarrow \text{glucose 1-phosphate}$$
$$+ \text{glycogen}_{(n-1 \text{ glucose residues})}$$

The reaction catalyzed by glycogen phosphorylase is physiologically irreversible and is the regulated step in glycogen mobilization. Different isozymes of glycogen phosphorylase are present in muscle and liver, permitting organ-specific regulation of glycogenolysis.

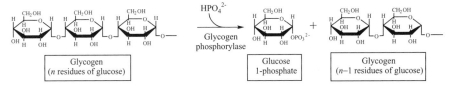

FIGURE 8-2 Glycogen phosphorylase catalyzes glycogenolysis.

8.4.1.2 Debranching Enzyme. Glycogen phosphorylase catalyzes the cleavage of α-1,4 glycosidic bonds but not of α-1,6 glycosidic bonds. Furthermore, glycogen phosphorylase is incapable of removing glucose residues that lie within four residues of a branch point. The debranching enzyme that solves this problem has two distinct catalytic activities (Fig. 8-3). First, the oligo-α-1,4 $\rightarrow$ α-1,4-glucanotransferase catalytic site transfers three glycosyl residues en bloc from the α-1,6 branch to an adjacent α-1,4 chain, leaving behind a single glucose unit in α-1,6 glucosidic linkage. The α-1,6-glucosidase catalytic site of the debranching enzyme then hydrolyzes the α-1,6 glycosidic bond of the remaining glucose unit, releasing free glucose and producing a long, unbranched α-1,4 glycosidic chain that is a suitable substrate for continued glycogenolysis by glycogen phosphorylase.

8.4.1.3 Significance of the Branched Structure of Glycogen. Given the complications involved in removing the branch-point glucose units from glycogen during glycogenolysis, it is reasonable to ask why these branches exist in the first place. There are two advantages to the highly branched structure of glycogen. First, the branched glycogen molecule provides numerous nonreducing termini that can serve as substrate for attack by multiple molecules of glycogen phosphorylase. This permits rapid mobilization of glycogen when the resulting glucose 1-phosphate is urgently needed either for muscle work or blood glucose homeostasis. The only alternative for providing multiple nonreducing ends of glycogen as available substrate for rapid mobilization by glycogen phosphorylase would be to have many smaller glucose polymers (oligosaccharides) instead of the larger glycogen granules. Given the substantial amounts of glycogen that can be stored in hepatocytes and muscle cells, having a comparable number of nonreducing ends on multiple linear (amylose) chains would substantially increase intracellular osmotic pressure and could cause cellular damage.

Second, the highly branched structure of glycogen permits formation of a compact glycogen granule. By contrast, long chains of relatively unbranched glucose polymers would also cause cellular damage. Evidence for this comes from the liver pathology associated with Anderson disease, a rare genetic disorder in which the ability to form branches in glycogen is impaired due to a deficiency of the branching enzyme. The presence of intracellular glycogen with very long outer branches leads to progressive cirrhosis of the liver and fatal liver failure.

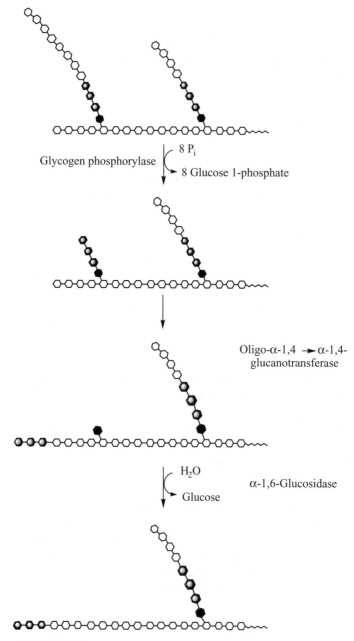

FIGURE 8-3   Role of the debranching enzyme in glycogenolysis.

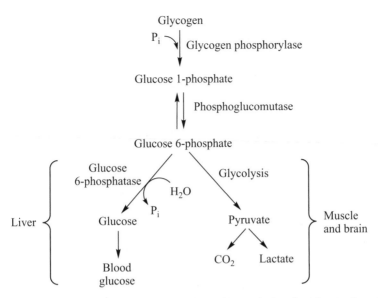

FIGURE 8-4   Alternate fates of glycogen-derived glucose 6-phosphate in muscle and liver.

8.4.1.4   What Is the Fate of Glucose 1-Phosphate in Skeletal Muscle?

One additional metabolic step is required before the glucose 1-phosphate generated in the glycogen phosphorylase reaction can be utilized as an energy source by muscle cells, and that is the transfer of the phosphate group from carbon-1 to carbon-6. Phosphoglucomutase catalyzes the following reaction:

$$\text{glucose 1-phosphate} \rightleftharpoons \text{glucose 6-phosphate}$$

This reaction is freely reversible, with the direction being determined by the relative concentrations of substrate and product. In muscle cells, the resulting glucose 6-phosphate molecule then enters the glycolytic pathway and is catabolized to provide energy for muscle contraction (Fig. 8-4).

8.4.1.5   What Is the Fate of Glucose 1-Phosphate in Hepatocytes?

Hepatocytes also contain phosphoglucomutase; however, the fate of glucose 6-phosphate is different in liver than in muscle (Fig. 8-4). The glucose 6-phosphate generated by glycogenolysis in liver is not catabolized within the hepatocyte but is, instead, released from the cell as free glucose. This irreversible dephosphorylation step is catalyzed by glucose 6-phosphatase:

$$\text{glucose 6-phosphate} + H_2O \rightarrow \text{glucose} + P_i$$

Glucose 6-phosphatase is an intrinsic enzyme embedded in the membrane of the endoplasmic reticulum (ER) (Fig. 8-5). Hydrolysis of glucose 6-phosphate occurs not

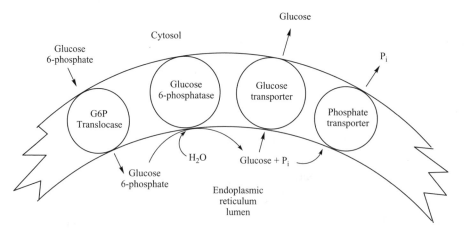

FIGURE 8-5 Role of transporters in the functioning of glucose 6-phosphatase.

in the cytosol but in the lumen of the endoplasmic reticulum. A specific transporter, glucose 6-phosphate translocase, is required to transport glucose 6-phosphate from the cytosol into the lumen of the ER; an anion transporter returns the resulting inorganic phosphate to the cytosol. Although the hepatic ER also contains a glucose transporter, most of the resultant glucose is not transported back into the cytosol of the hepatocyte; instead, the glucose is secreted to maintain blood-glucose homeostasis. Glucose 6-phosphatase is expressed in liver and the renal cortex, but not in muscle, and plays an important role in providing blood glucose from gluconeogenesis as well as glycogenolysis.

8.4.1.6 *What Is the Advantage of Intracellular Phosphorolysis Versus Hydrolysis of Glycogen?* Phosphorolysis is particularly advantageous for skeletal muscle and heart, which utilize the glucose units of glycogen for energy within the cell. As discussed in Chapter 4, phosphorylation of sugars serves both to activate them and to trap them within the cell. Mobilization of glycogen by phosphorolysis thus ensures that the exercising muscle can utilize the glycogen-derived glucose for its energy needs directly rather than allowing that glucose to be released into the bloodstream.

Another advantage of phosphorolytic degradation of glycogen as opposed to hydrolysis of glycogen is that generation of glucose phosphate instead of free glucose permits the glycogen-derived glucose to bypass the initial activation step of glycolysis: the hexokinase or glucokinase reaction. As also discussed in Chapter 4, each glucose molecule entering the glycolytic pathway requires 2 ATP to produce the activated glycolytic intermediate, fructose 1,6-bisphosphate. Subsequent glycolysis to lactate generates 4 ATP, for a net yield of 2 ATP/glucose. Phosphorolysis of glycogen, in effect, conserves some of the activation energy required for glycogen synthesis and only requires the input of one ATP to generate fructose 1,6-bisphosphate. This means that the net yield of ATP per glucose obtained from glycogen is 3 rather than 2; the 50% increase is clearly beneficial to strenuously exercising muscle.

It should be noted that not all of the glucose molecules of glycogen are mobilized as glucose 1-phosphate. As discussed above, during glycogenolysis the α-1,6-glycosidase component of the debranching enzyme releases a molecule of free glucose from each of the branch points of the glycogen molecule.

8.4.2 Glycogen Synthesis

As in the case of other anabolic pathways, the process of glycogen synthesis requires more energy input than is obtained during glycogen breakdown. The expenditure of an additional high-energy phosphate bond for each glucose molecule incorporated into glycogen during glycogen synthesis is the price the cell must pay for being able to store glucose in glycogen when fuel is plentiful and have that endogenous glucose available when exogenous fuels are scarce. The activated form of glucose that is used for glycogen synthesis is UDP-1-glucose.

8.4.2.1 *Pathway for Glycogen Synthesis from Free Glucose.* Following a meal, glucose is taken up by cells and phosphorylated by hexokinase (in muscle) or glucokinase (in liver) to generate glucose 6-phosphate:

$$glucose + ATP \rightarrow glucose\ 6\text{-phosphate} + ADP$$

When liver and muscle are poised to synthesize glycogen, phosphoglucomutase then shifts the phosphate group from C_6 to C_1:

$$glucose\ 6\text{-phosphate} \rightleftharpoons glucose\ 1\text{-phosphate}$$

This reversible reaction is catalyzed by the same enzyme that catalyzes the conversion of glucose 1-phosphate to glucose 6-phosphate during glycogenolysis.

8.4.2.2 *UDP-Glucose Pyrophosphorylase.* UDP-glucose pyrophosphorylase catalyzes the synthesis of UDP-1-glucose from glucose 1-phosphate (Fig. 8-6):

$$glucose\ 1\text{-phosphate} + UTP \rightleftharpoons UDP\text{-1-glucose} + PP_i$$

Although the enzyme is named for the reverse reaction (glucose 1-phosphate production), the reaction is driven to the right by the rapid hydrolysis of inorganic pyrophosphate to inorganic phosphate by pyrophosphatase. Note that the pathway of glycogen synthesis has thus far consumed two high-energy bonds (1 ATP and 1 UTP) per molecule of glucose activated to the level of UDP-1-glucose.

8.4.2.3 *Glycogen Synthase.* Glycogen synthase then transfers the activated glucose moiety of UDP-1-glucose to the 4-hydroxyl group of a glucose residue

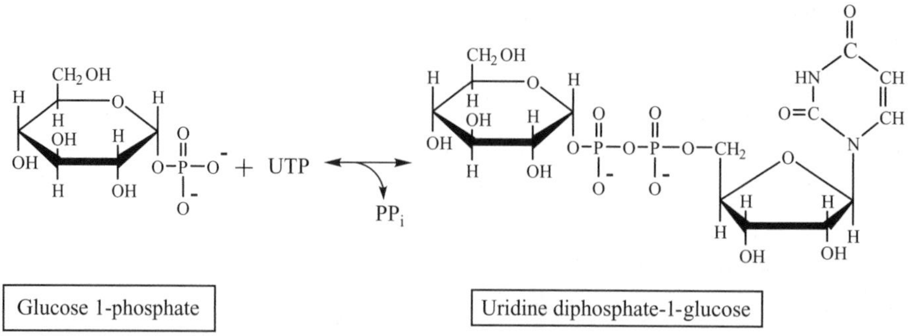

FIGURE 8-6   Synthesis of uridine diphosphate-1-glucose (UDP-1-glucose).

at one of the numerous nonreducing ends of the branched glycogen molecule, thereby forming a new α-1,4 linkage and extending the carbohydrate chain by one glucose unit.

8.4.2.4 Formation of Branches in the Growing Glycogen Molecule.
The formal name of the branching enzyme is amylo-(α-1,4 → α-1,6) transglycosylase. The enzyme transfers en bloc a six- to seven-residue oligosaccharide from the nonreducing end of a newly elongated α-1,4 glucose polymer to form a new α-1,6 glycosidic linkage (Fig. 8-7).

8.4.2.5 How Is the Synthesis of a New Molecule of Glycogen Initiated?
Glycogen synthase adds glucose moieties to an already existing α-1,4 glucose polymer. It is reasonable to ask about the origins of the initial glucose polymer. The answer is that a specialized protein called *glycogenin* serves as a primer for glycogen synthesis. Glycogenin is self-glucosylating and catalyzes both the covalent attachment of the initial glucose moiety to the protein and the subsequent addition of glucose units to form an α-1,4 glucose oligomer which can serve as substrate for glycogen synthase. The glycogenin molecule remains in the core of the resulting glycogen polymer, where it is attached to the reducing end of the initial glucose unit (Fig. 8-1).

8.5 REGULATION OF GLYCOGEN METABOLISM

As indicated above, glycogen synthesis occurs in the fed state, whereas glycogenolysis occurs both in the fasted state and in response to strenuous exercise. The two pathways—glycogen synthesis and glycogenolysis—are reciprocally regulated so that both processes are not active at the same time, thus preventing the operation of a futile, energy-wasting cycle. Glycogen metabolism is regulated by both allosteric and hormonal mechanisms. We first discuss the reciprocal hormonal regulation of the two pathways and then the specific allosteric mechanisms that apply to each of the key regulatory enzymes.

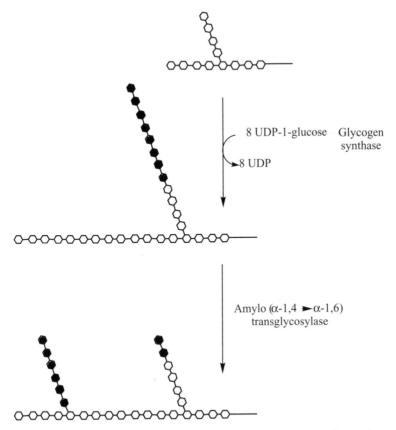

FIGURE 8-7   Glycogen synthesis and the role of amylo (α-1,4 $\rightarrow$ α-1,6)transglycosylase (the branching enzyme).

8.5.1 Hormonal Regulation

The major regulated steps of glycogenolysis and glycogen synthesis are catalyzed by glycogen phosphorylase and glycogen synthase, respectively. As discussed in Chapter 2, the signal transduction cascade initiated by glucagon or epinephrine acts through protein kinase A (PKA) to phosphorylate specific serine residues on certain key regulatory enzymes, including glycogen synthase and a highly specific protein kinase, phosphorylase kinase, which is dedicated to phosphorylating glycogen phosphorylase (Fig. 8-8). Protein phosphorylation activates glycogen phosphorylase and inactivates glycogen synthase, thereby stimulating glycogenolysis while concurrently inhibiting glycogen synthesis.

Binding of insulin to its receptor on hepatocytes and muscle cells leads to many downstream intracellular effects, one of which is activation of protein phosphatase 1. Protein phosphatase 1, in turn, reverses the effects of protein kinase A by dephosphorylating the enzymes glycogen synthase, glycogen phosphorylase, and phosphorylase

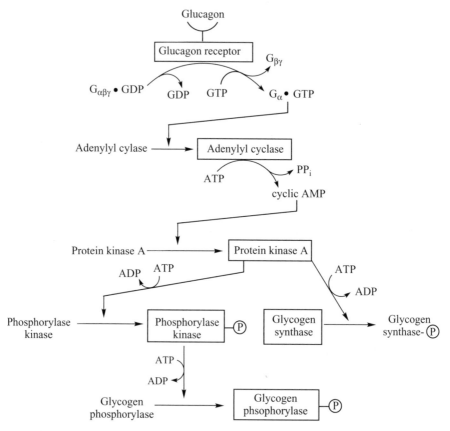

FIGURE 8-8   Role of glucagon in the coordinated regulation of glycogenolysis and glycogen synthesis. The active forms of the receptor and the enzymes in the signaling cascade are indicated in boxes.

kinase. Insulin thus acts to reverse the effects of glucagon and/or epinephrine on these enzymes, effectively activating glycogen synthesis while inactivating glycogen mobilization.

8.5.1.1 *Regulation of Glycogen Phosphorylase.* Glycogen phosphorylase

is a dimeric enzyme that can assume one of two states, R (relaxed) and T (tense or taut), that are in equilibrium with each other. In the R state, the active site of glycogen phosphorylase is accessible to the substrate (glycogen) and the enzyme is active; conversely, in the T state, the active site is inaccessible and the enzyme is inactive. The nonphosphorylated form of glycogen phosphorylase (referred to as *phosphorylase b*) exists primarily in the inactive T form. Phosphorylation of phosphorylase *b* (producing phosphorylase *a*) shifts the equilibrium in favor of the R form, thus increasing glycogen phosphorylase activity.

The equilibrium between the R and T forms of glycogen phosphorylase is also affected by allosteric regulators. Muscle and liver contain different isozymes of glycogen phosphorylase, thus permitting tissue-specific modulation of enzyme activity.

Muscle Phosphorylase. Phosphorylase *b* (the nonphosphorylated form of the enzyme) is inactive under most physiological conditions. Both ATP and glucose 6-phosphate are allosteric modulators that act to keep the glycogen phosphorylase primarily in the inactive T state. Thus, when fuel and energy are plentiful, glycogen phosphorylase is inactive. Exercise changes this situation by depleting supplies of both ATP and glucose 6-phosphate. The decreased energy charge of exercising muscle results in increased concentrations of AMP, which shift the equilibrium toward the R form, allosterically activating the otherwise inactive phosphorylase *b*.

Liver Phosphorylase. Glucose is an allosteric inhibitor of liver glycogen phosphorylase which acts to inhibit the otherwise active phosphorylase *a*. This mechanism serves to restrict glycogenolysis when it is not needed to maintain blood glucose.

8.5.1.2 Regulation of Phosphorylase Kinase.

Phosphorylase kinase is a serine–threonine protein kinase that phosphorylates and activates glycogen phosphorylase. Like glycogen phosphorylase, there are muscle and liver isozymes of phosphorylase kinase. Muscle phosphorylase kinase is only partially activated by phosphorylation. Phosphorylase kinase is also partially activated by an increase in the intracellular concentration of calcium ions that bind to calmodulin, a ubiquitous calcium-binding regulatory protein. Each molecule of phosphorylase kinase contains one molecule of calmodulin as its δ-subunit. Nerve impulses and muscle contraction increase cytosolic $[Ca^{2+}]$, thereby enhancing hormone-stimulated activation of phosphorylase kinase and subsequent mobilization of intramuscular glycogen. Full activation of phosphorylase kinase requires both PKA-catalyzed phosphorylation and binding of Ca^{2+} to the calmodulin subunit.

8.5.1.3 Regulation of Glycogen Synthase.

As described above, PKA-stimulated phosphorylation of glycogen synthase inactivates the enzyme. Several other protein kinases, including glycogen synthase kinase-3 (GSK3), can also phosphorylate specific serine residues of glycogen synthase and thus contribute to inactivation of the enzyme. By contrast, high concentrations of glucose 6-phosphate increase the activity of the otherwise inactive phosphorylated glycogen synthase in both muscle and liver. This allosteric regulation explains how blood glucose concentration modulates glycogen synthesis by the liver.

8.6 ABNORMALITIES OF GLYCOGEN METABOLISM

8.6.1 Glycogen Storage Diseases

There are a number of inborn errors of metabolism which result in excessive accumulation of intracellular glycogen; these are called *glycogen storage diseases*. Deficiencies in the activity of the liver isozymes of glycogen phosphorylase or phosphorylase kinase result in liver enlargement (hepatomegaly) and fasting hypoglycemia. The symptoms of a deficiency of glycogen phosphorylase or phosphorylase kinase are not as severe as those associated with a deficiency of glucose 6-phosphatase (von Gierke disease) where neither glycogenolysis nor gluconeogenesis is available to support blood glucose homeostasis because glucose 6-phosphate cannot be dephosphorylated and released from the hepatocytes.

By contrast, a genetic deficiency of the muscle isozyme of glycogen phosphorylase (McArdle disease) is accompanied by muscle-specific accumulation of glycogen and a limited ability to exercise strenuously. In the absence of glycogenolysis to support exercise, affected persons rapidly develop a depleted energy charge within the muscle that causes severe muscle pain and weakness.

8.6.1.1 Pompe Disease. During cellular recycling and tissue remodeling, most cells of healthy persons accumulate some glycogen within lysosomes. Catabolism of glycogen within lysosomes utilizes an alternative pathway in which lysosomal acid maltase hydrolyzes both α-1,4- and α-1,6-glycosidic bonds. Unlike the other glycogen storage diseases, deficiency of acid maltase results in accumulation of glycogen in lysosomes rather than as granules in the cytosol. Whereas many organs are affected in Pompe disease, the most serious problem is cardiac damage, which ultimately results in death due to cardiorespiratory failure.

8.6.2 Neonatal Hypoglycemia

At delivery, the maternal supply of glucose is cut off abruptly, resulting in a transient hypoglycemic state in the first 1 to 2 hours. In response to hypoglycemia, increased concentrations of epinephrine and glucagon in the blood stimulate the liver to initiate glycogenolysis, which is crucial to neonatal survival. However, some newborns lack adequate hepatic glycogen reserves to support blood glucose homeostasis in the first few hours after birth. During the third trimester, the fetal liver normally accumulates 10 to 12 g of glycogen, which provides the neonate with about a 12-hour supply of glucose. It is common for premature infants to be born before they have had time to accumulate a sufficient quantity of this crucial glucose reserve. Inadequate neonatal glycogen reserves can also be the result of in utero malnutrition. In either case, the result can be life threatening if intravenous glucose is not supplied in a timely manner.

A related problem can arise in an infant born to a mother who has any form of poorly controlled diabetes mellitus, including *gestational diabetes*, which is also called *glucose intolerance of pregnancy*. Although fetal exposure to a constant oversupply of glucose results in higher-than-normal glycogen stores in what are often large-for-gestational-age-infants, it also results in increased plasma insulin levels in the fetus. The persistence of an elevated insulin level in the infant after birth and the concomitant lower-than-normal concentration of glucagon act to suppress mobilization of glycogen from the neonatal liver. Thus, although these infants have ample glycogen reserves, hyperinsulinemia may prevent them from maintaining adequate circulating levels of blood glucose and put them at risk for developing life-threatening hypoglycemia.

8.6.3 Glycogen Loading

Although the preferred fuel of resting muscle is fatty acids, exercising muscle uses a mixture of glucose and fatty acids. With an increased rate of work, the percentage contribution of glucose to energy metabolism is increased. Sustained vigorous exercise normally depletes glycogen stores within 1 to 3 hours; since muscle glycogen is available only for the intracellular use of the muscle, glycogen stores are depleted only in those muscles that are actively exercising. Following exercise, repletion of the intramuscular glycogen stores usually occurs within 1 to 2 days, particularly if the diet provides at least 50 to 60% of total calories as carbohydrates. Interestingly, such repletion may increase glycogen stores to levels that are twice their original level. An athlete who trains vigorously two or three days before a competitive event and then consumes a high-carbohydrate diet can therefore enter the competition with higher-than-normal glycogen reserves and an increased capacity for sustained physical exertion. It should be noted that with training, the muscles also become less dependent on glucose, further increasing one's ability to go further on their glycogen reserves.

CHAPTER 9

GLUCONEOGENESIS

9.1 FUNCTION OF GLUCONEOGENESIS

In Chapter 3 we learned that some tissues, particularly the brain, red blood cells, and the renal medulla, depend heavily on glycolysis to satisfy their ATP needs. Yet, as discussed in Chapter 8, the glycogen stores of the body are limited and provide only about an 8- to10-hour supply of glucose to maintain the plasma glucose concentration within normal limits.

Gluconeogenesis is the process by which the body synthesizes glucose from endogenous noncarbohydrate precursors, primarily lactate and glycerol, and the carbon skeletons of the amino acids alanine and glutamine. This pathway is essential for maintaining the concentration of blood glucose in the fasted state. The term *gluconeogenesis* is used to indicate that this process is explicitly distinct from the interconversion of hexose sugars, through which glucose is generated from dietary fructose or galactose.

Unlike other fuel sources such as amino acids, lactate, and glycerol, the carbon skeletons of most fatty acids cannot be utilized for gluconeogenesis. Virtually all physiological fatty acids contain an even numbers of carbons (usually, C16 or C18) and their catabolism involves cleavage of the fatty acid chain into two-carbon acetyl-CoA units. Humans and other animals lack a pathway for converting acetyl-CoA to glucose. On the other hand, the oxidation of the relatively rare odd-chain fatty acids and branched methyl fatty acids that are present in human diets do generate small amounts of propionic acid that can be converted into glucose.

Medical Biochemistry: Human Metabolism in Health and Disease   By Miriam D. Rosenthal and Robert H. Glew
Copyright © 2009 John Wiley & Sons, Inc.

9.2   TISSUES IN WHICH GLUCONEOGENESIS IS ACTIVE

The liver has long been considered the major gluconeogenic organ. However, recent studies indicate that contributions of the renal cortex to gluconeogenesis have been underappreciated. Current estimates indicate that, on a wet weight basis, the renal cortex produces more glucose than the liver, much of it for use by the renal medulla. Indeed, renal gluconeogenesis protects the body from severe hypoglycemia under conditions of liver failure. It should be noted, however, that the kidney lacks significant glycogen stores and, unlike the liver, can contribute to glucose homeostasis via gluconeogenesis but not by way of glycogenolysis.

9.3   PHYSIOLOGICAL CONDITIONS IN WHICH GLUCONEOGENESIS IS ESPECIALLY ACTIVE

Gluconeogenesis is active primarily when a person is in the fasted state, when dietary carbohydrates have been utilized or stored as glycogen, and the plasma concentration of glucose has declined. The liver starts synthesizing glucose in response to the decreased insulin/glucagon ratio which occurs after postprandial processing of absorbed nutrients and increases the rate of gluconeogenesis as glycogen stores become depleted during a subsequent fasting period.

Gluconeogenesis also increases during prolonged physical exercise and serves to provide glucose for heart and active skeletal muscle. After exercise, the rate of gluconeogenesis remains elevated and contributes to modest replenishment of muscle glycogen stores prior to the availability of dietary glucose. During both fasting and recovery from prolonged exercise, the substantial energy cost of gluconeogenesis is met primarily by concurrent β-oxidation of fatty acids to acetyl-CoA in the liver.

Fasting gluconeogenesis is particularly important in the neonate. In the first few hours after delivery, the newborn experiences a period of transient hypoglycemia resulting from loss of the continuous glucose infusion that had been provided through the umbilical cord from the maternal circulation. Since glycogen stores in the newborn are insufficient to meet the resulting need for blood sugar, the healthy neonate responds by increasing the rate of gluconeogenesis.

9.4   REACTIONS OF GLUCONEOGENESIS

9.4.1   Enzymes Unique to Gluconeogenesis

The pathway for gluconeogenesis utilizes many, but not all, of the enzymes of glycolysis (Fig. 9-1). The reactions that are common to glycolysis and gluconeogenesis are the reversible reactions. This is best illustrated when one considers lactate as the starting point for gluconeogenesis. As discussed in Chapter 3, lactate is the endpoint of glycolysis when pyruvate cannot be metabolized through pyruvate dehydrogenase and the TCA cycle. Anaerobic glycolysis occurs in red blood cells, in the renal medulla, and

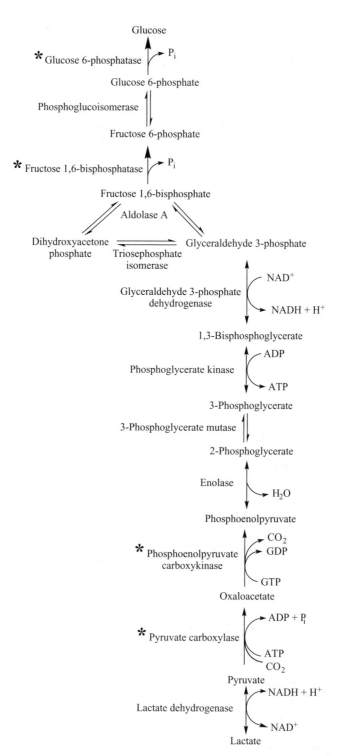

FIGURE 9-1 Pathway for gluconeogenesis from lactate. Asterisks indicate irreversible steps.

in skeletal muscle during strenuous exercise. Gluconeogenesis serves to regenerate glucose from lactate, thereby ensuring a constant supply of glucose for those cells and tissues that are highly dependent on glycolysis for their energy needs.

There are three physiologically irreversible steps in the glycolytic pathway that must be bypassed by gluconeogenesis-specific enzymes. Two of these irreversible steps are the two ATP-requiring activation reactions of glycolysis catalyzed by glucokinase and phosphofructokinase-1; they are bypassed by glucose 6-phosphatase and fructose 1,6-bisphosphatase, respectively. The third irreversible step of glycolysis is the second ATP-generating reaction, which is catalyzed by pyruvate kinase. (The other substrate-level phosphorylation reaction that generates ATP is catalyzed by phosphoglycerate kinase; it is reversible and is utilized in gluconeogenesis as well as glycolysis.) The gluconeogenesis pathway utilizes the reactions catalyzed by pyruvate carboxylase and phosphoenolpyruvate carboxykinase to bypass the irreversible pyruvate kinase reaction of glycolysis (Fig. 9-2).

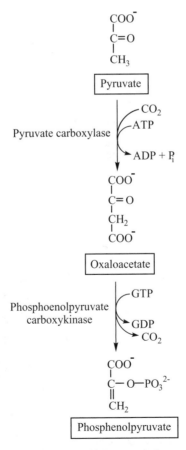

FIGURE 9-2 The two initial steps of gluconeogenesis.

Thus, the pathway for gluconeogenesis from lactate shown in Figure 9-1 has many reactions in common with the glycolytic pathway shown in Figure 3-3. The reactions catalyzed by enzymes such as phosphoglucoisomerase, aldolase A, glyceraldehyde 3-phosphate dehydrogenase, and phosphoglycerate kinase are reversible and are at near-equilibrium under the normal intracellular conditions of pH and concentrations of substrates and products. These reversible reactions can therefore be used for either glycolysis or gluconeogenesis, depending on the activity of each of the enzymes that catalyze the irreversible steps that regulate the availability of substrates for the two pathways.

Let us consider those steps that are unique to gluconeogenesis.

9.4.1.1 *Pyruvate Carboxylase.*

Carboxylation of pyruvate by pyruvate carboxylase generates oxaloacetate and requires input of energy in the form of ATP:

$$\text{pyruvate} + CO_2 + ATP \rightarrow \text{oxaloacetate} + ADP$$

The mitochondrial synthesis of oxaloacetate was discussed briefly in Chapter 4 in the context of anaplerotic reactions that serve to replenish the supply of citric acid cycle intermediates. Similar to other carboxylases, the catalytic domain of pyruvate carboxylase contains a molecule of biotin that is covalently attached to the enzyme by means of an amide bond between the carboxyl group of the biotin side chain and the ε-amino group of a lysine residue of the enzyme (Fig. 9-3). Biotin serves as a carrier of the activated CO_2 that is transferred to pyruvate to form oxaloacetate. This same CO_2 molecule is released during the next step in gluconeogenesis and does not appear in the resulting glucose molecule. Pyruvate carboxylase is activated by

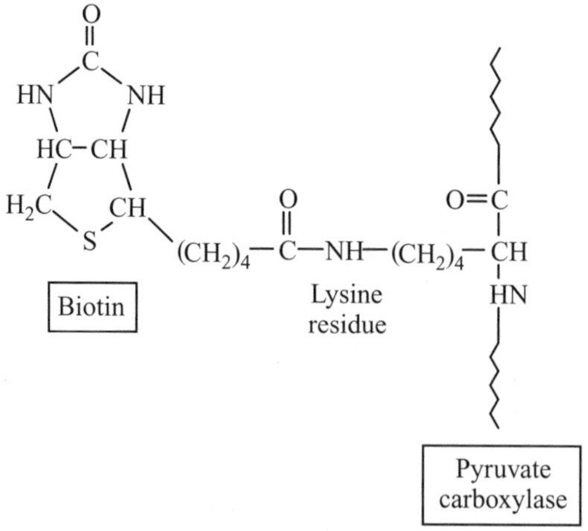

FIGURE 9-3   Covalent linkage of biotin to pyruvate-CoA carboxylase.

acetyl-CoA, ensuring that gluconeogenesis occurs only when there is sufficient fatty acid oxidation to provide the energy needed for glucose synthesis.

9.4.1.2 *Phosphoenolpyruvate Carboxykinase.* The next step in gluconeo-genesis involves the simultaneous decarboxylation and phosphorylation of oxalo-acetate. GTP provides the high-energy phosphate group that ends up in the product, phosphoenolpyruvate (PEP):

$$\text{oxaloacetate} + \text{GTP} \rightarrow \text{phosphoenolpyruvate} + CO_2 + \text{GDP}$$

By virtue of the phosphate group it contains, PEP is trapped in the cytosol and thus remains available for the next step in gluconeogenesis.

The sequence of reactions catalyzed by pyruvate carboxylase and pyruvate car-boxykinase serves to bypass the irreversible pyruvate kinase reaction of glycolysis. The utilization of these two reactions is not, however, without an extra energy cost. Whereas one ATP is generated during glycolysis by the conversion of PEP to pyruvate in the pyruvate kinase reaction, it takes two high-energy phosphate groups (one from ATP and one from GTP) to generate PEP from pyruvate during gluconeogenesis.

9.4.1.3 *Fructose 1,6-Bisphosphatase.* Hydrolysis of fructose 1,6-bisphos-phate is the next irreversible step of gluconeogenesis:

$$\text{fructose 1,6-bisphosphate} + H_2O \rightarrow \text{fructose 6-phosphate} + P_i$$

This reaction reverses the step in glycolysis that is catalyzed by phosphofructo-kinase-1, but it does not regenerate the ATP utilized in that reaction. Like phosphofructokinase-1, which is the main step that is regulated in glycolysis, fruc-tose 1,6-bisphosphatase is a critical allosteric enzyme that is involved in regulation of gluconeogenesis.

9.4.1.4 *Glucose 6-Phosphatase.* This enzyme, which is as integral protein of the endoplasmic reticulum, catalyzes the intraluminal hydrolysis of glucose 6-phosphate to generate free glucose:

$$\text{glucose 6-phosphate} + H_2O \rightarrow \text{glucose} + P_i$$

Glucose 6-phosphatase-catalyzed dephosphorylation of glucose 6-phosphate is essential for the release of glucose from the cell. Expression of glucose 6-phosphatase in liver and renal cortex permits these tissues to utilize gluconeogenesis to maintain blood glucose. As discussed in Chapter 8, hepatic glucose 6-phosphatase is also necessary for the release of glucose from the liver during glycogenolysis. In the absence of glucose 6-phosphatase, glucose 6-phosphate is trapped within the cell and either utilized for glycolysis or incorporated into glycogen.

9.4.2 Precursors for Gluconeogenesis

9.4.2.1 Lactate. The gluconeogenesis pathway described above started with lactate as the substrate and essentially reversed glycolysis to produce glucose. Gluconeogenesis thus provides a mechanism by which the liver and renal cortex can synthesize glucose from the lactate produced by skeletal muscle during strenuous exercise and continuously by red blood cells. The metabolic interchange between lactate-generating cells and gluconeogenic cells, called the *Cori cycle*, is illustrated in Figure 4-5.

The Cori Cycle Requires Net Energy Input. More energy is required to generate glucose from lactate in the liver than is obtained by oxidizing glucose in red blood cells. Glycolysis of glucose to lactate produces a net of two molecules of ATP per molecule of glucose oxidized. By comparison, gluconeogenesis from lactate requires 6 ATP equivalents (4 ATP, 2 GTP) to produce one molecule of glucose. The two-step sequence catalyzed by pyruvate carboxylase and pyruvate carboxykinase requires one ATP plus one GTP per molecule of pyruvate. An additional ATP is required for the phosphoglycerate kinase-catalyzed conversion of 3-phosphglycerate to 1,3-bisphosphoglycerate. Since two molecules of lactate are utilized for the synthesis of each molecule of glucose, the total energy cost is 2×3 or 6, high-energy bonds.

Is the Cori Cycle Just a Waste of Energy? It may appear that continued breakdown and resynthesis of glucose is wasteful. It is, however, the small energy cost paid by the liver and renal cortex to permit effective functioning of other cells. Since erythrocytes lack both mitochondria and a nucleus, they are smaller than most cells; their relatively small cell size permits easy passage of red cells through tiny capillaries but renders red blood cells completely dependent on glycolysis to lactate for their ATP. The conversion of lactate to glucose occurs in the liver, where ample ATP can be generated from the β-oxidation of long-chain fatty acids.

Glycolysis to lactate is also advantageous during strenuous exercise. Although the yield of ATP per glucose molecule metabolized is much lower than when glucose is oxidized all the way to CO_2 and water, the rate at which ATP can be generated by glycolysis is greater than the rate at which ATP can be produced by oxidative phosphorylation. Most of the lactate generated by muscle is secreted into the circulation and returned to the liver, where it is converted back into glucose by the process of gluconeogenesis. The energy required to convert lactate into glucose is derived from the mitochondrial β-oxidation of fatty acids and transfer of the reducing equivalents from NADH and $FADH_2$ into the ATP-generating oxidative phosphorylation system of mitochondria.

9.4.2.2 Alanine. Alanine is the major gluconeogenic amino acid substrate of the liver. In the fasted state, proteolysis of muscle proteins provides substrates for maintaining blood glucose homeostasis. However, not all amino acid carbon skeletons can be converted into glucose. In particular, muscle protein contains a significant percentage (approximately 20%) of branched-chain amino acids that are ketogenic or

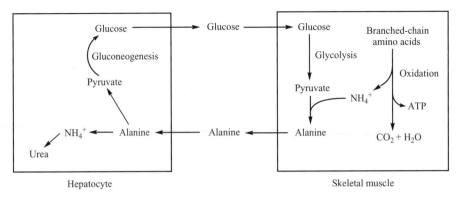

FIGURE 9-4 Alanine cycle.

mixed ketogenic and glucogenic. Oxidation of the carbon chains of branched-chain amino acids occurs primarily within muscle and serves as a significant energy source for muscle during fasting. Before branched-chain amino acids can be oxidized, the α-amino groups must be removed by transamination and exported from the muscle primarily as alanine and glutamine.

In the synthesis of alanine in muscle, pyruvate serves as the acceptor molecule for the α-amino groups transferred, with the pyruvate being derived from glycolysis. This means that muscle cells need a constant supply of glucose to sustain the net export of gluconeogenic precursors. That glucose supply is provided mainly by hepatic gluconeogenesis from alanine. The interorgan cycle of glucose catabolism in the muscle to generate alanine and the recycling of the carbon skeletons of alanine to glucose in the liver is called the *alanine cycle* (Fig. 9-4). Like the Cori cycle discussed above, the alanine cycle has a net energy cost. Nevertheless, the alanine cycle has significant advantages to the organism as a whole since it permits efficient catabolism of muscle proteins that provide substrates for gluconeogenesis.

The pathway for hepatic gluconeogenesis from alanine is similar to that from lactate in that both lactate and alanine are readily converted to pyruvate. In the case of alanine, the reaction involves transamination in which the α-amino group of an amino acid is transferred to α-ketoglutarate and subsequently excreted as urea:

$$\text{alanine} + \alpha\text{-ketoglutarate} \rightarrow \text{pyruvate} + \text{glutamate}$$

9.4.2.3 Glutamine. In the renal cortex, glutamine is the preferred substrate for gluconeogenesis. Like alanine, glutamine is synthesized by skeletal muscle in the fasted state as a means of exporting the amino groups of amino acids. In the kidney, the two amino groups of glutamine are removed by glutaminase and glutamate dehydrogenase, producing free ammonium ions and α-ketoglutarate. The ammonium ions serve to buffer acids excreted in the urine, while the α-ketoglutarate provides substrate for gluconeogenesis. As a result of the linkage between the generation of free ammonium ions and α-ketoglutarate, gluconeogenesis in the kidney increases

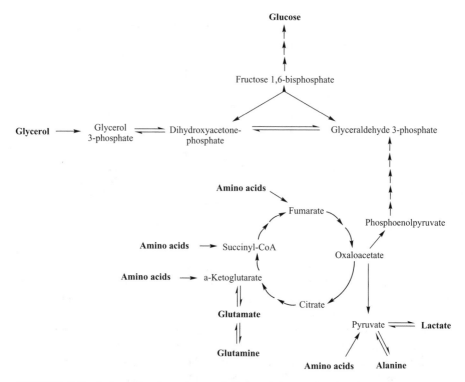

FIGURE 9-5 Pathway for gluconeogenesis from substrates other than lactate. Amino acids that generate pyruvate: alanine, cysteine, glycine, methionine, serine, threonine, and tryptophan. Amino acids that generate α-ketoglutarate: arginine, glutamate, glutamine, histidine, and proline. Amino acids that generate succinyl-CoA: isoleucine, threonine, and valine. Amino acids that generate fumarate: phenylalanine, tyrosine, and aspartate (via the urea cycle). Amino acids that generate oxaloacetate: asparagine and aspartate.

significantly during conditions of acidosis as well as fasting. Oxidation of α-ketoglutarate via the TCA cycle produces oxaloacetate (Fig. 9-5), which then enters the same pathway as that used to synthesize glucose from lactate (Fig. 9-1).

9.4.2.4 Many Other Amino Acids, Including Histidine, Proline, and Asparagine, Are Glucogenic.

A number of the other amino acids can contribute all or a part of their carbon skeletons to gluconeogenesis. In each instance, the carbon skeletons of these glucogenic amino acids are metabolized either to pyruvate or to one of the TCA-cycle intermediates, such as oxaloacetate, succinyl-CoA, or α-ketoglutarate (Fig. 9-5).

Not all amino acid carbon skeletons can be utilized for gluconeogenesis, because the catabolism of certain amino acids generates acetyl-CoA, and humans cannot convert acetyl-CoA into glucose. The pyruvate dehydrogenase reaction is irreversible, and animal cells lack an alternative pathway for utilizing acetyl-CoA for the net synthesis of TCA-cycle intermediates. Amino acids such as leucine which are catabolized to acetyl-CoA do not provide carbon skeletons that are suitable for glucose synthesis in humans. Still other amino acids are both glucogenic and ketogenic. For example, catabolism of tryptophan produces both pyruvate and acetyl-CoA: The pyruvate can be utilized for gluconeogenesis, but the acetyl-CoA cannot.

9.4.2.5   Glycerol. In the fasted state, mobilization of adipose triacylglycerols provides free fatty acids and glycerol. Although even-chain fatty acids are catabolized to acetyl-CoA and, like ketogenic amino acids, are not substrates for gluconeogenesis, the glycerol that is released during lipolysis can be a significant source of substrate for glucose synthesis.

Glycerol released from adipocytes is taken up by the liver and phosphorylated by glycerol kinase:

$$\text{glycerol} + \text{ATP} \rightarrow \text{glycerol 3-phosphate} + \text{ADP}$$

NAD^+-dependent glycerol 3-phosphate dehydrogenase then oxidizes the glycerol 3-phosphate to dihydroxyacetone phosphate, which enters the gluconeogenic pathway at the level of the aldolase A reaction (Fig. 9-5):

$$\text{glycerol 3-phosphate} + NAD^+ \rightarrow \text{dihydroxyacetone phosphate} + \text{NADH} + H^+$$

9.4.3   Localization of Gluconeogenesis

All of the enzymes of the glycolytic pathway are located in the cytosol. Although most of the enzymes of gluconeogenesis are also found in the cytosol, there are two exceptions: pyruvate carboxylase and glucose 6-phosphatase.

9.4.3.1   Pyruvate Carboxylase. Pyruvate carboxylase is a mitochondrial enzyme that provides oxaloacetate for the TCA cycle as well as for gluconeogenesis. Before oxaloacetate can be converted into glucose, it must be shuttled out of the mitochondrion. This is accomplished by first reducing oxaloacetate to malate inside the mitochondrion; following its transport into the cytosol, malate is then reoxidized to oxaloacetate. Oxaloacetate derived from α-ketoglutarate or other TCA-cycle intermediates is also transported out of the mitochondrion as malate. Since both the mitochondrial and cytosolic malate dehydrogenase isozymes are NAD^+-linked, the net effect of this process is the conversion of NADH to NAD^+ in the mitochondrion and the generation of NADH from NAD^+ in the cytosol (Fig. 9-6). Cytosolic NADH is required for the subsequent conversion of 1,3-bisphosphoglycerate to glyceraldehyde 3-phosphate (Fig. 9-1). It should be noted that the gluconeogenic pathway from lactate generates NADH in the initial lactate dehydrogenase reaction:

$$\text{lactate} + NAD^+ \rightarrow \text{pyruvate} + \text{NADH} + H^+$$

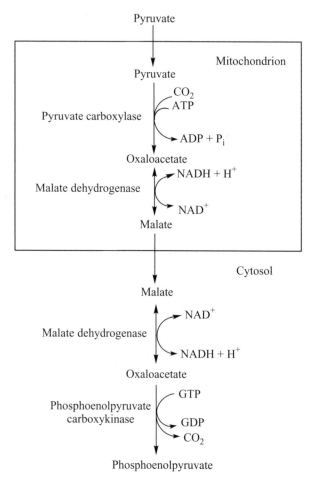

FIGURE 9-6   Intracellular localization of the initial steps of gluconeogenesis.

whereas gluconeogenesis from alanine does not:

$$\text{alanine} + \alpha\text{-ketoglutarate} \rightarrow \text{pyruvate} + \text{glutamate}$$

9.4.3.2 Glucose 6-Phosphatase. The hydrolysis of glucose 6-phosphate to glucose is required for the release of glucose from the cell and as such is the last step in both gluconeogenesis and glycogenolysis in the liver. As discussed in Chapter 8, the hydrolysis of glucose 6-phosphate to free glucose occurs not in the cytosol but in the lumen of the endoplasmic reticulum (Fig. 8-5). Although some glucose may be transported back into the cytosol, most of it remains extracellular, where it serves to maintain blood glucose homeostasis.

9.5 REGULATION OF GLUCONEOGENESIS

Glycolysis and gluconeogenesis in hepatocytes are reciprocally regulated in that physiological conditions that activate one pathway concurrently inactivate the other. As described in Chapter 3, the major regulated steps of glycolysis are those catalyzed by phosphofructokinase 1 (PFK-1) and pyruvate kinase. The major regulated steps of gluconeogenesis are the reactions that bypass these two irreversible reactions of glycolysis: those catalyzed by fructose 1,6-bisphosphatase, pyruvate carboxylase, and phosphoenolpyruvate carboxykinase (Fig. 9-7). Regulation of gluconeogenesis occurs on several levels, as described below.

9.5.1 Allosteric Regulation

9.5.1.1 *Pyruvate Carboxylase and Phosphoenolpyruvate Carboxykinase.* Both of these enzymes are inhibited when the energy charge of the cell

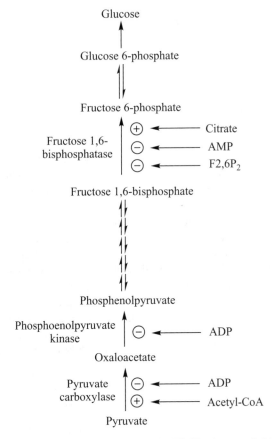

FIGURE 9-7   Regulation of gluconeogenesis. F2,6P$_2$, fructose 2,6-bisphosphate.

is low; in both cases the allosteric inhibitor is ADP rather than AMP. In addition, pyruvate carboxylase is stimulated by high concentrations of mitochondrial acetyl-CoA.

9.5.1.2 Fructose 1,6-Bisphosphatase. This enzyme is activated by citrate and inhibited by AMP. Both of these regulatory mechanisms serve to ensure that gluconeogenesis occurs only when sufficient energy is available for the synthesis of glucose. Citrate is exported from the mitochondrion during the process of shuttling the acetyl moiety of acetyl-CoA into the cytosol for the synthesis of cholesterol and fatty acids. Thus, the cytosolic citrate concentration increases when mitochondrial acetyl-CoA is present at a concentration which exceeds that required for ATP generation via the combined actions of the TCA cycle and the electron transport chain. Conversely, the high levels of AMP and low cytosolic citrate concentrations that occur when the energy charge of the cell is low act to inhibit gluconeogenesis. Since phosphofructokinase-1 is inhibited by citrate and stimulated by AMP, the low-energy charge of the cell also increases the rate of glycolysis.

9.5.2   Hormonal Regulation of Enzymatic Activity

Synthesis of glucose by means of gluconeogenesis is essential to blood glucose homeostasis in the fasted state. Glucagon and epinephrine stimulate gluconeogenesis in part by increasing substrate availability (e.g., glycerol and amino acids) and also through their effects on the activity of fructose 1,6-bisphosphatase. These two hormones also act to inhibit glycolysis in liver by inhibiting phosphofructokinase-1 (PFK-1). By contrast, insulin acts to inhibit fructose 1,6-bisphosphatase and stimulate PFK-1.

The mechanism of hormonal effects on both gluconeogenesis and glycolysis involves regulation of the concentration of fructose 2,6-bisphosphate, which is an allosteric inhibitor of the gluconeogenic enzyme fructose 1,6-bisphosphatase as well as an activator of the glycolytic enzyme phosphofructokinase-1. As described in Chapter 3 and illustrated in Figure 3-9, synthesis and degradation of fructose 2,6-bisphosphate are catalyzed by two enzyme activities, phosphofructokinase-2 and fructose 2,6-bisphosphatase, respectively, which are both contained on one bifunctional protein. Phosphorylation of the liver isozyme by protein kinase A inhibits phosphofructokinase-2 and activates fructose 2,6-bisphosphatase, thereby decreasing the intracellular concentration of fructose 2,6-bisphosphate and stimulating gluconeogenesis. By contrast, insulin initiates a signal transduction cascade, which results in the dephosphorylation of the bifunctional protein, increasing the intracellular concentration of fructose 2,6-bisphosphate and thus inhibiting gluconeogenesis while increasing the rate of glycolysis.

9.5.3   Transcriptional Regulation of Gene Expression

Expression of three of the four gluconeogenic enzymes—phosphoenolpyruvate carboxykinase, fructose 1,6-bisphosphatase, and glucose 6-phosphatase—is increased in response to a high glucagon/insulin ratio in the fasted state, with phosphoenolpyruvate carboxykinase being the major regulated enzyme. Although the transcriptional response is slower than the hormonal regulation involving enzyme phosphorylation, increases in enzyme activities occur within 40 minutes of the return to fasting blood glucose levels. Expression of gluconeogenic enzymes is also stimulated by other hormones, including hydrocortisone. Indeed, enhancement of glucose synthesis is one of the major physiological roles of the glucocorticoids.

Unlike the other enzymes required for gluconeogenesis but not glycolysis, pyruvate carboxylase is expressed constitutively. This is consistent with the dual role of pyruvate carboxylase: It is a gluconeogenic enzyme in the fasted state and an important anaplerotic enzyme in the fed state.

9.6 ABNORMAL FUNCTION OF THE GLUCONEOGENIC PATHWAY

9.6.1 Genetic Disease

Type I glycogen storage disease or *von Gierke disease* is actually a condition of impaired gluconeogenesis as well as glycogen metabolism. The reason for this is that the defective enzyme in von Gierke disease, glucose 6-phosphatase, is needed for the export of glucose derived from gluconeogenesis as well as that derived from hepatic mobilization of glycogen stores. Insufficient glucose 6-phosphatase activity results in accumulation of excess glucose 6-phosphate and consequent excessive glycogen storage in both liver and kidneys. Von Gierke disease can also be a result of defects in the transporter systems that transport glucose 6-phosphate from the cytosol to the lumen of the endoplasmic reticulum and return P_i and glucose to the cytosol (Fig. 8-5). Since neither gluconeogenesis nor glycogenolysis can provide glucose to the blood in the absence of glucose 6-phosphatase, the fasting hypoglycemia of von Gierke disease is more severe than that due to glycogen phosphorylase deficiency.

9.6.2 Poorly Controlled Diabetes Mellitus Type 1

Diabetes mellitus derives its name from the fact that the urine of patients with this disease has a characteristic sweetness caused by urinary excretion of glucose. The underlying pathology of type I diabetes mellitus is inadequate insulin production secondary to autoimmune damage to the β-cells of the pancreas. The result is a hormonal milieu in which the body perceives that it is starving even in the postprandial state. The liver responds to the elevated glucagon/insulin ratio by increasing gluconeogenesis, thus resulting in hyperglycemia and glycosuria.

9.6.3 Excess Ethanol Consumption

As discussed to a greater extent in Chapter 13, the major pathway that metabolizes ethanol involves two successive oxidation steps, both of which require NAD^+ and generate NADH. The resulting elevated $NADH/NAD^+$ ratio depletes gluconeogenic substrates by driving all of the following reactions to the right:

$$\text{pyruvate} + \text{NADH} + \text{H}^+ \rightarrow \text{lactate} + \text{NAD}^+$$

$$\text{dihydroxyacetone phosphate} + \text{NADH} + \text{H}^+ \rightarrow \text{glycerol 3-phosphate} + \text{NAD}^+$$

$$\text{oxaloacetate} + \text{NADH} + \text{H}^+ \rightarrow \text{malate} + \text{NAD}^+$$

Acute ethanol intoxication is thus often associated with severe, even life-threatening hypoglycemia, especially if the person consuming the ethanol is malnourished and has limited glycogen stores. A similar problem can occur when ethanol consumption follows strenuous, glycogen-depleting exercise. The remedy in either case is to provide the person with oral or, if necessary, intravenous glucose.

CHAPTER 10

FATTY ACID OXIDATION AND KETONES

10.1 FUNCTIONS OF FATTY ACID OXIDATION

10.1.1 Fatty Acid Oxidation Provides Energy for Cellular and Metabolic Work

Triacylglycerols (TAG) are the major energy store of the body and the major endogenous fuel in the fasted state. Triacylglycerols are not only a more concentrated energy source than glucose (or glycogen), generating 9 kcal/g compared to 4 kcal/g from glucose, but they can also be stored in a more compact, nonhydrated form. Current American diets typically contain 35 to 50% of calories as fat. In addition, after a meal, dietary carbohydrates in excess of immediate caloric needs are converted to fat and stored for future use.

Most of the fatty acids oxidized by the β-oxidation pathway are linear, unbranched molecules comprised of 16 or 18 carbon atoms. These long-chain fatty acids (LCFA) include the saturated fatty acids palmitic acid (16:0) and stearic acid (18:0), where the notation in parentheses indicates the number of carbons and, after the colon, the number of double bonds (Fig 10-1). The most common monounsaturated fatty acid is oleic acid (9c–18:1) in which the *cis* carbon–carbon double bond starts on carbon atom 9 from the carboxyl end of the molecule. The most common polyunsaturated fatty acid is linoleic acid, which is essential in the diet. Linoleic acid is an 18-carbon, diunsaturated fatty acid with carbon–carbon double bonds starting on carbon atoms 9 and 12 from the carboxyl end of the molecule (9c,12c–18:2). Linoleic acid may also be written as 18:2ω6 (18:2*n*-6) to indicate that that the first carbon–carbon double

Medical Biochemistry: Human Metabolism in Health and Disease   By Miriam D. Rosenthal and Robert H. Glew
Copyright © 2009 John Wiley & Sons, Inc.

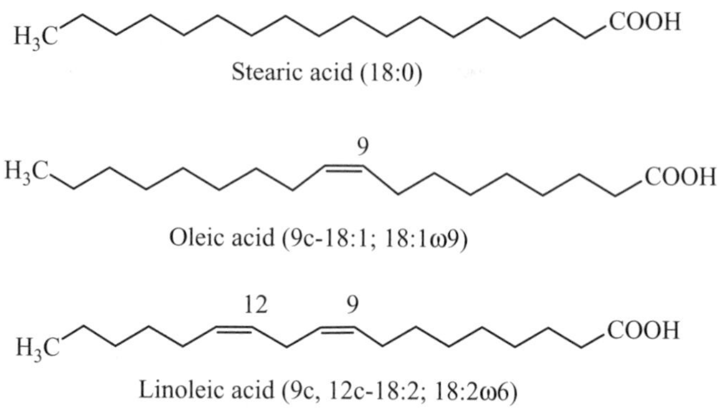

H_3C ⌒⌒⌒⌒⌒⌒⌒⌒COOH

Stearic acid (18:0)

H_3C ⌒⌒⌒⌒ = ⌒⌒⌒COOH

Oleic acid (9c-18:1; 18:1ω9)

H_3C ⌒⌒ = ⌒ = ⌒⌒COOH

Linoleic acid (9c, 12c-18:2; 18:2ω6)

FIGURE 10-1   Structures of some common fatty acids.

bond starts on the sixth carbon from the methyl or omega (ω) end of the molecule. Using the latter nomenclature, oleic acid would be written 18:1ω9.

β-Oxidation is the pathway by which the long-chain fatty acids from both dietary fat and adipose tissue TAG are oxidized to acetyl-CoA. The reducing equivalents released during fatty acid oxidation are captured in the form of $FADH_2$ and NADH, which are used to support oxidative phosphorylation. In most circumstances, the acetyl-CoA units generated by β-oxidation will subsequently be oxidized through the tricarboxylic acid (TCA) cycle, generating additional $FADH_2$ and NADH and ultimately, additional ATP. Particularly in the fasted state, many cells and tissues depend on β-oxidation of fatty acids to provide the ATP needed to maintain ion gradients and to support biosynthetic processes such as gluconeogenesis.

10.1.2   Fatty Acid Oxidation Provides Fuel to the Brain During Starvation

Since the brain does not utilize long-chain fatty acids as an energy source, it is normally dependent on the oxidation of glucose to meet its energy needs. However, during prolonged fasting or starvation the brain meets its energy needs by oxidizing ketones (ketone bodies) as well as glucose. The ketones of physiological significance are four-carbon anions (β-hydroxybutyrate and acetoacetate) produced from acetyl-CoA generated by the β-oxidation of long-chain fatty acids in the liver (Fig. 10-2). Oxidation of ketones by the brain reduces the brain's dependence on glucose and thus decreases the body's need to catabolize muscle proteins to provide amino acid carbon skeletons for gluconeogenesis.

10.1.3   Fatty Acid Oxidation Generates Heat

Brown fat is a specialized tissue that has a high metabolic rate although it does not produce very much ATP from the NADH and $FADH_2$ generated during the oxidation of fatty acids. As discussed in Chapter 6, the presence of thermogenin or "uncoupling protein" in the inner membrane of the mitochondria of brown fat results in the

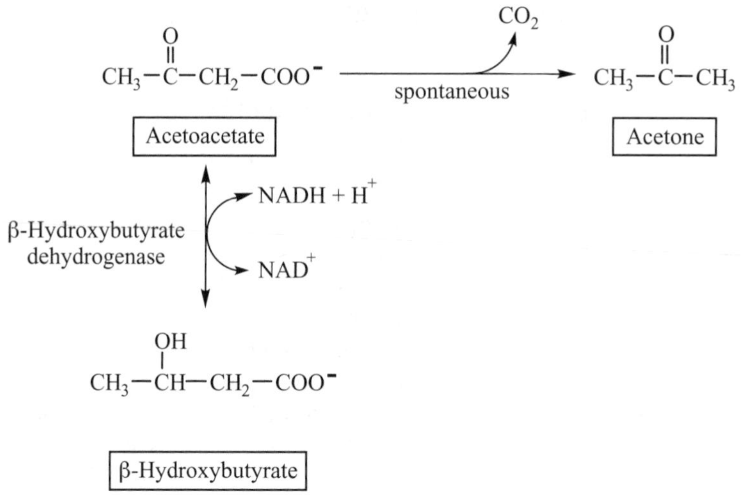

$$CH_3-\overset{\overset{\displaystyle O}{\|}}{C}-CH_2-COO^- \quad \xrightarrow[\text{spontaneous}]{\quad\quad CO_2\quad\quad} \quad CH_3-\overset{\overset{\displaystyle O}{\|}}{C}-CH_3$$

| Acetoacetate | | Acetone |

β-Hydroxybutyrate
dehydrogenase

NADH + H$^+$

NAD$^+$

$$\overset{\displaystyle OH}{CH_3-\overset{|}{C}H-CH_2-COO^-}$$

β-Hydroxybutyrate

FIGURE 10-2 Ketone bodies; acetoacetate is the precursor of both β-hydroxybutyrate and acetone.

generation of heat rather than ATP from electron flow through the electron-transport chain. This process contributes to nonshivering thermogenesis, which is particularly important for maintaining body temperature in newborns.

10.1.4 Fatty Acid Oxidation Provides a Pathway for Catabolism of Diverse Dietary Fatty Acids

The major fatty acids of both dietary TAG and adipose stores in the body contain 16 to 18 carbons and are oxidized through mitochondrial β-oxidation. Essentially the same pathway is utilized for short- (C4–C6) and medium-chain (C8–C12) fatty acids. Related peroxisomal pathways, which are discussed later in the chapter, oxidize the less common branched-chain and very long-chain ($\geq$C22) fatty acids.

10.2 TISSUES IN WHICH FATTY ACID OXIDATION IS ACTIVE

All cells and tissues except red blood cells and the brain oxidize fatty acids to generate ATP. The β-oxidation pathway is absent in red blood cells because they lack mitochondria. Although neuronal cells in the brain do contain mitochondria, there is only limited transport of fatty acids across the blood–brain barrier. This explains why fatty acids per se are not a significant fuel source for the brain.

Fatty acid oxidation is most active in tissues that are highly active metabolically. Thus, skeletal and heart muscle in particular have a large capacity for oxidizing fatty acids. Normally, 60 to 90% of the energy required for contraction of the heart is derived from the oxidation of fatty acids.

10.3   PHYSIOLOGICAL CONDITIONS IN WHICH FATTY ACID OXIDATION IS MOST ACTIVE

10.3.1   Fatty Acid Oxidation Increases in the Fasted State

In the fasted state, the plasma glucose concentration is depressed and tissues such as muscle rely primarily on oxidation of fatty acids to generate ATP. In addition, during a fast the liver has an increased requirement for ATP to provide the energy required for gluconeogenesis. The liver meets this requirement by β-oxidizing long-chain fatty acids to acetyl-CoA. The process of gluconeogenesis markedly reduces the intramitochondrial supply of oxaloacetate, thereby limiting subsequent oxidation of acetyl-CoA via the TCA cycle. Under these conditions, the two-carbon units of acetyl-CoA are utilized to synthesize four-carbon ketone bodies—acetoacetate and β-hydroxybutyrate—which leave hepatocytes and enter the blood. Oxidation of the ketone bodies to CO_2 and water occurs in tissues such as muscle and brain that do not carry out gluconeogenesis.

10.3.2   Fatty Acid Oxidation Increases During Exercise

Increased rates of fatty acid oxidation are required during exercise and provide a substantial fraction of the ATP required for muscle work. As discussed more fully in Chapter 25, the precise nature of the mix of glucose and fatty acids utilized by muscle depends on exercise intensity and duration and on a person's prior conditioning. It should be noted that, since substrate-level phosphorylation does not occur in the pathway of fatty acid oxidation, generation of ATP from β-oxidation is dependent on the availability of oxygen for oxidative phosphorylation. For this reason, strenuous exercise that depletes oxygen renders muscle more dependent on glycolysis to lactate than is the case when the work rate of muscle is lower.

10.4   PATHWAYS OF FATTY ACID OXIDATION

10.4.1   Transport of Fatty Acids

Since long-chain fatty acids are poorly soluble in aqueous media, they must be transported in the plasma complexed with albumin. When a fatty acid dissociates from albumin it is transferred from the capillary lumen through the capillary endothelium and interstitial space to the cells below. Long-chain (C16–C20) fatty acids enter cells both by simple diffusion and by carrier-mediated transport. The diffusion mechanism involves initial penetration of the outer leaflet of the plasma membrane by the hydrophobic tail of the fatty acid, followed by a "flip-flop" within the membrane. Thus, when the fatty acid emerges on the cytosolic side of the membrane, the carboxyl group of the fatty acid enters the cytosol ahead of the hydrocarbon tail.

There are three major fatty acid transporter proteins in the plasma membranes of human cells: fatty acid transporter (FAT/CD36), plasma membrane fatty acid

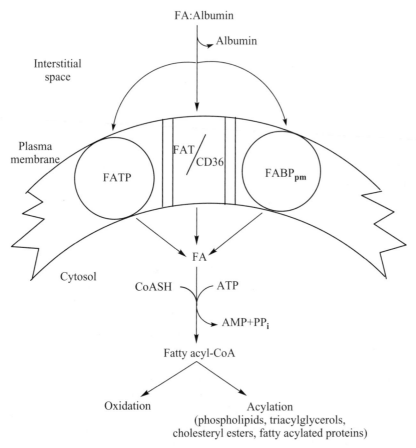

FIGURE 10-3   Fatty acid transporters in the plasma membrane. FATP, fatty acid transport protein; FABP$_{pm}$, fatty acid–binding protein of the plasma membrane.

transport protein (FABPpm), and fatty acid transport protein (FATP) (Fig. 10-3). Of these, FAT/CD36 is the major fatty acid transporter in heart muscle, skeletal muscle, adipocytes, and intestine. Intracellularly, fatty acids are bound to cytosolic fatty acid binding proteins (FABP), which deliver the fatty acids to the sites where they are metabolized.

10.4.2   Activation of Free Fatty Acids

Once inside cells, fatty acids must be activated before they can be metabolized. In contrast to glucose, which is activated and trapped within cells as glucose 6-phosphate, fatty acids are converted not to acyl phosphates but to thioesters of coenzyme A

(CoASH) by acyl-CoA synthetase:

$$CoASH + ATP + R-COOH \rightleftharpoons R-\overset{\overset{\displaystyle O}{\displaystyle \|}}{C}-SCoA + AMP + PP_i$$

This reversible reaction is pulled in the direction of acyl-CoA synthesis by pyrophosphatase, which hydrolyzes PP_i to $2P_i$.

Acyl-CoA synthetases are localized to three different sites in cells: the cytosolic face of the endoplasmic reticulum, the outer mitochondrial membrane, and the peroxisomal membrane. There are at least five genetically distinct acyl-CoA synthetase (ACS) isoforms, each having its own specificity with regard to the fatty acid substrate. For example, ACS4 prefers polyunsaturated fatty acids such as arachidonic acid (20:4ω6) and docosahexaenoic acid (DHA, 22:6ω3).

10.4.3 Mitochondrial β-Oxidation

10.4.3.1 The Carnitine–Fatty Acid Transport System. Long-chain fatty acids destined for β-oxidation are activated to their CoA forms primarily on the surface of the outer mitochondrial membrane. The inner mitochondrial membrane is, however, impermeable to long-chain fatty acyl-CoA molecules. Transport of fatty acids containing 16 to 20 carbon atoms across the inner mitochondrial membrane is facilitated by a fatty acid transport mechanism called the *carnitine translocase system.*

Carnitine is a quaternary amine that has a hydroxyl group to which a fatty acid can be attached (Fig. 10-4). The fatty acid is linked to carnitine by means of an oxygen ester bond that is unusual because it is energy-rich and highly reactive (unlike the more stable oxygen ester linkage between fatty acids and glycerol in TAG). Since carnitine can be synthesized in the liver and kidney from trimethyllysine, it is not usually considered an essential dietary nutrient. Synthesis of trimethyllysine occurs as a posttranslational modification of muscle proteins, with *S*-adenosylmethionine

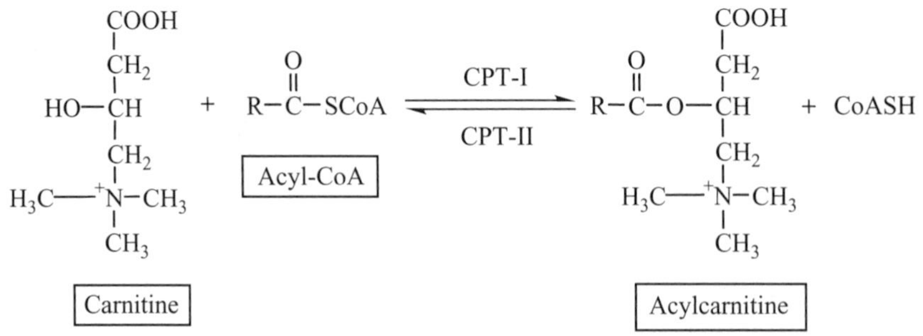

FIGURE 10-4 Transfer of acyl groups catalyzed by carnitine palmitoyltransferase I (CPT-I) and carnitine palmitoyltransferase II (CPT-II).

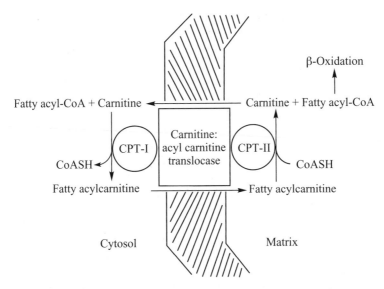

FIGURE 10-5 Transport of long-chain fatty acids into mitochondria by means of the carnitine–acylcarnitine translocase system. CPT-I, carnitine palmitoyltransferase I; CPT-II, carnitine palmitoyltransferase II.

(the activated form of methionine) serving as the methyl donor; when proteins turn over, trimethyllysine is released and made available for carnitine synthesis.

The crux of the system for transport of long-chain fatty acids into the mitochondrion is a carnitine translocase, which is embedded in the inner mitochondrial membrane (Fig. 10-5). This translocase transports fatty acylcarnitine into the mitochondria in exchange for free carnitine, which is concurrently exported from the mitochondrial matrix into the cytosol.

The activity of the carnitine translocase system is dependent on two enzymes, carnitine palmitoyltransferase I (CPT-I) and carnitine palmitoyltransferase II (CPT-II), both of which catalyze the reversible transfer of long-chain fatty acids between coenzyme A and carnitine:

$$\text{carnitine} + \text{R}-\overset{\overset{\displaystyle O}{\|}}{\text{C}}-\text{SCoA} \rightleftharpoons \text{R}-\overset{\overset{\displaystyle O}{\|}}{\text{C}}-\text{O}-\text{carnitine} + \text{CoASH}$$

CPT-I is localized to the mitochondrial outer membrane and acts to generate acylcarnitine. CPT-II, which is localized to the matrix face of the inner mitochondrial membrane, forms intramitochondrial acyl-CoA from CoASH and acylcarnitine. As shown in Figure 10-5, the net effect is the transfer of a long-chain fatty acid from an extramitochondrial CoASH molecule to an intramitochondrial CoASH molecule.

10.4.3.2 *Mitochondrial β-Oxidation Pathway.* The term β-*oxidation* is derived from the fact that the critical chemistry of the four core reactions that comprise the pathway takes place on the third carbon from the carboxyl end: that is, the β-carbon atom.

The first of the four core reactions of the β-oxidation pathway is irreversible and is catalyzed by acyl-CoA dehydrogenase (Fig. 10-6). Two hydrogen atoms are removed—one each from the α and β carbons—generating a carbon–carbon double bond between the α and β carbons of the fatty acyl-CoA chain. These hydrogen atoms are transferred to FAD to form FADH$_2$. Mitochondria contain a family of FAD-linked acyl-CoA dehydrogenases: a long-chain acyl-CoA dehydrogenase, which is specific for fatty acids containing 14 to 20 carbon atoms; a medium-chain acyl-CoA dehydrogenase, which oxidizes intermediate chain-length fatty acids (C8–C12); and a short-chain acyl-CoA dehydrogenase, which oxidizes C4 and C6 fatty acids.

The second step in β-oxidation involves hydration of the carbon–carbon double bond between the α- and β-carbons by enoyl-CoA hydratase (Fig. 10-6). The hydroxyl group is introduced onto the β-carbon. A second dehydrogenase, NAD$^+$-linked β-hydroxyacyl-CoA dehydrogenase, then oxidizes the hydroxyacyl-CoA molecule to generate a β-ketoacyl-CoA and a molecule of NADH + H$^+$. The fourth and final step in β-oxidation involves cleavage of the fatty acid chain with attachment of a second molecule of CoASH to the β-carbon and generation of one molecule of acetyl-CoA. The enzyme that catalyzes this reaction is called β-ketoacyl-CoA thiolase, reflecting the fact that the cleavage of the carbon–carbon bond involves a sulfhydryl group.

The net effect of the four steps in β-oxidation is the production of one molecule of acetyl-CoA and one fatty acyl-CoA molecule whose carbon chain is two carbons shorter than the original substrate. The four steps are then repeated, with successive chain shortening by two carbon atoms. The final thiolytic cleavage reaction converts the 4-carbon β-ketobutyryl-CoA (acetoacetyl-CoA) into two molecules of acetyl-CoA.

10.4.3.3 *Energy Yield from β-Oxidation.* Calculation of the amount of ATP that can be derived from the oxidation of one long-chain fatty acid such as palmitate (16:0) is a useful exercise for demonstrating the energy-dense nature of fatty acids. The two dehydrogenase reactions that are part of each round of mitochondrial β-oxidation produce one molecule of FADH$_2$ and one molecule of NADH. As discussed in Chapter 6, passage of the reducing equivalents from these reduced cofactors into the electron transport chain yields 2.5 ATP/NADH and 1.5 ATP/FADH$_2$. The seven cycles of β-oxidation that are required to oxidize a 16-carbon palmitic acid molecule to 8 molecules of acetyl-CoA thus generate 7 × 4, or 28, ATP. Subsequent oxidation of these acetyl-CoA molecules via the TCA cycle generates 10 ATP/acetyl-CoA or 80 ATP/8 acetyl CoA, for a total yield of 108 ATP per palmitoyl-CoA. Subtracting the two high-energy bonds expended in the fatty acid synthetase and pyrophosphatase reactions that converts palmitic acid into palmitoyl-CoA, each molecule of palmitic acid can thus generate 106 molecules of ATP.

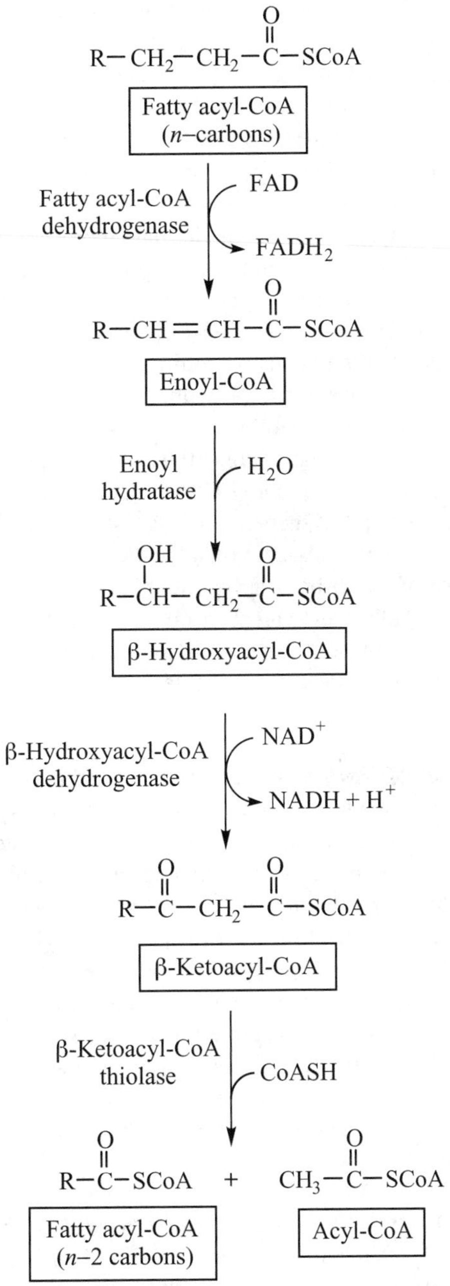

$$
\underset{\substack{\text{Fatty acyl-CoA} \\ (n\text{–carbons})}}{\boxed{}}\; R-CH_2-CH_2-\overset{\overset{\displaystyle O}{\|}}{C}-SCoA
$$

Fatty acyl-CoA
dehydrogenase $\quad\Bigg\langle\; \begin{array}{l} \text{FAD} \\ \\ FADH_2 \end{array}$

$$
R-CH = CH-\overset{\overset{\displaystyle O}{\|}}{C}-SCoA
$$

Enoyl-CoA

Enoyl
hydratase $\quad\Big\langle\; H_2O$

$$
R-\underset{\underset{\displaystyle OH}{|}}{CH}- CH_2-\overset{\overset{\displaystyle O}{\|}}{C}-SCoA
$$

β-Hydroxyacyl-CoA

β-Hydroxyacyl-CoA
dehydrogenase $\quad\Bigg\langle\; \begin{array}{l} NAD^+ \\ \\ NADH + H^+ \end{array}$

$$
R-\overset{\overset{\displaystyle O}{\|}}{C}-CH_2-\overset{\overset{\displaystyle O}{\|}}{C}-SCoA
$$

β-Ketoacyl-CoA

β-Ketoacyl-CoA
thiolase $\quad\Big\langle\; CoASH$

$$
R-\overset{\overset{\displaystyle O}{\|}}{C}-SCoA \quad + \quad CH_3-\overset{\overset{\displaystyle O}{\|}}{C}-SCoA
$$

Fatty acyl-CoA
(n–2 carbons) $\qquad\qquad$ Acyl-CoA

FIGURE 10-6 Mitochondrial β-oxidation pathway.

10.4.3.4 Ancillary Reactions to the Pathway of β-Oxidation.

β-Oxidation of saturated fatty acids generates an unsaturated intermediate with a $\Delta^{2,3}$-*trans* double bond, which is then hydrated by enoyl-CoA hydratase. The metabolism of unsaturated and polyunsaturated fatty acids requires additional enzymes to act on pre-existing *cis* double bonds. As illustrated in Figure 10-7, the oxidation of linoleic acid (c9,c12–18:2) proceeds in essentially the same manner as the β-oxidation of saturated fatty acids, with successive cleavage of two-carbon units as acetyl-CoA. However, after three cycles of β-oxidation, the chain-shortening process produces an acyl-CoA molecule that has a *cis*-3,4 double bond. At this point, an ancillary enzyme, Δ^3,Δ^2-enoyl-CoA isomerase, converts the $\Delta^{3,4}$-*cis* double bond to a $\Delta^{2,3}$-*trans* double bond, thus providing a suitable substrate for enoyl-CoA hydratase.

A slightly different situation arises when the chain-shortening process produces an acyl-CoA molecule that has a *cis*-4,5 double bond. Under these conditions, the acyl-CoA dehydrogenase step generates a $\Delta^{2,3}$-*trans*/$\Delta^{4,5}$-*cis* conjugated diunsaturated fatty acyl-CoA intermediate (Fig. 10-7). At this point a second ancillary enzyme, NADPH-dependent 2,4-dienoyl-CoA reductase, transfers hydrogen atoms from NADPH to carbons 4 and 5, generating a $\Delta^{3,4}$-*trans*-enoyl-CoA. Δ^3,Δ^2-Enoyl isomerase then converts $\Delta^{3,4}$-*trans*-enoyl-CoA to $\Delta^{2,3}$-*trans*-enoyl-CoA.

Note that by virtue of the carbon–carbon double bonds they contain, unsaturated fatty acids generate slightly less ATP than do saturated fatty acids. For example, β-oxidation of oleic acid bypasses one acyl-CoA dehydrogenase step and thereby generates one less $FADH_2$ molecule (and 1.5 fewer ATP) than its saturated counterpart, stearic acid.

10.4.3.5 Oxidation of Medium-Chain Fatty Acids.

Breast milk contains relatively large amounts of medium-chain (C8–C12) fatty acids that provide nursing infants with substantial amounts of energy. These shorter fatty acids are more soluble than are their more common C16–C20 counterparts and can enter the mitochondrion directly from the cytosol without need for the carnitine transporter system. The C8–C12 fatty acids are activated to their corresponding acyl-CoA derivatives within the mitochondrion and then undergo β-oxidation. The initial oxidation reaction is catalyzed by medium-chain acyl-CoA dehydrogenase (MCAD). Because medium-chain fatty acids bypass both the carnitine transporter system and the β-oxidation enzymes specific to long-chain acyl-CoAs, TAG comprised of C8–C12 fatty acids are sometimes used as nutrient supplements to treat persons who cannot oxidize long-chain fatty acids due to genetic deficiencies of proteins involved in either the transport or oxidation of long-chain fatty acids.

10.4.3.6 Oxidation of Fatty Acids Containing an Odd Number Of Carbon Atoms.

Dietary lipids often contain a small amount of odd-carbon fatty acids such as 17:0. Odd-chain fatty acids are oxidized by the normal mitochondrial β-oxidation process. However, the last thiolytic cleavage step produces one molecule of acetyl-CoA and one molecule of propionyl-CoA from the methyl end of the fatty acid. As indicated in Chapter 9, propionyl-CoA is gluconeogenic in a fasting person because it can be metabolized to glucose in the liver. The incorporation of propionyl-CoA

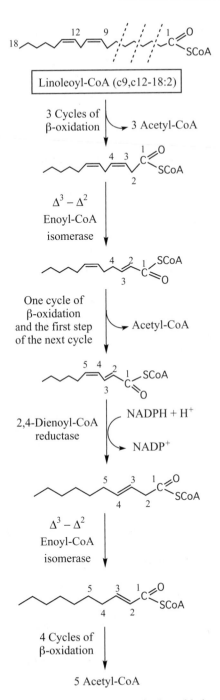

FIGURE 10-7 Role of ancillary enzymes in the oxidation of linoleic acid.

into mainstream metabolism involves carboxylation by propionyl-CoA carboxylase, followed by a vitamin B_{12}-requiring reaction that generates the TCA-cycle intermediate succinyl-CoA. It should be noted that only three of the carbons of an odd-chain fatty acid such as 17:0 are gluconeogenic; the other carbons, like those of even-chain fatty acids, generate only acetyl-CoA and cannot be utilized to synthesize glucose.

10.4.4   Ketone Bodies

The association of ketones (a.k.a., *ketone bodies*) with the ketoacidosis of diabetes has given these substances an undeservedly sinister connotation. However, ketones are normal metabolites that serve as circulating fuels, especially during periods of moderate (12 to 24 hours) or severe (>5 days) fasting. The two physiologically significant ketones are acetoacetate (β-ketobutyrate) and β-hydroxybutyrate (Fig. 10-2). Acetone is the product of the nonenzymatic decarboxylation of acetoacetate. Unlike hydrophobic long-chain fatty acids that require albumin for their transport in the plasma, ketone bodies are water-soluble and do not require a carrier protein for transport.

10.4.4.1   Ketone Synthesis.   Ketones are synthesized mainly in the liver, with a smaller contribution from the renal cortex. In both tissues, the substrate for ketogenesis is mitochondrial acetyl-CoA, which is derived from the β-oxidation of long-chain fatty acids and, to a lesser extent, from the oxidation of ketogenic amino acids (e.g., leucine). In the fasted state, much of the acetyl-CoA generated by β-oxidation cannot enter the TCA cycle because of a relative shortage of oxaloacetate which has been diverted to gluconeogenesis. The pathway of ketone body synthesis (Fig. 10-8) converts two acetyl-CoA molecules into one four-carbon acetoacetate molecule while releasing two free CoASH molecules, which are required for continued β-oxidation. Continued β-oxidation, in turn, provides $FADH_2$ and NADH substrate for oxidative phosphorylation.

The first step in acetoacetate synthesis is catalyzed by β-ketothiolase, which also catalyzes the last step in β-oxidation:

$$2\,\text{acetyl-CoA} \rightleftharpoons \text{acetoacetyl-CoA} + \text{CoASH}$$

This reversible reaction is driven to the right by a high concentration of acetyl-CoA arising from β-oxidation.

The acetoacetyl-CoA from the β-ketothiolase reaction is then combined with a third molecule of acetyl-CoA to form β-hydroxy-β-methylglutaryl-CoA (HMG-CoA) in the reaction is catalyzed by HMG-CoA synthase:

$$\text{acetoacetyl-CoA} + \text{acetyl-CoA} \rightarrow \text{β-hydroxy-β-methyl-glutaryl-CoA}$$

$$+ \text{CoASH}$$

Most cells contain a second HMG-CoA synthase that is localized to the cytosol, where it is involved in the pathway of cholesterol synthesis rather than ketogenesis.

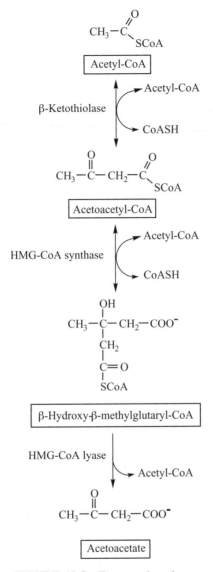

FIGURE 10-8   Ketogenesis pathway.

Mitochondrial HMG-CoA is then hydrolyzed by HMG-CoA lyase to produce acetoacetate plus acetyl-CoA:

β-hydroxy-β-methylglutaryl-CoA + H_2O → acetoacetate + acetyl-CoA

While about one-third of the acetoacetate produced by HMG-CoA lyase is secreted by the liver into the circulation, the other two-thirds is first reduced by mitochondrial β-hydroxybutyrate dehydrogenase and then secreted:

$$\text{acetoacetate} + \text{NADH} + \text{H}^+ \rightleftharpoons \beta\text{-hydroxybutyrate} + \text{NAD}^+$$

This reaction is driven in the direction of β-hydroxybutyrate synthesis by the relatively high mitochondrial ratio of NADH/NAD$^+$ generated by active β-oxidation of fatty acids. β-Hydroxybutyrate is more reduced and more energy-rich than acetoacetate.

10.4.4.2 Utilization of Ketones. Although the liver and red blood cells do not oxidize ketones, heart and skeletal muscle are capable of efficiently oxidizing ketones. Furthermore, after several days of fasting, the brain is also capable of utilizing ketones as an energy source. Ketone utilization is initiated by mitochondrial β-hydroxybutyrate dehydrogenase, which converts β-hydroxybutyrate back into acetoacetate:

$$\beta\text{-hydroxybutyrate} + \text{NAD}^+ \rightleftharpoons \text{acetoacetate} + \text{NADH} + \text{H}^+$$

Acetoacetate is then activated (and trapped within the cell) by one of two mitochondrial enzymatic reactions: The first trapping reaction is reversible and catalyzed by succinyl-CoA:β-ketoacid CoA-transferase:

$$\text{acetoacetate} + \text{succinyl-CoA} \rightleftharpoons \text{acetoacetyl-CoA} + \text{succinate}$$

The other trapping reaction is catalyzed by acetoacetyl-CoA synthetase:

$$\text{acetoacetate} + \text{ATP} + \text{CoASH} \rightarrow \text{acetoacetyl-CoA} + \text{AMP} + \text{PP}_i$$

β-Ketoacid CoA-transferase and acetoacetyl-CoA synthetase are both absent from hepatocytes, which accounts for the inability of liver to oxidize ketones.

As discussed above, acetoacetyl-CoA is the penultimate intermediate in the pathway of β-oxidation, and is cleaved by β-ketothiolase into two molecules of acetyl-CoA:

$$\text{acetoacetyl-CoA} + \text{CoASH} \rightarrow 2\ \text{acetyl-CoA}$$

Since tissues such as muscle that oxidize ketone bodies do not perform gluconeogenesis and thus do not deplete their supply of oxaloacetate in the fasted state, the acetyl-CoA molecules generated from acetoacetate and β-hydroxybutyrate are readily oxidized in the TCA cycle.

Overall, ketogenesis and ketone utilization constitute a multiorgan process that allows for the complete oxidation of long-chain fatty acids to CO_2 and water. Oxidation of acetyl-CoA derived from ketones can be a considerable source of ATP and is particularly significant in the brain during a period of prolonged fasting.

10.4.5 Peroxisomal Oxidation of Fatty Acids

10.4.5.1 Oxidation of Very Long-Chain Fatty Acids. The initial oxidation of very long-chain fatty acids (VLCFA) comprised of 22 carbon atoms or more is accomplished by a modified β-oxidation pathway that operates in peroxisomes (Fig. 10-9). One major difference between the mitochondrial and peroxisomal pathways is that, since peroxisomes lack an electron transport system, the reduced cofactors

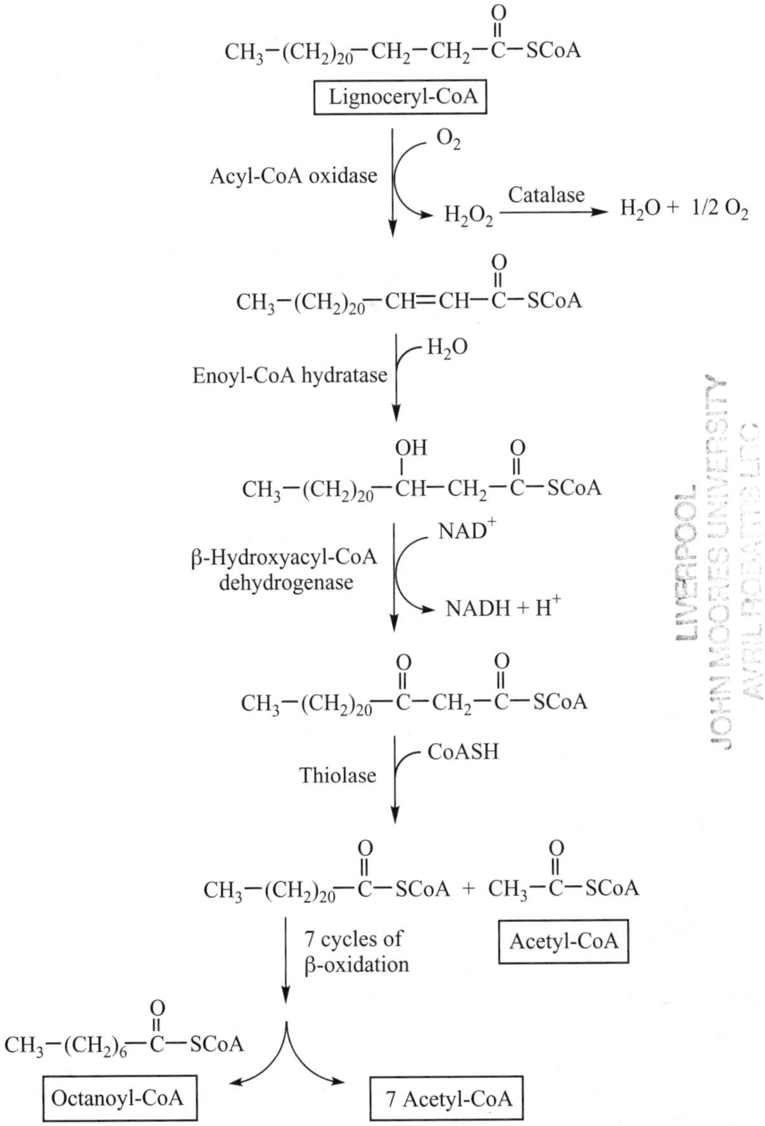

FIGURE 10-9 Pathway for oxidizing very long-chain fatty acids ($\geq$C22) in peroxisomes.

generated during peroxisomal β-oxidation are not channeled directly into oxidative phosphorylation.

The VLCFA are first activated to acyl-CoAs by a distinct acyl-CoA synthase called *lignoceroyl* (24:0) *ligase*:

$$\text{lignoceric acid} + \text{ATP} + \text{CoASH} \rightarrow \text{lignoceroyl-CoA} + \text{AMP} + \text{PP}_i$$

The first FAD-linked dehydrogenase step in the peroxisomal β-oxidation pathway is different from the corresponding step in standard mitochondrial β-oxidation. The peroxisomal FAD-linked dehydrogenase (called *acyl-CoA oxidase*) that removes two hydrogen atoms from the fatty acid chain transfers those hydrogens to molecular oxygen, thus producing H_2O_2.

$$R\text{--}CH_2\text{--}CH_2\text{--}\overset{\overset{\displaystyle O}{\|}}{C}\text{--}SCoA + FAD \rightarrow R\text{--}CH{=}CH\text{--}\overset{\overset{\displaystyle O}{\|}}{C}\text{--}SCoA + FADH_2$$

$$FADH_2 + O_2 \rightarrow FAD + H_2O_2$$

Catalase within the peroxisome then breaks down the hydrogen peroxide:

$$2H_2O_2 \rightarrow 2H_2O + O_2$$

The subsequent steps of the β-oxidation pathway in peroxisomes are similar to those that operate in mitochondrial β-oxidation. The reducing equivalents from the NADH generated by hydroxyacyl-CoA dehydrogenase are utilized for synthetic reactions within peroxisomes or shuttled out of the peroxisomes and eventually into mitochondria.

Once the peroxisomal β-oxidation pathway has reduced the very long-chain fatty acid chain to the level of an 8- or 10-carbon acyl-CoA molecule, the shortened fatty acid chain is transferred to mitochondria and further catabolized via the mitochondrial β-oxidation pathway. The peroxisomal acetyl-CoA units are probably hydrolyzed to acetate, which is subsequently oxidized in mitochondria.

10.4.5.2 α-Oxidation of Fatty Acids Containing a Branched Methyl Group.

The chemistry of the β-oxidation pathway entails removal of both hydrogen atoms from the β-carbon atom. Therefore, fatty acids that have a methyl group on C3 (the β-carbon) cannot be oxidized by regular β-oxidation and require a specialized pathway, which is called *α-oxidation*. One such branched-chain fatty acid is phytanic acid, which is derived from the phytanol side chain of chlorophyll. Phytanic acid has methyl groups on carbon atoms 3, 7, 11, and 15 (Fig. 10-10). Humans do not derive phytanic acid directly from dietary chlorophyll, but do obtain it from dietary dairy products, beef, and fatty fish.

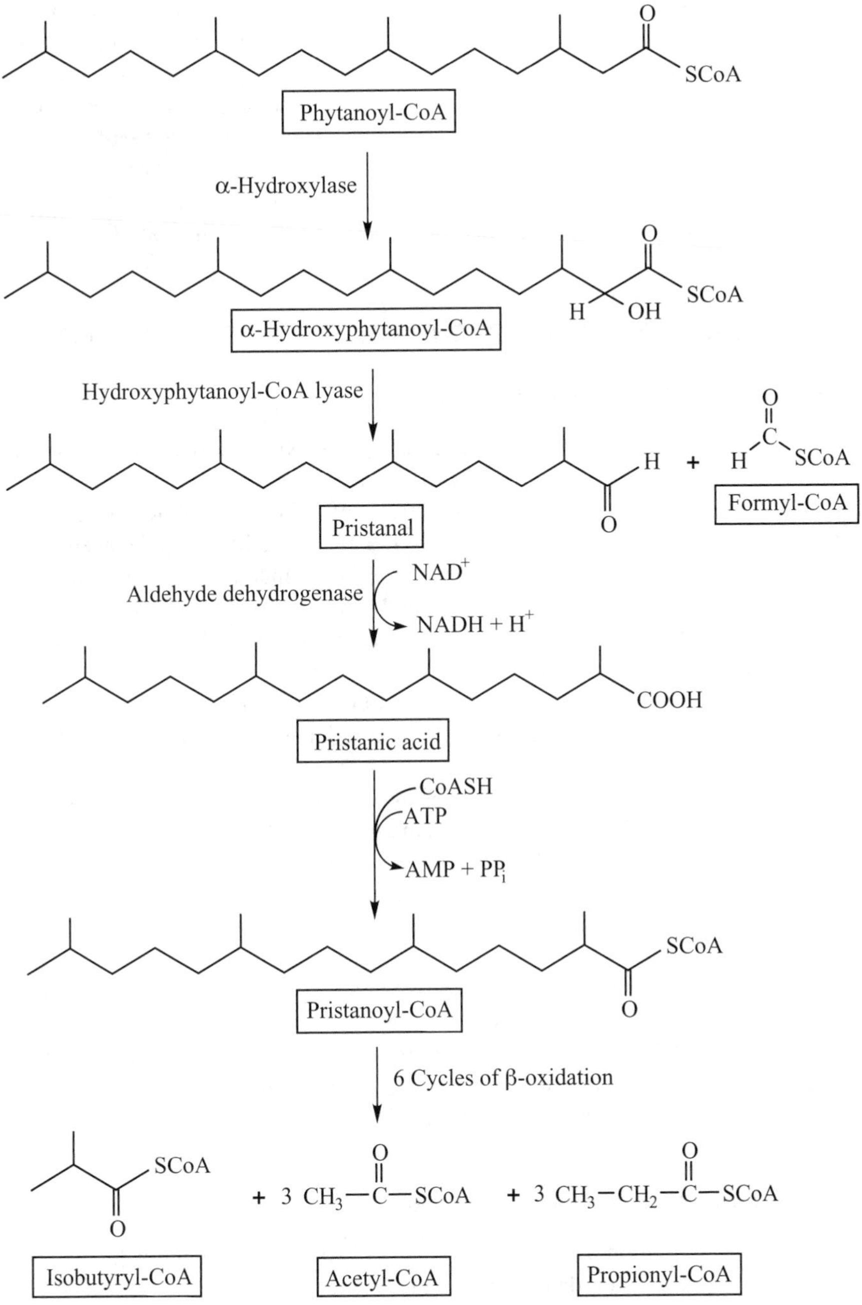

FIGURE 10-10 Peroxisomal α-oxidation of phytanic acid.

Peroxisomal α-oxidation of phytanoyl-CoA removes C1 of phytanic acid, thereby creating a shorter molecule in which the methyl group nearest the carboxyl is now on the α-carbon instead of the β-carbon; therefore, the product, pristanoic acid, is a suitable substrate for β-oxidation. Furthermore, the remaining methyl groups of the fatty acyl-CoA chain are now positioned on even-numbered carbon atoms and therefore do not present a problem for the enzymes of the standard β-oxidation pathway. Wherever a methyl group is attached to the α-carbon, cleavage of the carbon chain by β-ketothiolase will generate propionyl-CoA rather than acetyl-CoA.

10.5   REGULATION OF MITOCHONDRIAL FATTY ACID OXIDATION

10.5.1   Regulation by Energy Charge

The major site of regulation of the mitochondrial β-oxidation pathway is carnitine palmitoyltransferase-I (CPT-I), which controls the entry of long-chain (C16–C20) fatty acids into the mitochondrion. The activity of CPT-I is inhibited by malonyl-CoA, the product of the key regulatory enzyme of fatty acid synthesis: acetyl-CoA carboxylase. In the fed state, inhibition of CPT-I by malonyl-CoA prevents fatty acid oxidation when glucose is plentiful and when acetyl-CoA is being directed toward fatty acid synthesis. When a cell is actively synthesizing fatty acids de novo, the malonyl-CoA concentration in the cytosol increases. Subsequent inhibition of CPT-1 by malonyl-CoA decreases import of long-chain fatty acids into mitochondria, thereby preventing a futile cycle of simultaneous fatty acid synthesis and β-oxidation.

Conversely, when the energy charge of the cell is low, the increased concentration of AMP activates AMP-activated protein kinase (AMPK), which phosphorylates acetyl-CoA carboxylase, thereby inhibiting the enzyme so that it no longer produces malonyl-CoA. Thus, the effect of AMP activation of AMPK is to permit transport of fatty acids into the mitochondrion and ultimately increase the rate of β-oxidation.

β-Oxidation of fatty acids within the mitochondrion is also regulated by the energy charge of the cell. A high ATP/ADP ratio inhibits entry of reducing equivalents from NADH and $FADH_2$ into the electron-transport chain. The resulting increased concentrations of these reduced cofactors in turn prevent the two dehydrogenases of β-oxidation from acting when further generation of ATP is not required.

10.5.2   Regulation by Gene Transcription

Peroxisome proliferation-activator receptor-α (PPAR-α) is a ligand-activated transcription factor that stimulates fatty acid oxidation in liver and muscle. Ligands for PPAR-α include certain prostaglandins as well as some nonsteroidal anti-inflammatory drugs (e.g., indomethacin, ibuprofen). Ligand-activated PPAR-α induces the synthesis of many different genes, including members of the family of enzymes and proteins involved in β-oxidation.

10.6   DISEASES RELATED TO FATTY ACID OXIDATION

10.6.1   Medium-Chain Acyl-CoA Dehydrogenase Deficiency (MCADD)

The most common genetic defect in fatty acid oxidation is the one that affects the medium-chain acyl-CoA dehydrogenase. A deficiency in medium-chain acyl-CoA dehydrogenase activity is associated with high concentrations of both C8–C12 fatty acids and C8–C12 acylcarnitines in the plasma and urine of affected persons. Partial oxidation of these intermediate-chain-length fatty acids also generates unusual dicarboxylic fatty acids whose presence in body fluids is diagnostic of MCADD. These medium-chain dicarboxylic acids arise by a process called ω-*oxidation*, which takes place in the endoplasmic reticulum and involves oxidation of a fatty acid from its methyl end.

MCADD causes fasting hypoglycemia and muscle weakness. Limited utilization of fatty acids as fuels results in an increased dependence on glucose for muscle work. At the same time, gluconeogenesis is impaired because of the limited production of both ATP and NADH substrates needed to drive hepatic gluconeogenesis. Treatment of persons with MCADD involves avoiding periods of fasting that would tend to produce hypoglycemia. Patients with MCADD are advised to take frequent small meals that are relatively high in carbohydrates. They also benefit from consuming uncooked starch, which is digested and absorbed more slowly than cooked starch, thereby reducing the tendency toward hypoglycemia.

10.6.2   Genetic Defects in Long-Chain Fatty Acid Utilization

Genetic defects in many of the other proteins required for mitochondrial β-oxidation have also been documented. They include deficiencies in the genes encoding CPT-I, CPT-II, carnitine translocase, acyl-CoA dehydrogenase, and β-hydroxyacyl-CoA dehydrogenase. In all of these cases, the clinical manifestations include muscle weakness and fasting hypoglycemia, similar to those observed in patients with MCADD. Unlike the situation with MCADD, people with deficiencies in enzymes that metabolize long-chain fatty acids do benefit from diets that contain TAG composed primarily of medium-chain fatty acids. The utilization of these medium-chain fatty acids is not dependent on the palmitoylcarnitine translocase system or the β-oxidation enzymes that are specific for long-chain acyl-CoAs.

10.6.3   Systemic Carnitine Deficiency

Since mitochondrial oxidation of long-chain fatty acids depends on carnitine, anything that depletes the body of carnitine or which impairs intracellular carnitine availability will compromise a cell's capacity to carry out β-oxidation. Kidneys contain a carnitine transporter (the sodium-dependent organic cation transporter-2 or OCTN-2), which recovers 95% of filtered carnitine. The presence of OCTN-2 is also required for uptake of carnitine into peripheral tissues such as heart and skeletal muscle. Thus, genetic loss of the function of this renal carnitine transporter results in both carnitine wastage and impaired β-oxidation of long-chain fatty acids.

Carnitine deficiency can also occur in newborns who have a limited capacity for carnitine synthesis. It is not uncommon for underweight or premature newborns to be born with relatively low stores of carnitine. Since human milk contains relatively low concentrations of carnitine, premature infants may benefit from carnitine supplementation.

10.6.4 Hypoglycin

There are also environmental factors that can reduce a person's ability to oxidize fatty acids. One such factor is hypoglycin, a substance that is present in the unripe fruit of the tropical akee tree. Ingested hypoglycin inhibits the mitochondrial acyl-CoA dehydrogenase responsible for oxidizing short- and medium-chain acyl-CoAs, thus causing severe, life-threatening hypoglycemia.

10.6.5 Impaired β-Oxidation of VLCFA

X-linked adrenoleukodystrophy (ALD) is a relatively common (1/20,000) genetic disease characterized by elevated levels of C26:0 and an elevated C26:0/C22:0 ratio in plasma. Pathology results from the accumulation of cholesteryl esters of VLCFA, particularly in the central nervous system, the adrenal glands, and the testes, with adverse effects on membrane structure and steroidogenesis. The genetic defect lies not in any of the enzymes of activation or β-oxidation of VLCFA, but rather in a gene (ALDP for adrenoleukodystrophy protein) that is a member of the adenosine triphosphate–binding cassette (ABC) family of transporters, and which appears to be involved in the activation of VLCFA-CoA synthetase.

Impaired β-oxidation of VLCFA is also observed in patients with peroxisomal biogenesis disorders such as Zellweger syndrome and neonatal ALD. These persons have a defect in one or more of the *PEX proteins* that are required to import enzymes into the peroxisome. Cells of people with peroxisomal biogenesis disorders are essentially devoid of peroxisomes and exhibit deficits in multiple peroxisomal metabolic pathways, including synthesis of unsaturated ether lipids, α-oxidation of phytanic acid, and processing of bile acid intermediates as well as β-oxidation of VLCFA.

10.6.6 Refsum Disease

This genetic disease is caused by a lack of the α-hydroxylase required for α-oxidation of fatty acids, such as phytanic acid, that have a methyl group on an odd-numbered carbon. Accumulation of large quantities of phytanic acid in the nervous tissue and liver results in chronic polyneuropathy and cerebellar disfunction.

10.6.7 Ketosis

The condition in which the blood and urine concentrations of ketones are markedly elevated is called *ketosis* or *ketoacidosis*. Ketosis occurs when hepatic gluconeogenesis is especially active and ketone production exceeds oxidation of ketones by muscle and other tissues. Children are more susceptible to ketosis than adults because of their higher metabolic rate, lower glycogen stores, and higher brain weight/liver weight ratio. Children may develop ketosis as a result of infections that induce anorexia and vomiting. Ketosis is also commonly seen in patients with untreated type 1 diabetes, where insulin insufficiency results in increased fat mobilization, gluconeogenesis, and ketone synthesis.

Since both acetoacetate and β-hydroxybutyrate are organic acids, ketosis is a form of metabolic acidosis. In order to be excreted in the urine, the anionic metabolic acids in the urine must be counterbalanced by equivalent numbers of cations. Therefore, ketosis may result in depletion of body stores of sodium and potassium and in some instances even in the loss of divalent cations such as calcium and magnesium.

CHAPTER 11

FATTY ACID SYNTHESIS

11.1 FUNCTIONS OF FATTY ACID SYNTHESIS

Fatty acid synthesis serves two main functions: One is to convert dietary carbohydrates and the carbon skeletons of excess amino acids into triacylglycerols (TAG) that can be stored until needed during periods of fasting. The other function is to produce a variety of fatty acids, which are components of the complex lipids of biological membranes and the precursors of the eicosanoid lipid hormones.

The major pathway of fatty acid synthesis converts acetyl-CoA molecules derived from dietary carbohydrates and amino acids into the long-chain fatty acid palmitic acid (16:0). Additional enzymes elongate and desaturate both endogenous palmitate and dietary fatty acids to produce a number of other fatty acids, of which the most common are stearic acid (18:0) and oleic acid (c9–18:1) (see Fig. 10.1).

Two fatty acids, linoleic acid (c9,c12–18:2ω6) and α-linolenic acid (c9,12, 15–18:3ω3), are essential fatty acids in the sense that they cannot be synthesized by humans, and as such must be obtained from the diet (Fig. 11–1). Although neither linoleic acid nor α-linolenic acid can be synthesized by humans, these dietary fatty acids can be elongated and further desaturated to produce 20- and 22-carbon polyunsaturated fatty acids such as arachidonic acid (20:4ω6) and docosahexaenoic acid (22:6ω3).

Medical Biochemistry: Human Metabolism in Health and Disease   By Miriam D. Rosenthal and Robert H. Glew
Copyright © 2009 John Wiley & Sons, Inc.

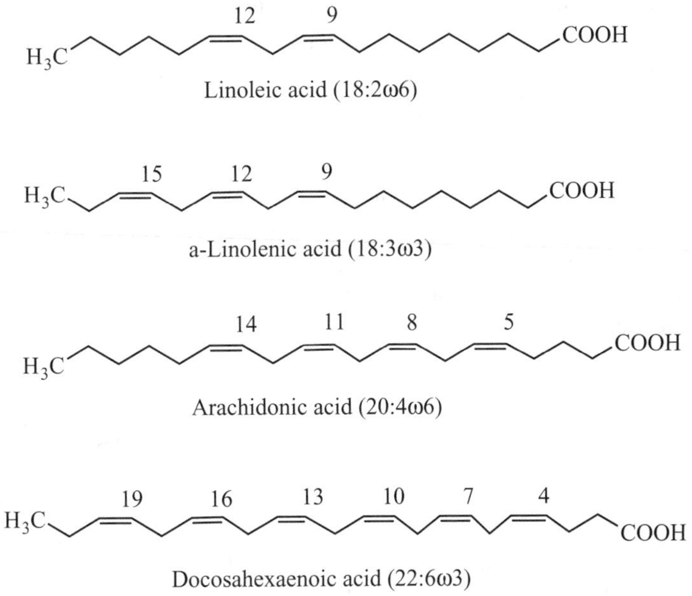

Linoleic acid (18:2ω6)

α-Linolenic acid (18:3ω3)

Arachidonic acid (20:4ω6)

Docosahexaenoic acid (22:6ω3)

FIGURE 11-1   Structures of some common polyunsaturated fatty acids.

11.2 LOCALIZATION OF FATTY ACID SYNTHESIS

Fatty acid synthesis takes place in the cytosol of most cells and tissues; however, hepatocytes and adipocytes are endowed with an especially high capacity for de novo fatty acid synthesis. In the case of fat cells, the fatty acids are esterified to glycerol and stored in the form of TAG. In the fasted state, the TAG in adipocytes are hydrolyzed sequentially by the triacylglycerol lipase desnutrin, hormone-sensitive lipase, and monoacylglycerol lipase, and the free fatty acids are released from adipocytes and transported through the blood bound to albumin.

Although the liver is the primary site of fatty acid synthesis in humans, hepatocytes do not normally accumulate TAG. Instead, the TAG are packaged into very low density lipoproteins (VLDL) and secreted into the circulation. In fact, the accumulation of extensive amounts of triacylglycerol in the liver is pathologic and can ultimately result in cirrhosis.

11.3 CONDITIONS WHEN FATTY ACID SYNTHESIS OCCURS

Fatty acid synthesis is most active following a meal. In the first few hours after foods containing carbohydrates such as starch and sucrose have been digested and absorbed, the body experiences a period of transient hyperglycemia, which triggers insulin secretion from the β-cells of the pancreas and suppresses the secretion of glucagon.

The resulting high insulin/glucagon ratio signals hepatocytes and adipocytes to take up glucose from the circulation and convert it into fatty acids, and ultimately into TAG. Fatty acid synthesis is thus greater when a person is consuming a high-carbohydrate diet than a diet that is relatively low in carbohydrates.

The rate of de novo fatty acid synthesis is very high during embryogenesis and in fetal lungs when there is a need for palmitic acid to support the synthesis of dipalmitoylphosphatidylcholine-rich pulmonary surfactant. Fatty acid synthesis is also greatly increased in cancer cells. Whereas normal cells obtain most of the fatty acids they need for membrane phospholipid synthesis from extracellular sources (e.g., plasma fatty acids and lipoprotein-associated triglycerides), cancer cells derive most of their fatty acids by means of de novo fatty acid synthesis.

11.4   REACTIONS THAT SYNTHESIZE AND MODIFY FATTY ACIDS

The pathway for de novo fatty acid synthesis generates palmitate (16:0) by the sequential addition of two-carbon units derived from acetyl-CoA to the growing fatty acid chain. These acetyl-CoA units are first activated by addition of CO_2 (HCO_3^{-1}) to form malonyl-CoA (see below). Synthesis of fatty acids also requires reducing equivalents that are provided by NADPH. The overall process of fatty acid synthesis can be summarized as follows:

$$8 \, acetyl\text{-}CoA + 7ATP + 14NADPH + H^+ + H_2O \rightarrow$$

$$palmitate + 7ADP + 7P_i + 14NADP^+ + 8CoASH$$

Palmitate (as palmitoyl-CoA) can be elongated to form stearoyl-CoA, and both palmitoyl-CoA and stearoyl-CoA can be desaturated to generate palmitoleoyl-CoA (c9–16:1) and oleoyl-CoA (c9–18:1), respectively.

11.4.1   Sources of Cytosolic Acetyl-CoA

The major source of acetyl-CoA for fatty acid synthesis is glucose. Acetyl-CoA is also generated from oxidation of the carbon skeletons of excess dietary amino acids and from ethanol. In all cases, acetyl-CoA that is not needed for the immediate generation of ATP via the TCA cycle and electron-transport system is routed to the synthesis of fatty acids.

Generation of acetyl-CoA from glucose involves metabolism of glucose via glycolysis to pyruvate in the cytosol, and formation of acetyl-CoA from pyruvate by means of the pyruvate dehydrogenase reaction that occurs in mitochondria. Since the inner mitochondrial membrane is impermeable to acetyl-CoA, fatty acid–synthesizing cells transport acetyl-CoA equivalents from the mitochondrial matrix into the cytosol in the form of citrate. This process is accomplished as follows. First, acetyl-CoA condenses with oxaloacetic acid in the citrate synthase reaction that normally functions

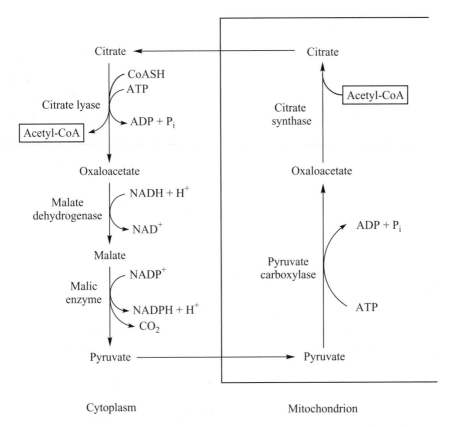

FIGURE 11-2   Pathway by which mitochondrial acetyl-CoA is transported into the cytosol and NADPH is concurrently generated from the reducing equivalents of NADH.

to introduce acetyl-CoA into the TCA cycle (Fig. 11-2):

$$oxaloacetate + acetyl\text{-}CoA \rightarrow citrate + CoASH$$

After being transported into the cytosol, citric acid is cleaved into acetyl-CoA and oxaloacetic acid by citrate lyase:

$$citrate + CoASH + ATP \rightarrow acetyl\text{-}CoA + oxaloacetate + ADP + P_i$$

This reaction serves two purposes: First, it provides acetyl-CoA substrate for fatty acid synthesis; second, it provides oxaloacetate for the NADPH-generating trans-hydrogenation pathway described below.

11.4.2 Sources of NADPH

11.4.2.1 Pentose Phosphate Pathway. Most of the NADPH that supplies the reducing equivalents for fatty acid synthesis is derived from the pentose phosphate pathway. The oxidative branch of this pathway utilizes two successive dehydrogenase reactions (catalyzed by glucose 6-phosphate dehydrogenase and 6-phosphogluconate dehydrogenase) to generate two molecules of NADPH for each molecule of glucose 6-phosphate oxidized to ribulose 5-phosphate.

11.4.2.2 Transhydrogenation Pathway. The net effect of the transhydrogenation pathway is the ATP-driven transfer of reducing equivalents from NADH to $NADP^+$ to produce NADPH. The pathway also serves to replenish the mitochondrial oxaloacetate pool needed to transport acetyl-CoA out of the mitochondrion in the form of citrate. The transhydrogenation pathway consists of three reactions (Fig. 11-2):

(1) $\qquad$ oxaloacetate $+ NADH + H^+ \rightarrow$ malate $+ NAD^+$

(2) $\qquad$ malate $+ NADP^+ \rightarrow$ pyruvate $+ CO_2 + NADPH + H^+$

(3) $\qquad$ pyruvate $+ ATP + CO_2 \rightarrow$ oxaloacetate $+ ADP + P_i$

net reaction : $NADH + NADP^+ + ATP \rightarrow NAD^+ + NADPH + ADP + P_i$

Reaction (1) is catalyzed by cytosolic malate dehydrogenase. The oxaloacetate in this reaction comes from the citrate lyase reaction described above. The malic enzyme, which catalyzes reaction (2), converts malate to pyruvate by means of $NADP^+$-dependent, oxidative decarboxylation. The pyruvate produced by the malic enzyme enters the mitochondrion, where it is carboxylated in reaction (3) by pyruvate carboxylase.

The NADH substrate for reaction (1) is derived from the glyceraldehyde 3-phosphate dehydrogenase step in glycolysis. Since glycolysis and fatty acid synthesis usually operate at the same time, metabolism of glucose thus provides reducing equivalents as well as acetyl-CoA for fatty acid synthesis.

11.4.3 Generation of Malonyl-CoA

Acetyl-CoA is the immediate donor of the two carbons at the methyl end of a newly synthesized fatty acid. Malonyl-CoA serves as the high-energy, highly reactive donor of the additional acetyl units used during the process of fatty acid synthesis. Acetyl-CoA carboxylase, the cytosolic, biotin-containing enzyme that catalyzes the synthesis of malonyl-CoA, is the rate-limiting step of fatty acid synthesis (Fig. 11-3):

$$\text{acetyl-CoA} + ATP + HCO_3^- \rightarrow \text{malonyl-CoA} + ADP + P_i$$

As previously described for pyruvate carboxylase (Fig. 9-3), biotin is covalently attached to a lysine residue of acetyl-CoA carboxylase.

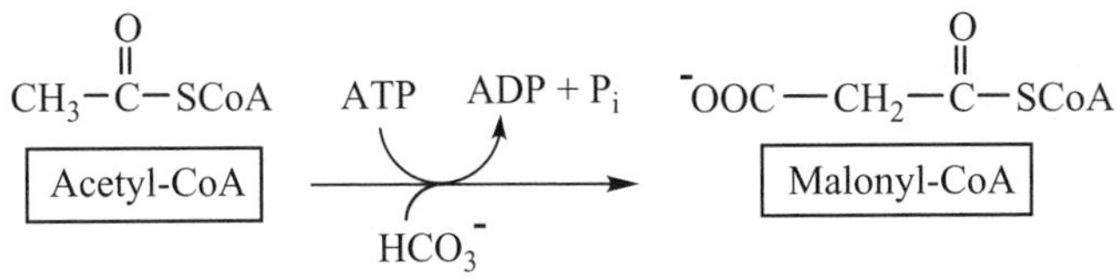

$$CH_3-\overset{\overset{\displaystyle O}{\|}}{C}-SCoA \quad ATP \quad ADP + P_i \quad {}^-OOC-CH_2-\overset{\overset{\displaystyle O}{\|}}{C}-SCoA$$

Acetyl-CoA ⟶ Malonyl-CoA

$$HCO_3^-$$

FIGURE 11-3 The acetyl-CoA carboxylase reaction.

Malonyl-CoA is an energy-rich compound that has a considerable fraction of the energy of the ATP molecule incorporated into its structure. During the process of fatty acid synthesis, release of the ionized carboxyl group of malonyl-CoA as CO_2 drives the formation of carbon–carbon bonds.

11.4.4 Fatty Acid Synthase Complex

Fatty acid synthase (FAS) is a multienzyme complex comprised of seven enzymes and one nonenzyme protein called *acyl carrier protein* (ACP). In humans, all seven enzymes and ACP occur as elements of a single large polypeptide. During fatty acid synthesis, all of the metabolic intermediates remain attached to the multienzyme fatty acid synthase complex.

The ACP domain of the FAS complex is similar to coenzyme A in that it contains a phosphopantothene group, which is composed of thioethanol amine in amide linkage to the vitamin pantothenic acid. However, unlike coenzyme A, the phosphopantothene group is esterified to the hydroxyl group of a serine residue of ACP rather than to adenosine 3′,5′-bisphosphate (Fig. 11-4). As its name indicates, the function of ACP is to carry the growing fatty acid chain during the process of fatty acid synthesis.

11.4.4.1 *Initial Charging of the FAS Complex.* The first two steps in fatty acid synthesis charge the FAS complex with acetyl and malonyl moieties from acetyl-CoA and malonyl-CoA, respectively. The first of the two reactions catalyzed by the malonyl/acetyl transferase component of the FAS affects the transfer of the acetyl unit of acetyl-CoA to the sulfhydryl group of ACP (Fig. 11-5):

acetyl-CoA + ACP-SH → acetyl-S-ACP + CoASH

As soon as the acetyl unit has become attached to ACP, it is transferred to the catalytically active sulfhydryl group of the condensing enzyme (CE-SH):

acetyl-S-ACP + CE-SH → acetyl-S-CE + ACP-SH

Shifting the acetyl group from ACP to the condensing enzyme frees up the sulfhydryl group of ACP so that it can accept a malonyl unit from malonyl-CoA. The transfer

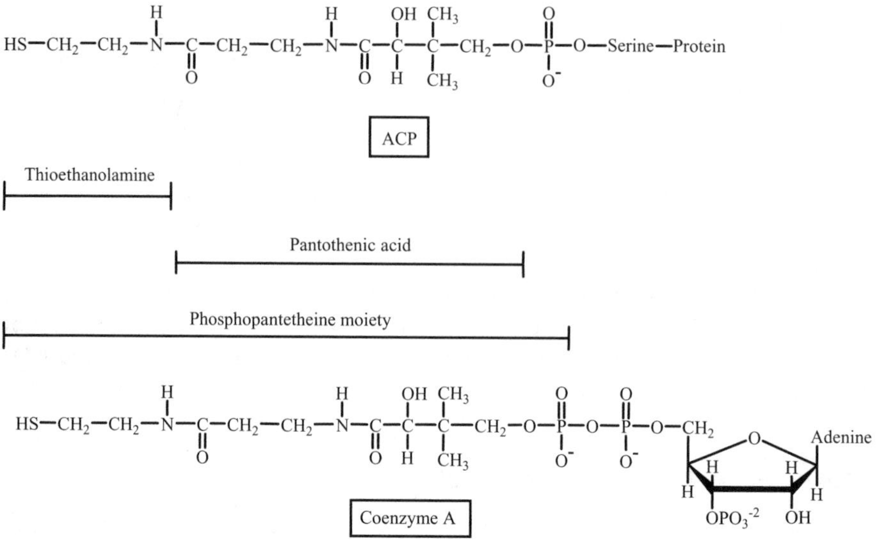

FIGURE 11-4   Comparison of the structures of acyl carrier protein (ACP) and coenzyme A (CoA).

of the malonyl moiety of malonyl-CoA onto ACP is accomplished by malonyl/acetyl transferase:

$$\text{malonyl-CoA} + \text{ACP-SH} \rightarrow \text{malonyl-S-ACP} + \text{CoASH}$$

With one acetyl and one malonyl moiety attached to the FAS complex, the stage is now set for the condensation step of the pathway.

11.4.4.2 *Condensation Reaction.* This bond-forming reaction is catalyzed by the condensing enzyme, β-ketoacyl-ACP synthase:

$$\text{acetyl-S-CE} + \text{malonyl-ACP} \rightarrow \beta\text{-ketobutyryl-ACP} + \text{CE-SH} + CO_2$$

Release of the high-energy carboxyl group of the malonyl moiety of malonyl-*S*-ACP as CO_2 pulls the reaction to the right.

11.4.4.3 *Reduction Sequence.* Next, there is a sequence of three reactions: two reduction reactions, both of which utilize NADPH as a source of reducing equivalents, and an intervening dehydration reaction (Fig. 11-6). The chemistry of the reduction reactions catalyzed by enzymes of the FAS complex resembles that of β-oxidation except that the fatty acid synthesis pathway operates in the opposite direction. Furthermore, both reduction reactions of fatty acid synthesis utilize NADPH, whereas the two oxidation reactions of β-oxidation generate $FADH_2$ and NADH.

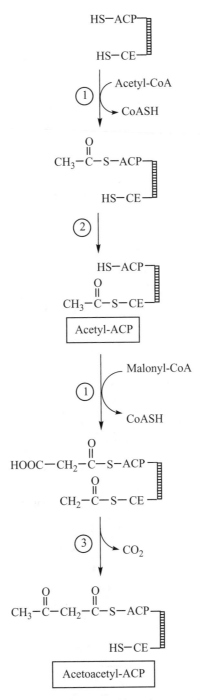

FIGURE 11-5 Reactions that load the fatty acid synthase complex and condense the two initial substrates. ①, ACP-malonyl/acetyl transferase; ②, acyl-group transfer within the fatty acid synthase complex; ③, β-ketoacyl synthase.

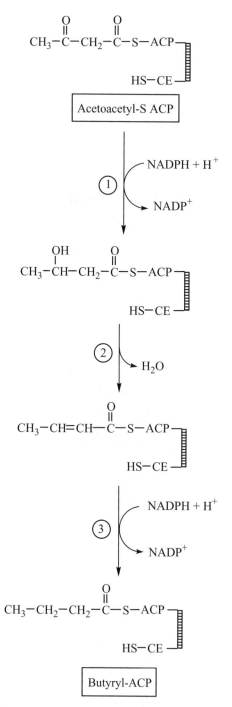

FIGURE 11-6   Reduction sequence of fatty acid synthesis: ①, β-ketoacyl-ACP reductase; ②, hydroxyacyl-ACP dehydratase; ③, enoyl-ACP reductase.

The first reduction reaction of fatty acid synthesis is catalyzed by β-ketoacyl-ACP reductase:

$$\beta\text{-ketoacyl-ACP} + NADPH + H^+ \rightarrow \beta\text{-hydroxyacyl-ACP} + NADP^+$$

The next step involves dehydration and is catalyzed by β-hydroxyacyl-ACP dehydratase:

$$\beta\text{-hydroxyacyl-ACP} \rightarrow trans\text{-enoyl-ACP} + H_2O$$

A second NADPH-dependent reductase, enoyl-ACP reductase, reduces the carbon–carbon double bond of enoyl-ACP:

$$enoyl\text{-ACP} + NADPH + H^+ \rightarrow butyryl\text{-ACP} + NADP^+$$

11.4.4.4 *Further Rounds of Fatty Acid Synthesis.*

Synthesis of butyryl-*S*-ACP completes one round of fatty acid synthesis. However, before another round of two-carbon addition to the growing fatty acid chain can occur, the butyryl moiety attached to ACP must first be shifted onto the sulfhydryl group of the condensing enzyme, which is the same –SH group that accepted the initial acetyl unit:

$$butyryl\text{-ACP} + CE\text{-SH} \rightarrow butyryl\text{-CE} + ACP\text{-SH}$$

This frees up the sulfhydryl group of ACP to accept a second malonyl unit and sets the stage for another round of reduction, dehydration, and reduction reactions. Additional rounds or cycles of the FAS complex take place, all with the growing acyl chain attached to ACP.

11.4.4.5 *Chain Termination.*

Once the acyl chain has reached 16 carbons, the palmitoyl chain is released from the complex by the thioesterase component of the FAS complex:

$$palmitoyl\text{-ACP} + H_2O \rightarrow palmitate + ACP\text{-SH}$$

The one exception to the generation of palmitate as the product of fatty acid synthesis in human cells occurs during the synthesis of milk fat in the mammary gland. During lactation, some of the fatty acids that comprise the TAG of breast milk are derived from de novo fatty acid synthesis in mammary epithelial cells. De novo fatty acid synthesis in lactating mammary glands is especially active in women whose diets are based largely on cereal staples such as maize, rice, or millet. Under these circumstances, the mammary gland synthesizes mainly medium-chain-length (C8–C12) fatty acids. This occurs because the mammary epithelium expresses a specialized thioesterase called decanoyl-ACP thioesterase, which terminates fatty acid synthesis when the FAS complex has generated acyl chains comprised of 8 to 12 carbon atoms.

11.4.5 Modification Reactions

Most cells have the ability to increase the chain length and degree of unsaturation of long-chain fatty acids. Modification of both dietary-derived fatty acids and the palmitate synthesized de novo in the body accounts for the great diversity of fatty acids in membrane lipids and those involved in signaling (e.g., eicosanoids).

11.4.5.1 Fatty Acid Chain Elongation.
Elongation of fatty acids occurs primarily in the endoplasmic reticulum and utilizes malonyl-CoA to add two-carbon units to long-chain fatty acyl-CoAs. There is a minor, secondary chain elongation system (elongase) in mitochondria that utilizes acetyl-CoA as the two-carbon donor and it appears to be involved primarily in the synthesis of lipoic acid, a cofactor for pyruvate dehydrogenase and α-ketoglutarate dehydrogenase.

The elongation system is comprised of a condensing enzyme that adds two carbons to a molecule of fatty acyl-CoA, and three additional enzyme activities, β-ketoacyl-CoA reductase, β-hydroxyacyl-CoA dehydratase, and enoyl-CoA reductase, whose activities are similar to the enzymes of FAS that catalyze the reduction sequence (Fig. 11-6).

The overall elongation reaction of palmitoyl-CoA that occurs in the endoplasmic reticulum is thus

$$\text{palmitoyl-CoA} + 2\text{NADPH} + \text{H}^+ + \text{malonyl-CoA} \rightarrow$$

$$\text{stearoyl-CoA} + 2\text{NADP}^+ + \text{CO}_2 + \text{CoASH}$$

There may be multiple elongation systems in the endoplasmic reticulum, with different specificities for the chain length and degree of unsaturation of the acyl chain substrate.

11.4.5.2 Δ⁹-Desaturation of Endogenously Synthesized Fatty Acids.
The major desaturase in human cells is stearoyl-CoA desaturase (Δ^9 desaturase), which introduces a double bond at carbon 9 from the carboxyl end of the fatty acid chain of fatty acyl-CoA. The enzyme desaturates both stearic acid (18:0) to oleic acid (*cis*9–18:1, *n*-9) and palmitic acid (16:0) to palmitoleic acid (*cis*9–16:1, *n*-7). Stearoyl-CoA desaturase is a mixed-function oxidase that utilizes molecular oxygen to oxidize both the long-chain fatty acyl-CoA and NADH:

$$\text{stearoyl-CoA} + \text{NADH} + \text{H}^+ + \text{O}_2 \rightarrow \text{oleoyl-CoA} + \text{NAD}^+ + 2\text{H}_2\text{O}$$

The desaturation complex includes the actual desaturase enzyme, cytochrome b_5 which serves as an electron acceptor, and NADH-cytochrome b_5 reductase, which contains FAD as a prosthetic group (Fig. 11-7).

11.4.5.3 Modification of Essential Fatty Acids.
Human cells cannot introduce double bonds beyond carbon 9 from the carboxyl end of long-chain fatty acids. For this reason, linoleate (c12–18:2; 18:2ω6) and α-linolenate (c9,c12,c15–18:3;

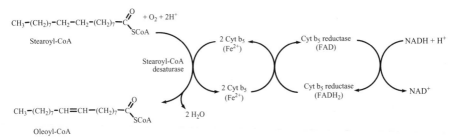

FIGURE 11-7   Desaturation of stearoyl-CoA.

18:3ω3) are essential dietary components that are obtained largely from vegetable oils. Both linoleate and α-linolenate can be modified by alternating desaturation and elongation steps to generate longer-chain polyunsaturated fatty acids. This process utilizes the fatty acid elongation system and fatty acid desaturases, which are specific for the Δ^5 and Δ^6 carbons of fatty acids. There does not appear to be a Δ^4 desaturase enzyme per se in humans; instead, Δ^4 desaturation is accomplished by a more complex, multistep process involving elongation and a subsequent β-oxidation atep.

The two parallel pathways by which the omega-6(ω6) fatty acid linoleic acid is converted into arachidonic acid (c5,c8,c11,c14–20:4) and the ω3 fatty acid α-linolenic acid is converted into docosahexaenoic acid (DHA, c4,c7,c10,c13,c16,c19–22:6) are shown in Figure 11-8. Note that although both ω6 and ω3 fatty acids such as linoleic acid and α-linolenic acid can be elongated and desaturated to generate longer-chain, more unsaturated ω6 and ω3 fatty acids, respectively, an ω3 fatty acid cannot be converted to an ω6 fatty acid, or vice versa.

11.5   REGULATION OF FATTY ACID SYNTHESIS

Fatty acid synthesis is regulated over the long term by insulin and in the short term by several different intracellular effectors: most important, citrate and long-chain fatty acyl-CoAs.

11.5.1   Regulation of the Activity of Acetyl-CoA Carboxylase

Acetyl-CoA carboxylase, the enzyme that catalyzes the rate-limiting step in fatty acid synthesis, is regulated both by allosteric modulators and by phosphorylation/dephosphorylation. The main allosteric activator of acetyl-CoA carboxylase is citrate. By contrast, palmitoyl-CoA and other long-chain fatty acyl-CoAs inhibit acetyl-CoA carboxylase; this phenomenon is an example of feedback inhibition.

Citrate activates acetyl-CoA carboxylase by inducing formation of an active, filamentous polymer from relatively inactive enzyme dimers. Activation of acetyl-CoA carboxylase by citrate reflects the energy status of the cell. In the fed state, when tissues are energy-replete, the high mitochondrial concentration of ATP inhibits

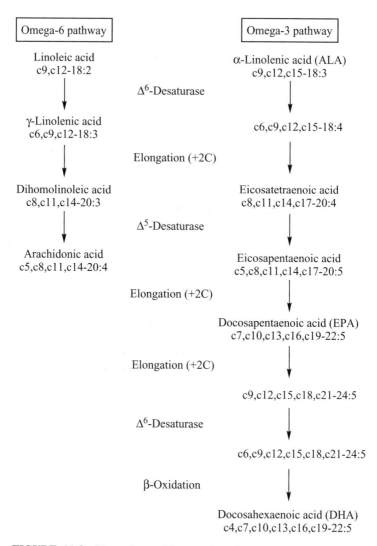

FIGURE 11-8 Elongation and desaturation of polyunsaturated fatty acids.

the TCA cycle by inhibiting isocitrate dehydrogenase. Since the citrate/isocitrate equilibrium favors citrate, it is citrate rather than isocitrate that accumulates within the mitochondrial matrix when the ATP concentration is high. As the mitochondrial concentration of citrate rises, citrate is transported into the cytosol, where it serves as a signal for the cell to synthesize fatty acids from excess acetyl-CoA.

Acetyl-CoA carboxylase activity is also regulated by protein phosphorylation/dephosphorylation. In the fasted state when the insulin/glucagon ratio is low, the intracellular concentration of cAMP increases, and cAMP-activated protein kinase A (PKA) phosphorylates and inhibits acetyl-CoA carboxylase. Acetyl-CoA carboxylase is also inhibited by the action of a second protein kinase, AMP-activated protein kinase (AMPK), whose activity reflects depletion of intracellular ATP. AMPK is an energy sensor that responds to decreases in the cellular energy level as reflected by a high AMP/ATP ratio. AMPK phosphorylates and inactivates acetyl-CoA carboxylase, thereby decreasing the concentration of malonyl-CoA and diminishing fatty acid synthesis. As discussed in Chapter 10, a decrease in the concentration of malonyl-CoA also results in enhanced β-oxidation of long-chain fatty acids. In the fed state, insulin activates the protein phosphatase, which dephosphorylates acetyl-CoA carboxylase, thus increasing the activity of the carboxylase.

11.5.2   Regulation of Enzyme Synthesis

Insulin is a powerful anabolic signal, particularly in hepatocytes and adipocytes, where it induces synthesis of the lipogenic family of enzymes, which includes acetyl-CoA carboxylase, citrate lyase, the malic enzyme, glucose 6-phosphate dehydrogenase, pyruvate kinase, and the FAS complex. The mechanism underlying this action by insulin involves activation of the sterol regulatory element–binding protein-1 (SREBP-1), a membrane-bound transcription factor that enhances transcription of the genes encoding proteins required for fatty acid synthesis. Glucagon, on the other hand, represses de novo synthesis of these enzymes in adipocytes and liver, and stimulates degradation of the lipogenic family of enzyme proteins. AMP-activated protein kinase also suppresses expression of fatty acid synthase, acetyl-CoA carboxylase, and citrate lyase.

11.6   ABNORMAL FUNCTION OF FATTY ACID SYNTHESIS

There are no known diseases resulting from deficiencies of enzymes in the fatty acid synthesis pathway. However, an increased rate of fatty acid synthesis sufficient to damage hepatocytes may occur in the liver of a person who chronically consumes large amounts of ethanol. This pathological condition is known as *alcoholic liver disease*.

11.6.1   Essential Fatty Acid Deficiency

Synthesis of 20- and 22-carbon polyunsaturated fatty acids requires adequate dietary intake of both ω6 and ω3 fatty acids. Essential fatty acid (EFA) deficiency is primarily a deficiency of ω6 fatty acids, which are required in substantially larger quantities than are the ω3 fatty acids. EFA deficiencies are rare now that lipid emulsions are utilized clinically as a component of parenteral nutrition solutions when it is necessary to bypass the gut. The major clinical symptoms of EFA deficiency are skin rash and alopecia. Biochemical evidence of EFA deficiency appears before clinical symptoms and involves the elongation of oleate (18:1 ω9) to produce the abnormal polyunsaturated fatty acid c5,c8,c11–20:3 by the enzymes that normally elongate and desaturate linoleate and α-linolenate.

11.6.2 Deficiency of Omega-3 Fatty Acids

Synthesis of phospholipids containing the highly polyunsaturated fatty acid docosa-hexaenoic acid (22:6 ω3 or DHA) is critical for normal brain development and retinal function. This requirement is specific for ω3 fatty acids and cannot be met by members of the ω6 polyunsaturated fatty acid family. DHA is the most abundant polyunsat-urated fatty acid in the central nervous systems and has been shown to modulate phosphatidylserine biosynthesis and neuronal signaling. Indeed, excess consumption of ω6 fatty acids such as linoleate actually exacerbates the deficiency of DHA by competing with α-linolenic acid as a substrate for the desaturation/elongation path-way. Premature infants who must sustain rapid brain growth and are born with limited fat stores are particularly at risk of DHA deficiency. Since prematurity is often asso-ciated with immature liver function and inadequate ability to elongate and desaturate polyunsaturated fatty acids, infant formulas are now supplemented with DHA and with arachidonic acid rather than with α-linolenate and linoleate.

CHAPTER 12

TRIACYLGLYCEROL TRANSPORT AND METABOLISM

12.1 FUNCTIONS OF TRIACYLGLYCEROLS

Both the triacylglycerols (TAG; also designated *triglycerides* or TG) stored in the body and most of the dietary triacylglycerols are comprised of three long-chain fatty acids (usually C16–C20) esterified to a molecule of glycerol (Fig. 12-1). The three hydroxyls are designated *sn* (stereospecific numbering)-1, *sn*-2, and *sn*-3. Fatty acids in the *sn*-1 and *sn*-3 positions tend to be long-chain saturated fatty acids (e.g., palmitic acid, stearic acid) or monounsaturated fatty acids (e.g., oleic acid), whereas those in the *sn*-2 position tend to be polyunsaturated fatty acids (e.g., linoleic acid, α-linoleic acid, arachidonic acid). The three most abundant fatty acids in the TAG of adipose tissue and plasma lipoproteins are palmitic acid, oleic acid, and linoleic acid.

12.1.1 Dietary Source of Energy

Populations in most Western countries consume 50 to 100 g of dietary fat per day, which amounts to 35 to 50% of their total daily energy intake. Dietary fat also contributes to the palatability of the diet.

12.1.2 Energy Storage

The main function of TAG in the body is to provide a compact and relatively unlimited means for storing energy. At 9 kcal/g, the energy content of TAG is more than

---

Medical Biochemistry: Human Metabolism in Health and Disease   By Miriam D. Rosenthal and Robert H. Glew
Copyright © 2009 John Wiley & Sons, Inc.

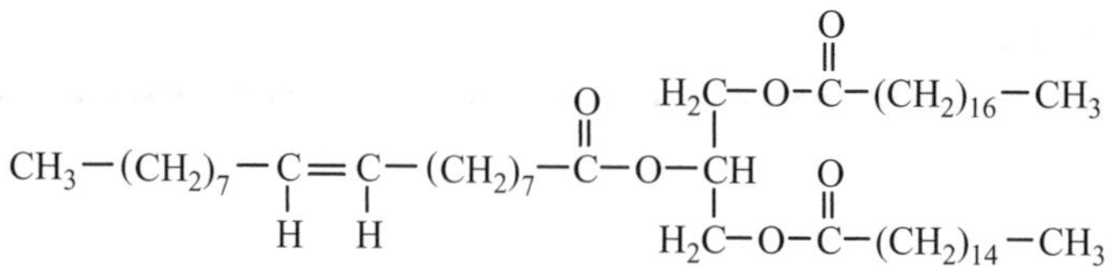

$$CH_3-(CH_2)_7-\underset{H}{\overset{}{C}}=\underset{H}{\overset{}{C}}-(CH_2)_7-\overset{O}{\overset{\|}{C}}-O-\underset{\substack{| \\ H_2C-O-\overset{O}{\overset{\|}{C}}-(CH_2)_{16}-CH_3 \quad sn\text{-}1 \\ | \\ \quad\quad\quad\quad\quad\quad\quad O \quad\quad sn\text{-}2 \\ \quad\quad\quad\quad\quad\quad\quad \| \\ H_2C-O-C-(CH_2)_{14}-CH_3 \quad sn\text{-}3}}{CH}}$$

FIGURE 12-1 Structure of a typical triacylglycerol molecule illustrating the stereospecific numbering (*sn*) system of glycerol: 1, stearoyl; 2, oleoyl; 3, palmitoyl.

twice that of the other major form of energy that humans store, namely glycogen (4.1 kcal/g). The fatty acids that comprise TAG are highly reduced. Except for the carboxyl group, most of the carbon atoms of a fatty acid have two hydrogen atoms attached to them and are bonded to another carbon atom. It is the energy that is released during the oxidation of these C−H and C−C bonds that ultimately supports the synthesis of ATP by the oxidative phosphorylation apparatus of mitochondria. Furthermore, whereas glycogen binds more than twice its weight in water, TAG are hydrophobic, such that only about 15% of the mass of adipose tissue is water. This contrast in water-binding capacity between fat and glycogen means that on a weight basis, it is more economical to store energy in the form of fat than in the form of glycogen.

12.1.3   Physical Functions of Triacylglycerols

Fat acts as a cushion, protecting certain organs (e.g., the kidneys) from physical injury. Adipose layers in the skin provide thermal insulation.

12.1.4   Nonshivering Thermogenesis

Because of their larger skin surface area relative to their small body mass, newborn infants are especially prone to heat loss, which can result in hypothermia. About 5% of the body mass of newborns is attributable to brown fat, which is located in the neck, midline of the upper back, mediastinum, and other organs. Since newborns do not shiver, the main function of brown fat is to generate heat to maintain body temperature. Brown fat is a metabolically active tissue rather than primarily a fat store; its characteristic color comes from the pigmented cytochromes in the abundant mitochondria contained therein. The presence of a specialized uncoupling protein called *thermogenin* prevents the potential energy that is released during fatty acid oxidation from being captured as ATP. Instead, as described in Chapter 6, the energy provided by electron flow through the electron-transport chain in the mitochondria of brown fat is released as heat.

12.1.5 TAG Carry Fat-Soluble Vitamins

Dietary fats not only provide energy and essential fatty acids; they also serve as a carrier for the intestinal absorption of the fat-soluble vitamins A, D, E, and K and carotenoids.

12.2 SITES OF TAG STORAGE

The adipocyte depot is the major site of triacylglycerol storage. Small amounts of TAG are also found in other tissues, including muscle, pancreas, and liver. TAG are also found in the blood in the form of lipoproteins.

12.3 CONDITIONS UNDER WHICH TRIACYLGLYCEROL METABOLISM IS ACTIVE

12.3.1 TAG Metabolism in the Fed State

Dietary TAG are partially hydrolyzed during digestion, reassembled into TAG in the enterocyte, and transported from the small intestine in lymph in the form of lipoprotein complexes called *chylomicrons*. The lymph empties into the bloodstream, where it carries TAG to other organs. Excess dietary carbohydrates and proteins are catabolized and converted to fat, mainly in the liver; however, some conversion of glucose to fat also occurs in adipose tissue. Under normal circumstances, TAG synthesized in the liver are not stored there but are released into the blood as a component of VLDL (very low density lipoprotein). The TAG in circulating lipoproteins (both diet-derived and endogenously synthesized) are hydrolyzed in the capillaries of skeletal muscle, adipose, and other tissues by lipoprotein lipase, and the resulting free fatty acids are taken up by adipocytes and muscle cells.

Intravenous fat emulsions are a common component of parenteral nutrition formulas, where they provide a concentrated source of calories plus essential fatty acids. They provide a substrate for lipoprotein lipase in the capillaries, thereby making free fatty acids available for cellular metabolism.

12.3.2 TAG Metabolism in the Fasted State

Between meals or during a fast when glycogen stores have been depleted, TAG in adipocytes are hydrolyzed to free fatty acids and glycerol. These fatty acids circulate bound to albumin rather than as components of triacylglycerol-containing lipoproteins. Formation of noncovalent complexes between fatty acids and albumin solubilizes the long-chain fatty acids. Transport on albumin also minimizes the potential damage to membranes by the detergent activity of free fatty acid ions.

12.3.3 TAG Synthesis During Lactation

During lactation, mammary glands synthesize TAG for secretion into breast milk. The fatty acids can be synthesized de novo within the mammary gland or derived from fatty acids released from VLDL triacylglycerols by lipoprotein lipase.

12.4 PATHWAYS OF TRIACYLGLYCEROL METABOLISM

12.4.1 Synthesis of TAG

The pathway for triacylglycerol synthesis in most tissues, including liver and adipocytes, utilizes glycerol 3-phosphate and fatty acyl-CoA (Fig. 12-2). The activated fatty acids (i.e., fatty acids attached to coenzyme A) are derived either from endogenous, de novo fatty acid synthesis or from dietary fats. By contrast, triacylglycerol synthesis in the small intestine begins with 2-monoacylglycerol.

12.4.1.1 Generation of Glycerol 3-Phosphate. There are two sources of glycerol 3-phosphate: the glycerol kinase and glycerol 3-phosphate dehydrogenase reactions. Glycerol kinase is expressed primarily in liver and catalyzes the following reaction:

$$\text{glycerol} + \text{ATP} \rightarrow \text{glycerol 3-phosphate} + \text{ADP}$$

In the fed state, the major source of glycerol substrate for the glycerol kinase reaction is glycerol released from circulating lipoprotein TAG by lipoprotein lipase. Small amounts of glycerol can also be obtained from the digestion of dietary glycerolipids (e.g., TAG, phospholipids).

Nonhepatic tissues obtain glycerol 3-phosphate from dihydroxyacetone phosphate (DHAP), an intermediate in both glycolysis and partial gluconeogenesis (a.k.a., glyceroneogenesis). The reaction is catalyzed by glycerol 3-phosphate dehydrogenase:

$$\text{dihydroxyacetone phosphate} + \text{NADH} + \text{H}^+ \rightleftharpoons \text{glycerol 3-phosphate} + \text{NAD}^+$$

12.4.1.2 Synthesis of Phosphatidic Acid. The first step in triacylglycerol synthesis is the trapping and activation of fatty acid inside the lipid-synthesizing cell by fatty acid synthetase:

$$\text{fatty acid} + \text{CoASH} + \text{ATP} \rightarrow \text{fatty acyl-CoA} + \text{AMP} + \text{PP}_i$$

Phosphatidic acid is then generated by the transfer of two molecules of fatty acid from their coenzyme A derivatives to glycerol 3-phosphate. The two reactions are catalyzed by glycerol 3-phosphate acyltransferase and 1-acylglycerol 3-phosphate

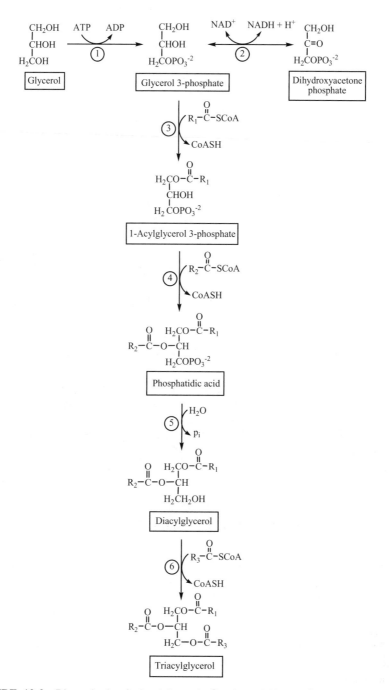

FIGURE 12-2 Biosynthesis of triacylglycerol: ①, glycerol kinase; ②, glycerol 3- phosphate dehydrogenase; ③, glycerol 3-phosphate acyltransferase; ④, 1-acylglycerol 3-phosphate acyltransferase; ⑤, phosphatidic acid phosphatase; ⑥, diacylglycerol acyltransferase.

acyltransferase, respectively (Fig. 12-2):

glycerol 3-phosphate + fatty acyl-CoA → 1-acylglycerol 3-phosphate + CoASH

1-acylglycerol 3-phosphate + fatty acyl-CoA → phosphatidic acid + CoASH

The metabolic intermediate in this reaction sequence, 1-acyl glycerol 3-phosphate, is commonly called *lysophosphatidic acid* since it can also be generated by hydrolysis of the *sn*-2 fatty acid of phosphatidic acid.

12.4.1.3 *Generation of TAG from Phosphatidic Acid.* Phosphatidic acid is hydrolyzed by phosphatidic acid phosphatase to generate 1,2-diacylglycerol. Pathways for phospholipid synthesis also utilize phosphatidic acid phosphatase to generate 1,2-diacylglycerol:

phosphatidic acid + H_2O → diacylglycerol + P_i

Addition of a third fatty acid group to the *sn*-3 hydroxyl by diacylglycerol acyltransferase yields triacylglycerol:

diacylglycerol + acyl-CoA → triacylglycerol + CoASH

12.4.1.4 *Intestinal TAG Synthesis.* As discussed in Chapter 3, the products of the intestinal digestion of dietary fat, free fatty acids and 2-monoacylglycerol, are taken up by enterocytes. Intracellularly, the fatty acids are activated to their respective acyl-CoA form by fatty acid synthetases. Reassembly of fatty acids and 2-monoacylglycerol into triacylglycerol involves successive addition of two acyl groups to 2-monoacylglycerol in reactions catalyzed by acyl transferases located in the endoplasmic reticulum.

12.4.2 Lipoproteins Transport TAG in the Blood

Plasma contains a class of macromolecular aggregates called *lipoproteins* that disperse and transport otherwise highly water-insoluble lipids—cholesteryl esters and TAG in particular—in the circulation. Lipoproteins also play a key role in the metabolism of these lipids and facilitate the two-way exchange of TAG between tissues and the blood. In addition, cholesteryl esters and TAG undergo rapid exchange between the various lipoproteins.

Figure 12-3 illustrates the key structural features of a plasma lipoprotein. The surface of a lipoprotein particle is coated with proteins, phospholipids, and free (nonesterified) cholesterol. The polar ends of the amphipathic lipids face the surface of the lipoprotein, whereas the hydrophobic portions are oriented toward the center of the particle. The core of the lipoproteins is composed of highly nonpolar lipids such as TAG and cholesteryl esters.

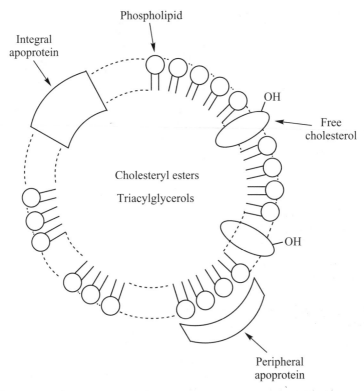

FIGURE 12-3 Generic structure of a plasma lipoprotein.

The proteins (or *apoproteins* as they are called when separated from the lipids) serve a number of functions, including stabilization of the lipoprotein's structure, recognition sites for lipoprotein receptors on cell membranes, and catalysis. For example, apo B100 is recognized by the LDL (low-density lipoprotein) receptor, whereas apo C2 activates the triacylglycerol-hydrolyzing enzyme, lipoprotein lipase.

For historical reasons, the lipoproteins are differentiated on the basis of density (Table 12-1). The most dense lipoprotein class is called high-density lipoprotein (HDL) (1.063 to 1.210 g/dL), followed by low-density lipoprotein (LDL) (1.016 to 1.063 g/dL), very low-density lipoprotein (VLDL) (0.95 to 1.006 g/dL), and chylomicrons (<0.95 g/dL). The major triacylglycerol-transporting lipoproteins are VLDL and chylomicrons. LDL and HDL transport primarily cholesteryl esters. HDL also contains some triacylglycerol. (See Chapter 17 for further discussion of lipoprotein metabolism.)

12.4.2.1 VLDL Transports Endogenous TAG. VLDL is synthesized and secreted mainly by the liver, with a small contribution by the small intestine following a meal. Its main function is to transport TAG that are synthesized in the liver. Functionally, the most prominent apoprotein component of VLDL is apo B100.

TABLE 12-1   Characteristics of the Major Lipoproteins

Lipoprotein	Density (g/mL)	Diameter (nm)	Major Apoprotein	Major Lipid
Chylomicrons	<0.95	75–1200	apo B48	Exogenous TAG
VLDL	0.95–1.006	30–80	apo B100	Exogenous TAG
LDL	1.019–1.063	12–25	apo B100	Cholesteryl ester
HDL2	1.063–1.12	10–20	apo A1	Cholesteryl ester
HDL3	1.12–1.21	7.5–10	apo A1	Cholesteryl ester

HDL, high-density lipoprotein, LDL, low-density lipoprotein, VLDL, very low-density lipoprotein, TAG, triacylglycerols.

While in the circulation, VLDL-triacylglycerols are hydrolyzed by lipoprotein lipase (see below), making free fatty acids available to adipose tissue for storage and to other peripheral tissues for energy production or biosynthetic purposes (e.g., phospholipid synthesis). As the TAG are hydrolyzed, the lipoprotein particles becomes smaller and less buoyant, passing through an intermediate-density lipoprotein (IDL) stage on their way to becoming low-density lipoproteins (LDL).

12.4.2.2 Chylomicrons Transport Exogenous TAG. Chylomicrons are transient in the sense that they are present in the circulation for only several hours following consumption of a fat-containing meal. The major apoprotein component of chylomicrons is apo B48, which contains only 48% of the full amino acid sequence of apo B100; its synthesis by enterocytes is the result of an RNA editing process in which a specific cytidine residue is deaminated to uridine, thus generating a stop codon within the RNA sequence.

Chylomicrons are synthesized by brush-border cells of the small intestine and secreted into the lymphatic system. Once in the blood, the TAG in chylomicrons are hydrolyzed by lipoprotein lipase. As they lose TAG, the chylomicrons shrink and become chylomicron remnants, which are eventually taken up by the liver.

12.4.3   Extracellular Hydrolysis of TAG

The enzymes that hydrolyze the ester bonds linking fatty acids to glycerol are classified as esterases and called *lipases*. The human body contains a number of lipases with distinctly different functions. The role of pancreatic lipase in the digestion of dietary TAG was discussed in Chapter 3. Lipases are also required to hydrolyze lipoprotein-associated TAG so that the constituent fatty acids can be taken up by cells. The lipases that act on TAG in lipoproteins include the three described below.

12.4.3.1 Lipoprotein Lipase (LpL). This glycoprotein is the main enzyme that hydrolyzes TAG in the blood. LpL is synthesized mainly in adipocytes, cardiac muscle, and lactating mammary glands. The enzyme is transported to the endothelial lining of capillaries, where it attaches to endothelial cells through noncovalent bonds to heparan sulfate proteoglycan. LpL catalyzes the hydrolysis of ester bonds in the

1- and 3-positions of the TAG in chylomicrons, VLDL and IDL:

$$\text{triacylglycerol} + 2H_2O \rightarrow 2 \text{ fatty acids} + 2\text{-monoacylglycerol}$$

Some of the monoacylglycerol generated in this reaction is internalized by vascular cells. The remainder isomerizes spontaneously to 1-monoacylglycerol, which is then hydrolyzed by either LpL or a monoacylglycerol lipase in plasma:

$$1\text{-monoacylglycerol} + H_2O \rightarrow \text{fatty acid} + \text{glycerol}$$

Activation of lipoprotein lipase requires a cofactor or activator protein, apo C2, which is synthesized by the liver. Newly synthesized chylomicrons and VLDL contain apo B48 and apo B100, respectively, but do not contain apo C2; instead, as they circulate, they acquire apo C2 from HDL. As the chylomicrons and VLDL gradually become delipidated of TAG, apo C2 returns to HDL and is thus recycled.

12.4.3.2 Hepatic Lipase (HL). Hepatic lipase is synthesized by hepatocytes and tethered to the surface of liver capillaries by heparan sulfate proteoglycans. It hydrolyzes phospholipids as well as TAG contained in high-density lipoprotein (HDL). Hepatic lipase also plays a role in internalizing lipoprotein-associated lipids by hepatocytes.

12.4.3.3 Endothelial Lipase. Endothelial lipase is synthesized by various cell types, including vascular endothelial cells. It also cleaves fatty acids from TAG associated with HDL. Like lipoprotein lipase and hepatic lipase, endothelial lipase has a high affinity for cell-surface heparan sulfate proteoglycan.

12.4.4 Intracellular Lipases

12.4.4.1 Adipocyte Lipases. Most of the body's fat is stored in adipocytes. During a fast or when there is a rapid and critical need for energy, adipocytes hydrolyze some of their triacylglycerol stores and release free fatty acids and glycerol into the circulation.

Hormone-sensitive lipase is a key enzyme involved in the hydrolysis of adipocyte triacylglycerol; its name reflects the fact that the enzyme is activated by the signal transduction cascade involving cAMP and protein kinase A (PKA). In adipocytes, activation of PKA is initiated primarily by epinephrine and to a lesser extent by glucagon and growth hormone. Hormone-sensitive lipase catalyzes the hydrolysis of the fatty acids at positions 1 and 3 of triacylglycerol molecules to produce 2-monoacylglycerol (2-MAG).

For decades, hormone-sensitive lipase was considered to be the key regulatory enzyme in the lipolysis pathway of adipose tissue. However, it is now believed that a recently discovered lipase called *desnutrin* or *adipose triglyceride lipase* (ATGL)

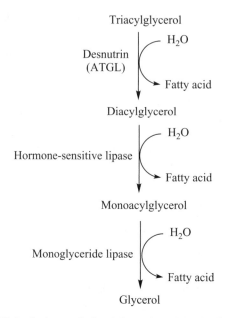

FIGURE 12-4 Pathway of triacylglycerol mobilization in adipocytes.

catalyzes the first step in triacylglycerol hydrolysis. Desnutrin catalyzes the hydrolysis of TAG to diacylglycerols (DAG) and is the rate-limiting step in triacylglycerol hydrolysis. Hormone-sensitive lipase is more active against DAG than against TAG. Thus, the current hypothesis regarding the lipolytic cascade in adipocytes involves three esterases acting sequentially: desnutrin hydrolyzes the first ester bond in TAG generating DAG; then DAG are hydrolyzed by hormone-sensitive lipase to produce 2-monoacylglycerols; and finally, monoacylglycerol lipase removes the third fatty acid to produce glycerol (Fig. 12-4).

12.4.4.2 Lipases in Muscle. Desnutrin and hormone-sensitive lipase are also present in those muscle fibers that have high aerobic oxidative capacity. These two lipases act to hydrolyze the normally modest intramuscular stores of TAG to provide energy during muscle work. Lipolysis in muscle is activated by muscular contraction as well as by epinephrine-induced signaling.

12.4.4.3 Lysosomal Acid Lipase. This lipase, which has a pH optimum near 5, catalyzes the hydrolysis of both LDL-derived triacylglycerols and cholesteryl esters that have been transported into lysosomes via the LDL receptor/endocytic pathway (see Chapter 17). The TAG are hydrolyzed to free fatty acids and monoacylglycerols which exit the lysosomes, whereupon these products can be used for cellular energy production or for phospholipid synthesis.

12.5 REGULATION OF TAG METABOLISM

Insulin is the most important regulator of triacylglycerol metabolism. Insulin enhances the rate of hydrolysis of lipoprotein-associated TAG by stimulating the synthesis and secretion of lipoprotein lipase by adipocytes and myocytes. Insulin also promotes TAG storage in adipocytes and TAG synthesis and VLDL export from hepatocytes. Simultaneously, insulin inhibits the breakdown of TAG, in fat cells. By contrast, hydrocortisone, epinephrine, and growth hormone oppose the action of insulin, inhibiting the synthesis of fatty acids in both hepatocytes and adipocytes, and promoting lipolysis in adipocytes in times of energy need, such as fasting and exercise.

12.5.1 Regulation of TAG Synthesis

Insulin stimulates dephosphorylation and activation of acetyl-CoA carboxylase, the enzyme that catalyzes the rate-limiting step in de novo fatty acid synthesis. Insulin also promotes fatty acid synthesis by inducing enzymes of the fatty acid synthesis family. In addition, insulin stimulates the catabolism of excess dietary carbohydrates, thereby increasing the supply of acetyl-CoA substrate for fatty acid and thus triacylglycerol synthesis.

12.5.2 Regulation of Adipocyte TAG Mobilization

Lipolysis in adipocytes is under tight hormonal control. As described above, desnutrin/ATGL is the enzyme that catalyzes the initial hydrolysis step involved in adipocyte triacylglycerol mobilization. The synthesis of desnutrin is induced by hydrocortisone and inhibited by insulin. Hormone-sensitive lipase is activated by epinephrine via a mechanism involving cAMP-dependent phosphorylation. By contrast, insulin acts to dephosphorylate hormone-sensitive lipase, thereby inhibiting lipolysis.

Adipocytes store TAG in the form of lipid droplets surrounded by a protein called *perilipin*. Like hormone-sensitive lipase, perilipin is phosphorylated by cAMP-dependent protein kinase A. In its unphosphorylated state, perilipin acts as a barrier that limits access of lipases to their substrates, thus maintaining a low rate of basal triacylglycerol hydrolysis. Phosphorylation of perilipin causes the lipid droplets to fragment and disperse, permitting efficient hydrolysis of adipocyte TAG.

12.5.3 Regulation of Lipoprotein Lipase Activity

Lipoprotein lipase (LpL) is active in both the fasted and fed states. In the fasted state, it plays an important role in making fatty acids from VLDL triacylglycerols available to cardiac and skeletal muscles, where the fatty acids serve as fuel. In contrast, in the fed state, LpL directs fatty acids from both chylomicrons and VLDL to adipocytes for storage in the form of TAG. Adipocyte LpL expression is reduced during fasting, while its expression in muscle is up-regulated. Conversely, adipocyte LpL expression is up-regulated in the fed state.

In addition, muscle- and adipocyte-specific forms of LpL have different kinetic properties, with the muscle enzyme having a lower K_m for triacylglycerol than the adipocyte enzyme. Thus, the active site of the LpL enzyme, which is localized to the surface of muscle capillaries, is saturated even during the fasted state, when circulating triacylglycerol-containing lipoprotein levels are low. By contrast, the activity of LpL associated with adipose tissue capillaries increases in the fed state, when the levels of triacylglycerol-rich lipoproteins are relatively high.

12.6   DISEASES INVOLVING ABNORMALITIES IN TRIACYLGLYCEROL METABOLISM

12.6.1   Obesity

Obesity is defined as excess adiposity, which refers to the storage of excess TAG in fat depots. A common practice for estimating obesity utilizes the *body mass index* (BMI), which is calculated as weight (kg)/height (m)^2. Obesity is defined as a body mass index > 30 kg/m^2, or, for example, a weight of 175 pounds or more for a woman who is 5 ft 4 in. tall. Obesity is epidemic in the United States and in many other countries as well. The primary metabolic factors that cause obesity involve overconsumption of food (carbohydrates and protein as well as fats) and insufficient physical activity.

12.6.2   Hypertriglyceridemia

Hypertriglyceridemia is defined as an abnormally high triglyceride concentration in the blood. A normal fasting plasma triglyceride concentration is considered below 150 mg/dL, borderline high at 150 to 199 mg/dL, high at 200 to 499 mg/dL, and very high at 500 mg/dL or above. A high plasma triglyceride level is associated with increased risk for cardiovascular disease, especially myocardial infarction. Hypertriglyceridemia may be caused by a genetic defect or secondarily by acquired factors, such as obesity, physical inactivity, ethanol consumption, diabetes mellitus, hypothyroidism, and drugs that either stimulate triacylglycerol synthesis or retard triacylglycerol catabolism.

12.6.2.1   Type II Diabetes. People with type II are insulin resistant. One effect of this condition is increased lipolysis in adipocytes, which increases the supply of free fatty acids to the liver. Since free fatty acid uptake by hepatocytes is directly proportional to the plasma free fatty acid concentration, the increased free fatty acid flux stimulates triacylglycerol synthesis in the liver. Increased hepatic triacylglycerol synthesis, in turn, leads to increased synthesis and secretion of VLDL, resulting in an elevated plasma triglyceride concentration.

12.6.3 Fatty Liver

The term *nonalcoholic fatty liver disease* refers to a disease spectrum ranging from asymptomatic triacylglycerol accumulation in hepatocytes to nonalcoholic steato-hepatosis (NASH), to cirrhosis (irreversible scarring of the liver) in persons who do not necessarily consume excessive amounts of ethanol. There is a strong association between insulin resistance and hepatic steatosis in such people.

As described above, insulin resistance results in increased uptake of fatty acids and triacylglycerol synthesis in the liver. Incorporation of TAG into VLDL depends on the synthesis of protein (primarily apo B100) and phospholipids (mainly phosphatidyl-choline). Apolipoprotein B100 production is stimulated by insulin and elevated free fatty acid levels. Phosphatidylcholine synthesis depends on an adequate supply of choline and methionine. Conditions such as type II diabetes may result in a supply of free fatty acids for triacylglycerol synthesis that overwhelms the ability of the liver to assemble and secrete VLDL. As a result, there is abnormal accumulation of TAG in the liver, which eventually results in liver disease.

12.6.4 Triacylglycerol Accumulation in Other Organs

Small amounts of TAG are normally found in many tissues, including cardiac muscle, skeletal muscle, and the pancreas. Increased accumulation of TAG in skeletal muscle occurs in obesity and is associated with insulin resistance and type II diabetes. It is not yet clear whether the impaired insulin signalling in such persons is the result of accumulation of TAG per se or of elevated levels of intermediates in the triacylglycerol synthesis pathway.

Paradoxically, TAG also accumulate in the skeletal muscle of trained athletes, but not because of insulin resistance. Instead, exercise training appears to increase the ability of skeletal muscle to mobilize intramuscular TAG and oxidize the resulting fatty acids. Exercise also increases the expression of enzymes of triacylglycerol synthesis (e.g., diacylglycerol acyltransferase) in muscle.

12.6.5 Single Gene Defects That Impair Triacylglycerol Metabolism

12.6.5.1 Chylomicronemia. People with this rare disorder are characterized by extremely high fasting plasma triglyceride concentrations (> 1000 mg/dL). The main causes of chylomicronemia are genetic deficiencies in lipoprotein lipase activity or apo C2, which is a cofactor of this lipase that normally increases the rate of lipolysis. Patients with chylomicronemia usually present in childhood with recurrent attacks of pancreatitis. It is postulated that the chylomicrons impair circulation in pancreatic capillaries thus leading to inflammation. Cell damage then leads to release and acti-vation of proteolytic enzymes in the pancreas rather than in the intestine, resulting in autodigestion of the pancreas. Treatment and prevention of chylomicronemia re-quire restricting dietary fat. Patients deficient in apo C2 often benefit from plasma infusions, which provide an exogenous source of this apoprotein.

12.6.5.2   *Abetalipoproteinemia.* Abetalipoproteinemia is a rare genetic disease characterized by the absence of all apo B–containing lipoproteins (chylomicrons, VLDL, LDL). The underlying defect is in the gene for the microsomal triglyceride transfer protein (MTTP), which facilitates the uptake of TAG by both apo B100 and apo B48. In the absence of MTTP, lipoprotein assembly is impaired in both hepatocytes and intestinal cells. Lack of chylomicron formation results in malabsorption of dietary fat and steatorrhea, and deficiencies of fat-soluble vitamins. At the same time, lack of hepatic VLDL synthesis results in fatty liver. Malabsorption of vitamin E, a fat-soluble vitamin, and the marked impairment of the interorgan transport of vitamin E result in demyelination and serious neurological impairment.

12.6.5.3   *Congenital Generalized Lipodystrophy.* One form of this rare disorder results from loss of function of AGPAT2, the adipocyte isozyme of 1-acylglycerol-3-phosphate acyltransferase, which is required for the biosynthesis of both TAG and phospholipids. Affected persons present at birth or in early infancy with a marked deficit of adipose tissue and compensatory accumulation of TAG in other organs, such as liver and skeletal muscle. They also exhibit extreme insulin resistance, hypertriglyceridemia, hepatic steatosis, and early onset of diabetes.

CHAPTER 13

ETHANOL

13.1 FUNCTION OF ETHANOL METABOLISM

It is believed that the original physiological function of alcohol dehydrogenase was to remove ethanol formed by microorganisms in the intestinal tract. Ethanol is also a common component of wine, beer, and distilled spirits. Since there is no significant renal or pulmonary excretion of ethanol and no storage of ethanol in the body, whatever ethanol is consumed must be disposed of through metabolism.

Ethanol can be a significant source of energy for people who consume large quantities of alcoholic beverages. The caloric content of ethanol is approximately 7 kcal/g, which is intermediate between those of glucose (4 kcal/g) and fat (9 kcal/g).

13.2 LOCATION OF ETHANOL METABOLISM

The major site of alcohol metabolism is the liver, which contains both alcohol dehydrogenase and the microsomal ethanol-oxidizing system (MEOS), the two enzymes most responsible for ethanol metabolism. However, alcohol dehydrogenase activity is also present in the gastric mucosa (more so in men than women), and to a lesser extent in other organs, including the kidneys, lungs, and small intestine.

Medical Biochemistry: Human Metabolism in Health and Disease   By Miriam D. Rosenthal and Robert H. Glew
Copyright © 2009 John Wiley & Sons, Inc.

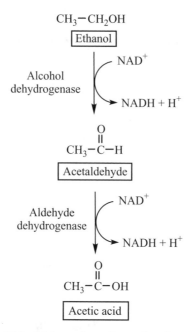

FIGURE 13-1   Major pathway from ethanol to acetyl-CoA.

13.3 PATHWAYS OF ETHANOL METABOLISM

The overall pathway for metabolizing ethanol involves oxidation of the alcohol to acetaldehyde, which is then oxidized to acetate (Fig. 13-1). The acetate derived from ethanol oxidation is activated to acetyl-CoA by acetate thiokinase (see below). The resulting acetyl-CoA can be metabolized through the TCA cycle or utilized for fatty acid synthesis. There are three enzymes or enzyme systems that convert ethanol to acetaldehyde: alcohol dehydrogenase, MEOS, and catalase.

13.3.1 Alcohol Dehydrogenase

In liver and stomach, ethanol metabolism is initiated mainly by NAD^+-dependent alcohol dehydrogenase (ADH), which is a cytosolic enzyme:

$$ethanol + NAD^+ \rightarrow acetaldehyde + NADH + H^+$$

There are several alcohol dehydrogenase isozymes that can oxidize ethanol. The gastric isozyme of ADH has a much higher K_m for ethanol than do the three ADH isozymes in liver. The NADH produced by alcohol dehydrogenase can be shuttled into the mitochondrion and utilized for ATP synthesis.

13.3.2 Microsomal Ethanol-Oxidizing System

The liver has a second pathway for oxidizing ethanol, which even though it can oxidize a variety of compounds in addition to ethanol, is designated the microsomal ethanol-oxidizing system (MEOS). Other substrates for MEOS include fatty acids, steroids, and barbiturates. MEOS oxidizes ethanol to acetaldehyde:

$$\text{ethanol} + NADPH + H^+ + O_2 \rightarrow \text{acetaldehyde} + NADP^+ + 2H_2O$$

MEOS is a mixed-function oxidase that oxidizes both ethanol and NADPH. In many ways MEOS is similar to stearoyl-CoA desaturase (see Chapter 11) in its use of a microsomal electron transport chain involving flavin nucleotides. In addition to its alcohol dehydrogenase activity per se, MEOS also contains cytochrome P450 and NADPH-dependent cytochrome P450 reductase. Since the major ethanol dehydrogenase component of MEOS, CYP2E1, has a much higher K_m for ethanol (11 mM) than the low-K_m forms of alcohol dehydrogenase (0.05 to 4 mM), MEOS functions only at relatively high concentrations of ethanol. To the extent that MEOS consumes NADPH, this reaction is a more wasteful way to metabolize ethanol than alcohol dehydrogenase, which generates NADH rather than consuming NADPH.

13.3.3 Catalase

Catalase, a ubiquitous enzyme, is also capable of oxidizing ethanol; however, its contribution to ethanol metabolism is minimal. Oxidation of ethanol by catalase utilizes hydrogen peroxide:

$$\text{ethanol} + H_2O_2 \rightarrow \text{acetaldehyde} + 2H_2O$$

The H_2O_2 required for catalase-catalyzed ethanol oxidation is derived mainly from the xanthine oxidase reaction of purine catabolism:

$$\text{hypoxanthine} + H_2O + O_2 \rightarrow \text{xanthine} + H_2O_2$$

and from the sequential actions of NADPH oxidase (1) and superoxide dismutase (2):

$$(1)\ 2O_2 + NADPH+ \rightarrow 2O_2^{\bullet -} + NADP^+ + H^+$$
$$(2)\ 2O_2^{\bullet -} + 2H^+ \rightarrow H_2O_2 + O_2$$

13.3.4 Metabolism of Acetaldehyde

Acetaldehyde generated by alcohol dehydrogenase, MEOS, or catalase is oxidized to acetate by NAD^+-dependent aldehyde dehydrogenase. Although the major isozyme of aldehyde dehydrogenase is located in the mitochondria, there is also a cytosolic aldehyde dehydrogenase isozyme. Both isozymes of aldehyde dehydrogenase catalyze the reaction

$$\text{acetaldehyde} + NAD^+ \rightarrow \text{acetate} + NADH + H^+$$

13.3.5   Metabolic Fate of the Acetate Derived from Ethanol

The acetate produced by the oxidation of ethanol is activated by acetyl-CoA synthetase (a.k.a., acetate thiokinase):

$$\text{acetate} + CoASH + ATP \rightarrow \text{acetyl-CoA} + AMP + PP_i$$

The major liver isozyme of acetate thiokinase is cytosolic, and the acetyl-CoA it generates is used for fatty acid and cholesterol synthesis. However, when these two pathways are inactive (due primarily to a high ratio of glucagon to insulin), acetate will diffuse out of the hepatocytes and be taken up and oxidized by heart and skeletal muscle which have high concentrations of mitochondrial acetyl-CoA synthetase. Thus, if ethanol is consumed along with significant amounts of carbohydrate, the acetate generated from ethanol will be used mainly as a substrate for hepatic fatty acid synthesis. If, however, ethanol is consumed in the absence of carbohydrate, the acetate derived from the oxidation of ethanol will be used mostly as fuel and oxidized to CO_2 and water.

13.4   REGULATION OF ETHANOL METABOLISM

Chronic consumption of ethanol can increase hepatic levels of CYP2E1 many fold. Ethanol also increases the expression of other cytochrome P450 genes. When induction of MEOS increases the rate of metabolism of ethanol, the increased production of acetaldehyde may exceed the ability of the acetaldehyde dehydrogenases to further oxidize acetaldehyde. The resultant accumulation of acetaldehyde increases the risk of liver damage.

Gender differences and genetic variants in the enzymes responsible for metabolizing ethanol may account for some of the individual variation in tolerance to ethanol. As noted earlier, premenopausal women normally have lower levels of gastric alcohol dehydrogenase than men. The lower level of gastric ADH activity in women, as well as gender-based differences in body size and total body-water space, are believed to account for the lower tolerance to ethanol in women relative to men.

A number of genetic polymorphisms in ethanol-metabolizing enzymes have been characterized. The inducibility of CYP2E1 can vary as much as 10-fold between persons with polymorphisms in the 5′-flanking region of the gene. Similarly, many persons of East Asian descent have an inactive or less active form of ALDH2, the hepatic mitochondrial isozyme of acetaldehyde dehydrogenase. When people with a mutation in ALDH2 consume ethanol, they are more susceptible to flushing, headache, and nausea, apparently because of acetaldehyde accumulation. The drug Antabuse, which is used to discourage alcoholics from drinking, acts by inhibiting acetaldehyde dehydrogenase; people who consume ethanol while taking Antabuse develop symptoms similar to those of people who have a genetic lack of ALDH2 activity.

13.5 METABOLIC ABNORMALITIES ASSOCIATED WITH ETHANOL METABOLISM

13.5.1 Alcoholic Hypoglycemia, Acidosis, and Ketoacidosis

Collectively, the successive reactions catalyzed by alcohol dehydrogenase and acetaldehyde dehydrogenase generate 2 mol of NADH per mole of ethanol oxidized. Metabolism of large or even moderate quantities of ethanol thus causes the $NADH/NAD^+$ ratio in the liver to increase markedly. Lack of NAD^+ inhibits lactate dehydrogenase and the entry of lactate into gluconeogenesis; the resulting increase in the plasma lactate concentration results in metabolic acidosis. A lack of NAD^+ also slows the action of other key enzymes required for gluconeogenesis, including glycerol 3-phosphate dehydrogenase, malate dehydrogenase, and glutamate dehydrogenase which is important in the removal of amino groups and subsequent entry of the carbon skeletons of amino acids into the gluconeogenesis pathway.

Metabolism of large quantities of ethanol can also lead to ketoacidosis, particularly when the plasma insulin concentration is depressed. Although much of the acetate derived from ethanol metabolism escapes the liver and is metabolized by other tissues, some of the acetate is activated by acetyl-CoA synthetase in hepatocytes. The high $NADH/NAD^+$ ratio slows entry of acetyl-CoA into the TCA cycle by decreasing the activity of NAD^+-linked malate dehydrogenase, thereby limiting the availability of oxaloacetate. The acetyl-CoA from ethanol metabolism is, instead, shunted to the synthesis of acetoacetate, which in turn is reduced to β-hydroxybutyrate, further exacerbating the metabolic acidosis. Alcoholic ketoacidosis is accompanied by dehydration, which results from a combination of vomiting, restricted fluid intake, and inhibition of antidiuretic hormone secretion by ethanol. Dehydration, in turn, impairs renal excretion of ketones.

13.5.2 Alcohol-Induced Fatty Liver

Chronic ethanol use perturbs normal hepatic metabolism, resulting in the accumulation of intracellular triacylglycerols. First, the high $NADH/NAD^+$ ratio decreases the activity of β-hydroxyacyl-CoA dehydrogenase, thus inhibiting β-oxidation of fatty acids. Second, metabolism of ethanol enhances fatty acid synthesis. Suppression of TCA-cycle activity leads to the accumulation of citrate in the cytosol, which stimulates acetyl-CoA carboxylase, the first enzyme in the pathway of fatty acid synthesis. At the same time, high concentrations of NADH increase the rate of production of NADPH (the reductant required for fatty acid synthesis) by means of the NADH/NADPH transhydrogenation pathway (Chapter 10). Accumulation of fatty acids in the liver is further enhanced by hormonal factors, particularly glucagon and hydrocortisone, which stimulate lipolysis in adipocytes, and provides the liver both with additional free fatty acids and with the glycerol needed for their esterification into triacylglycerols.

13.5.3 Acetaldehyde Toxicity

Accumulation of acetaldehyde, produced both by alcohol dehydrogenase and MEOS, is believed to be responsible for most of the alcohol-induced liver damage known as *cirrhosis*. By virtue of its aldehyde group, acetaldehyde is a highly reactive molecule

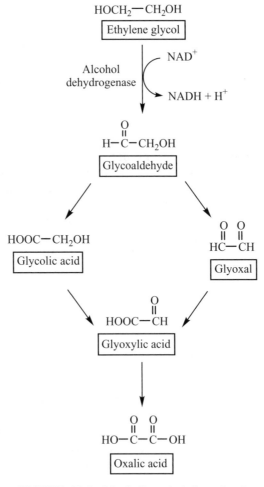

FIGURE 13-2 Metabolism of ethylene glycol.

that can form adducts with many different intracellular proteins. In particular, reaction of acetaldehyde with tubulin impairs secretion of serum proteins from hepatocytes, damaging these cells further. Increased oxidative stress, resulting from production of free radicals by CYP2E1, also contributes to liver damage in chronic alcoholics.

Alcohol dehydrogenase also plays a role in methanol and ethylene glycol toxicity. Methanol is oxidized by ADH to toxic formaldehyde, which in turn is oxidized to formic acid. ADH is also a key enzyme in the pathway that oxidizes ethylene glycol to three organic acids—glycolic acid, glyoxylic acid, and oxalic acid (Fig. 13-2)—which can cause life-threatening metabolic acidosis. Methanol and ethylene glycol poisoning can be treated by administration of the drug fomepizole (4-methylpyrazole) which inhibits ADH activity, thereby allowing these alcohols to be eliminated by the kidney and preventing formation of their more toxic metabolites.

Ethanol has also been utilized to treat ethylene glycol poisoning. Binding of ethanol to the catalytic site of alcohol dehydrogenase displaces ethylene glycol which can then be excreted by the kidneys instead of being oxidized. Since ethanol is a potent inhibitor of gluconeogenesis, concurrent provision of glucose is recommended to reduce the risk of inducing hypoglycemia, which could lead to brain damage.

13.5.4 Wernicke–Korsakoff Syndrome

People who chronically consume excessive amounts of ethanol are at risk for developing acute encephalopathy, peripheral nerve disfunction, and chronic impairment of short-term memory. This condition is known as *Wernicke–Korsakoff syndrome*. The

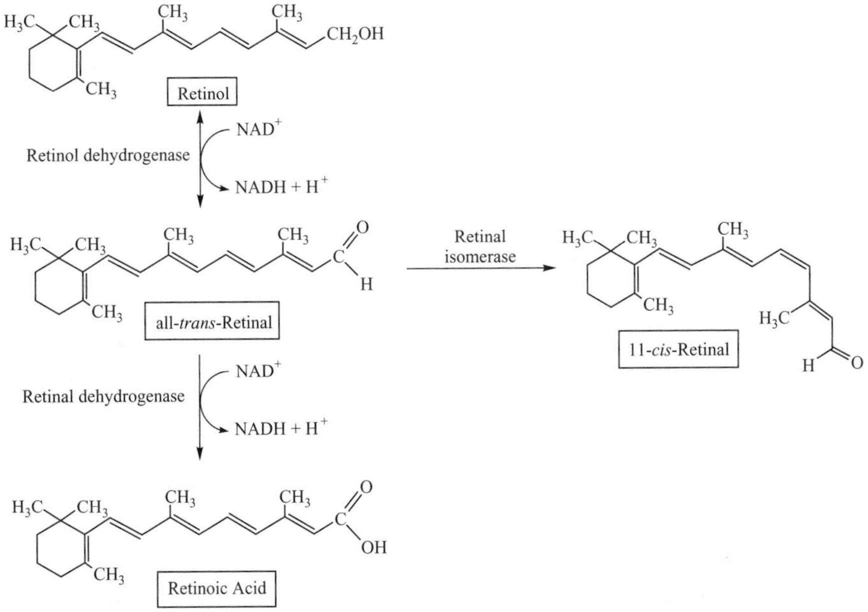

FIGURE 13-3 Metabolism of retinol.

underlying biochemical problem is a deficiency of thiamine (vitamin B_1), and early cases can be treated successfully with high doses of the vitamin. Thiamine pyrophosphate is the required cofactor for transketolase, which catalyzes two steps of the nonoxidative phase of the pentose phosphate pathway as well as for both pyruvate dehydrogenase and α-ketoglutarate dehydrogenase in the TCA cycle (Chapter 5), and for the α-keto acid dehydrogenase, which is involved in the catabolism of the carbon skeletons of branched-chain amino acids (Chapter 20). Dietary deficiency of thiamine is rare in Western societies except in people with alcoholism, who typically have poor diets. In addition, consumption of ethanol decreases thiamine absorption in the gut. Alcoholic cirrhosis impairs formation of the active cofactor thiamine pyrophosphate and contributes to excess excretion of thiamine.

13.5.5   Vitamin A Deficiency

Ethanol interferes with normal metabolism of vitamin A in two ways. First, ethanol is a competitive inhibitor of retinol dehydrogenase, which converts retinol to *all-trans*-retinal. Synthesis of retinal is essential for the formation of the visual pigment rhodopsin, which contains 11-*cis*-retinal. *All-trans*-retinal is also the precursor of retinoic acid, the active hormonal form of the vitamin (Fig. 13-3). Second, ethanol also induces the MEOS system, which enhances catabolism of retinol.

CHAPTER 14

PHOSPHOLIPIDS AND SPHINGOLIPIDS

14.1 FUNCTIONS OF PHOSPHOLIPIDS

Phospholipids are a heterogeneous class of molecules that are amphipathic; that is, they have both hydrophobic and hydrophilic domains. Most of the phospholipids are glycerophospholipids, in which the hydrophobic domain is 1,2-diacylglycerol. Phosphatidylcholine is an example of diacylglycerol phospholipid (Fig. 14-1). Other glycerophospholipids are ether lipids, in which the long-chain hydrocarbon in the 1-position is in ether linkage rather than ester linkage to glycerol. By contrast, sphingomyelin is a phospholipid that contains ceramide (N-acylsphingosine) in place of diacylglycerol (Fig. 14-1). The hydrophilic portion of all glycerophospholipids consists of a phosphodiester bridge that links the hydrophobic domain to the hydroxyl group of a small molecule such as choline, ethanolamine, serine, or inositol.

14.1.1 Membrane Structure

Most of the phospholipids in humans occur as structural elements of membranes where they separate the cytosol from the extracellular space or provide for intra-cellular compartmentalization (i.e., mitochondrial membranes). Phosphatidylcholine is the most abundant glycerophospholipid in membranes, and sphingomyelin is the most abundant ceramide-based lipid. The plasma membrane is asymmetric with respect to lipid composition in that phosphatidylcholine and sphingomyelin tend to be concentrated in the outer leaflet of the membrane bilayer, whereas

Medical Biochemistry: Human Metabolism in Health and Disease   By Miriam D. Rosenthal and Robert H. Glew
Copyright © 2009 John Wiley & Sons, Inc.

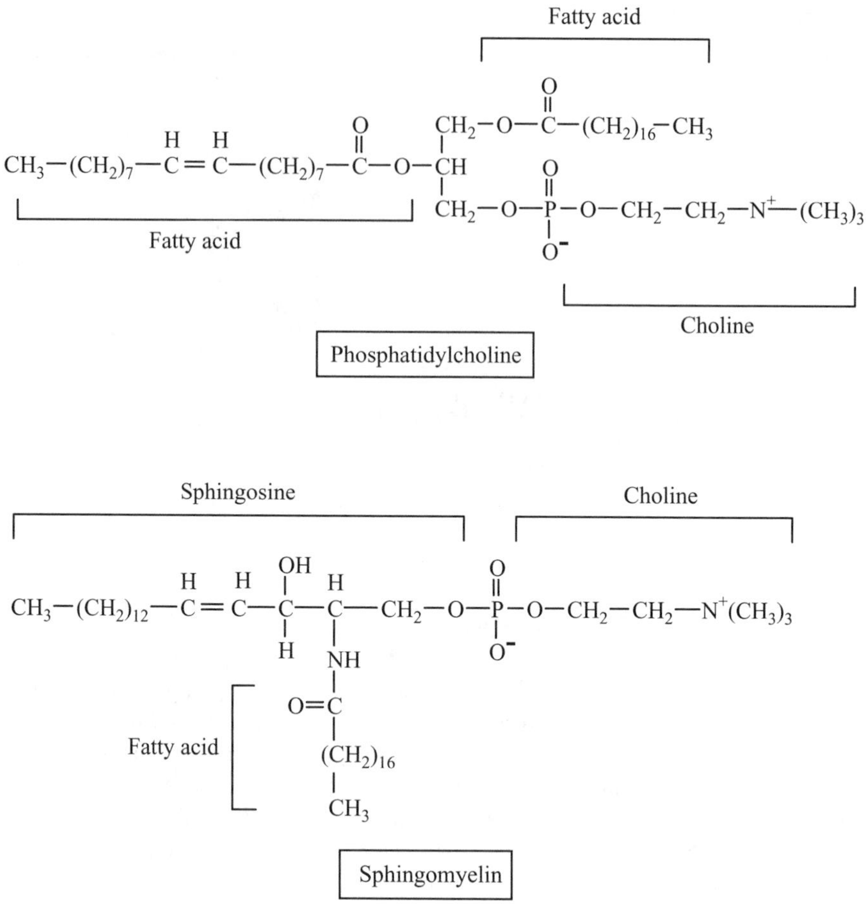

FIGURE 14-1 Structures of two common phospholipids: phosphatidylcholine and sphingomyelin.

phosphatidylethanolamine, phosphatidylserine, and phosphatidylinositol tend to be concentrated in the membrane's inner leaflet, facing the cytosol. Membranes of both myelin and gray matter of nervous tissue are particularly enriched in sphingomyelin.

Although phospholipids do move freely in the plane of the membrane, movement from one membrane leaflet to the other is limited because the polar head groups do not pass readily through the hydrophobic center of the membrane. Transfer of newly synthesized phospholipids from the cytosolic to the external face of the membrane and maintenance of appropriate membrane asymmetry requires the ATP-dependent action of transfer proteins known as *flippases*.

14.1.2 Phospholipids Are Emulsifiers

Phospholipids are amphipathic molecules that can disperse otherwise insoluble mixtures of hydrophobic molecules (e.g., triacylglycerols, cholesteryl esters) during digestion and during transport of lipids in the circulation. Phospholipids in bile serve at least two functions. First, in the intestine, phospholipids aid in the dispersion of dietary triacylglycerols and cholesteryl esters, thereby promoting their digestion and absorption. Second, phospholipids solubilize cholesterol in bile (most of which is unesterified), thereby minimizing precipitation of cholesterol and gallstone formation in the biliary tract. Phospholipids are also critical components of plasma lipoproteins that coat the surface of lipoprotein particles (e.g., chylomicrons, VLDL, LDL, HDL), which transport triacylglycerols and cholesteryl esters in the circulation.

14.1.3 Surfactant

Pulmonary surfactant is the layer of lipid (90%) and protein (10%) that coats the alveolar surface of the lung. Surfactant, which is produced by type II pneumocytes, lowers the surface tension across the air–water interface of alveoli, thereby preventing collapse of terminal respiratory chambers and conducting airways. About 75% of the surfactant lipid is phosphatidylcholine; of this about half is a specialized phosphatidylcholine, dipalmitoylphosphatidylcholine, which contains two molecules of palmitic acid, a 16-carbon saturated fatty acid. Two acidic phospholipids, phosphatidylinositol and phosphatidylglycerol, together account for about 15% of the total phospholipid content of surfactant.

14.1.4 Protein Anchors

Phosphatidylinositol (PI) functions as a tethering mechanism that anchors certain proteins to the external leaflet of the plasma membrane of many different types of cells. Glycosylphosphatidylinositol, commonly called the *GPI anchor*, consists of amphipathic PI linked through an ethanolamine-containing oligosaccharide to the carboxyl group of the C-terminal amino acid of a mature glycoprotein (Fig. 14-2). Acetylcholinesterase, alkaline phosphatase, and $5'$-nucleotidase are examples of proteins that are PI-anchored to membranes.

14.1.5 Activators of Enzymes

There are many instances where phospholipids function to promote enzyme activity. For example, glucocerebrosidase, a membrane-bound lysosomal enzyme that hydrolyzes glucosylceramide to glucose and ceramide, is activated by phosphatidylserine. Similarly, the activity of β-hydroxybutyrate dehydrogenase, which is involved in ketone body synthesis in the liver and β-hydroxybutyrate utilization in peripheral tissues (e.g., muscle, brain), is dependent on phosphatidylcholine. The activities of several of the proteases involved in blood coagulation require phosphatidylserine as well as calcium.

14.1.6 Precursors of Signaling Molecules

Hydrolysis of membrane phospholipids generates a variety of molecules that are involved in intracellular and cell–cell signaling. Diacylglycerol, ceramide,

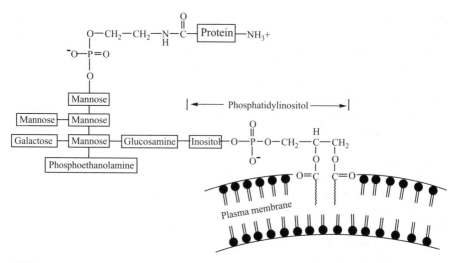

FIGURE 14-2 Generalized structure of a glycosylphosphatidylinositol-anchored membrane protein.

lysophosphatidic acid, and phosphatidic acid all activate various protein kinases, while sphingosine 1-phosphate and lysophosphatidic acid are extracellular signaling molecules whose respective roles include promoting angiogenesis and mitogenesis. Alkyl-linked choline phospholipids are the precursor for platelet-activating factor (PAF), which is an active hypotensive and inflammatory agent. Membrane phospholipids also provide a store of arachidonic acid, which is the precursor of prostaglandins and other autocoid eicosanoids.

14.1.7 Phospholipids as Scavengers of Free Radicals

Plasmalogens, a subclass of ether-linked glycerophospholipids that contain a vinyl ether double bond ($-CH_2-O-CH=CH-$) at the 1-position of the glycerol backbone, are present in high concentrations in the heart. By scavenging free radicals, plasmalogens may protect other membrane lipids from oxidative damage.

14.2 TISSUES IN WHICH PHOSPHOLIPIDS ARE SYNTHESIZED AND MODIFIED

All cells, with the possible exception of mature red blood cells, are capable of synthesizing one or more glycerophospholipids. Most of the reactions involved in phospholipid synthesis occur on the cytosolic face of the endoplasmic reticulum and Golgi complex.

The liver is a major site of phospholipid synthesis. In addition to making phospholipids for its own cellular needs, the liver generates phospholipids for secretion into bile and for coating plasma lipoproteins (e.g., VLDL). Two other tissues with a high capacity for phospholipid synthesis are intestinal enterocytes, which reesterify lysophospholipids produced from biliary and dietary phospholipids during digestion, and type II cells of the lung, which synthesize pulmonary surfactant.

14.3 METABOLIC PATHWAYS OF PHOSPHOLIPID METABOLISM

14.3.1 Biosynthesis of the Lipid Backbone

14.3.1.1 Synthesis of Diacylglycerol. The pathway for synthesis of glycerophospholipids, like that for triacylglycerols, starts with glycerol 3-phosphate. Phosphatidic acid is then generated by two successive acyltransferase-catalyzed transfers of fatty acid from their CoA derivatives to glycerol 3-phosphate (see Fig. 12-2). The specificity of these two acyltransferases is not very strict: The first one, the one that attaches a fatty acid to the 1-position of glycerol, is selective for the 16-carbon saturated fatty acyl-CoA, palmitoyl-CoA, and other saturated long-chain fatty acids; the second acyltransferase is selective for monounsaturated and polyunsaturated fatty acyl-CoAs, particularly linoleoyl-CoA. Pathways that generate phospholipids involve either phosphatidic acid or diacylglycerol as intermediates. Generation of diacylglycerol is catalyzed by phosphatidic acid phosphatase:

$$\text{phosphatidic acid} + H_2O \rightarrow \text{diacylglycerol} + P_i$$

14.3.1.2 Synthesis of Ether Lipids. The pathway for synthesis of ether lipids starts with acylation of the glycolytic intermediate dihydroxyacetone phosphate by dihydroxyacetone phosphate acyltransferase (Fig. 14-3):

$$\text{dihydroxyacetone phosphate} + \text{acyl-CoA} \rightarrow \text{1-acyl dihydroxyacetone phosphate}$$
$$+ \text{CoASH}$$

1-Alkyl dihydroxyacetone phosphate synthase then catalyzes the addition of a long-chain fatty alcohol group (ROH) in exchange for the fatty acid:

$$\text{1-acyl dihydroxyacetone phosphate} + \text{ROH} \rightarrow$$
$$\text{1-alkyl dihydroxyacetone phosphate} + \text{RCOOH}$$

This is followed by NADPH-dependent reduction of the C2 keto group by acyl/alkyl-DHAP reductase and acylation of the resulting hydroxyl group to generate the alkyl analog of phosphatidic acid, which then serves as a substrate for ether phospholipid synthesis.

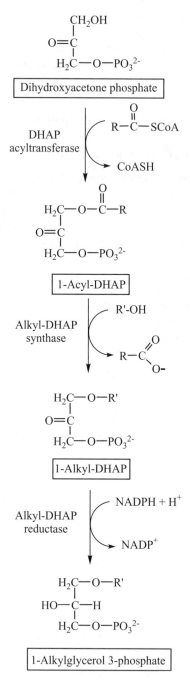

FIGURE 14-3 Synthesis of the 1-alkylglycerol 3-phosphate backbone of ether lipids. DHAP, dihydroxyacetone 3-phosphate; R'-OH, long-chain fatty alcohol.

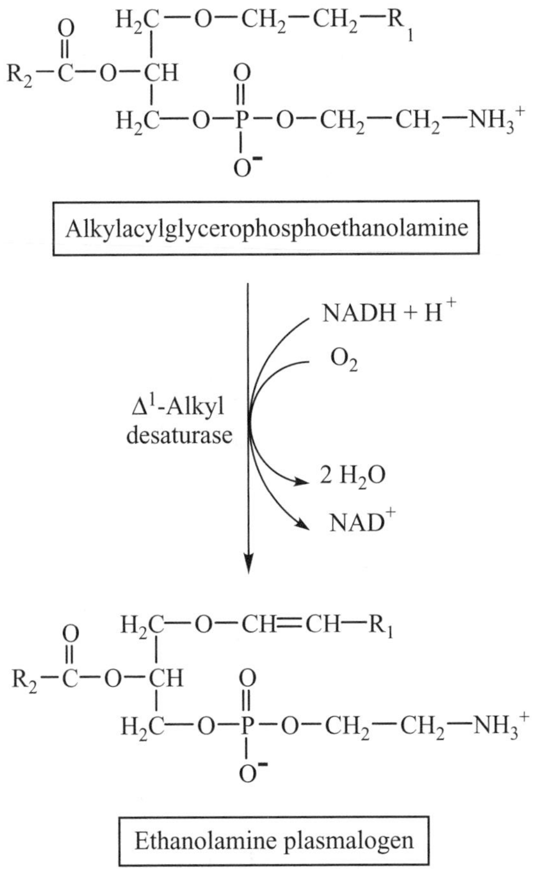

$$\begin{array}{c} \text{H}_2\text{C}-\text{O}-\text{CH}_2-\text{CH}_2-\text{R}_1 \\ \text{O} \quad\quad | \\ \| \quad\quad\quad \\ \text{R}_2-\text{C}-\text{O}-\text{CH} \quad\quad \text{O} \\ | \quad\quad\quad \| \\ \text{H}_2\text{C}-\text{O}-\text{P}-\text{O}-\text{CH}_2-\text{CH}_2-\text{NH}_3^+ \\ | \\ \text{O}^- \end{array}$$

| Alkylacylglycerophosphoethanolamine |

Δ^1-Alkyl
desaturase

NADH + H$^+$

O$_2$

2 H$_2$O

NAD$^+$

$$\begin{array}{c} \text{H}_2\text{C}-\text{O}-\text{CH}=\text{CH}-\text{R}_1 \\ \text{O} \quad\quad | \\ \| \quad\quad\quad \\ \text{R}_2-\text{C}-\text{O}-\text{CH} \quad\quad \text{O} \\ | \quad\quad\quad \| \\ \text{H}_2\text{C}-\text{O}-\text{P}-\text{O}-\text{CH}_2-\text{CH}_2-\text{NH}_3^+ \\ | \\ \text{O}^- \end{array}$$

| Ethanolamine plasmalogen |

FIGURE 14-4   Synthesis of ethanolamine plasmalogen from the 1-alkyl ether lipid.

Plasmalogen Synthesis. Plasmalogens are ether lipids with a double bond between C1 and C2 of the alkyl chain ($-\text{CH}_2-\text{O}-\text{CH}=\text{CH}-$). Generation of the double bond occurs after completion of 1-alkyl-2-acylphospholipid synthesis and is catalyzed by Δ^1-alkyl desaturase, which is a peroxisomal NADH-dependent mixed function oxidase (Fig. 14-4).

14.3.1.3 Ceramide Synthesis. The sphingosine backbone is synthesized from palmitoyl-CoA and serine. In the initial reaction, catalyzed by pyridoxal phosphate-dependent serine-palmitoyl transferase, serine is decarboxylated, and the resulting amine-containing two-carbon fragment condenses with palmitoyl-CoA (Fig. 14-5):

palmitoyl-CoA + serine $\rightarrow$ 3-ketosphinganine + CO$_2$ + CoASH

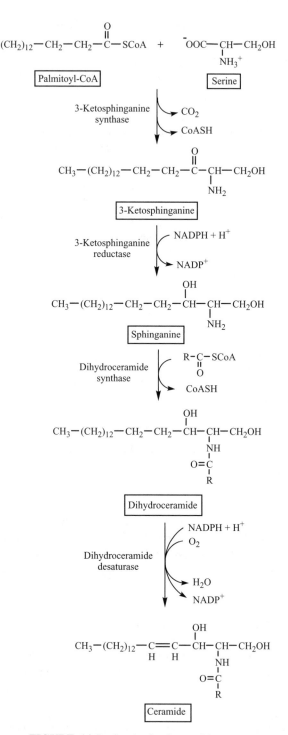

FIGURE 14-5 Synthesis of ceramide.

Next, 3-ketosphinganine reductase uses NADPH to reduce the keto group of 3-ketosphinganine, thereby generating *sphinganine* (also called *dihydrosphingosine*):

$$3-\text{ketosphinganine} + \text{NADPH} + \text{H}^+ \rightarrow \text{sphinganine} + \text{NADP}^+$$

Dihydroceramide synthase then attaches a long-chain fatty acid to the amino group of sphinganine to form dihydroceramide (*N*-acylsphinganine), which is then reduced to ceramide by *N*-acylsphinganine dehydrogenase.

14.3.2 Attachment of Polar Head Groups

14.3.2.1 *Synthesis of Phosphatidylcholine and Phosphatidylethanolamine.* In all human cells except hepatocytes, the major pathway of phosphatidylcholine (PC) synthesis starts with the activation of free choline. The initial step, the trapping of choline inside cells, is accomplished by the enzyme choline kinase (Fig. 14-6):

$$\text{choline} + \text{ATP} \rightarrow \text{phosphocholine} + \text{ADP}$$

Phosphocholine in turn is activated by CTP : phosphocholine cytidylyltransferase:

$$\text{phosphocholine} + \text{CTP} \rightarrow \text{CDP-choline} + \text{PP}_i$$

CDP-choline is then condensed with diacylglycerol in a reaction catalyzed by CDP-choline:diacylglycerol phosphocholine transferase:

$$\text{CDP-choline} + \text{diacylglycerol} \rightarrow \text{phosphatidylcholine} + \text{CMP}$$

An analogous pathway using CDP-ethanolamine is used to synthesize phosphatidylethanolamine (PE): namely, ethanolamine $\rightarrow$ phosphoethanolamine $\rightarrow$ CDP-ethanolamine $\rightarrow$ PE.

14.3.2.2 *Phosphatidylinositol Synthesis.* In contrast to the pathways of PC and PE synthesis, where their respective polar head groups, choline and ethanolamine, are activated at the expense of CTP, synthesis of PI involves CTP-dependent activation of the hydrophobic diacylglycerol moiety of phosphatidic acid, followed by condensation of CDP-diacylglycerol with inositol:

$$\text{phosphatidic acid} + \text{CTP} \rightarrow \text{CDP-diacylglycerol} + \text{PP}_i$$

$$\text{CDP-diacylglycerol} + \text{inositol} \rightarrow \text{phosphatidylinositol} + \text{CMP}$$

The inositol required for phosphatidylinositol synthesis is either obtained in the diet or synthesized by cyclization of glucose 6-phosphate. Major sites of inositol synthesis

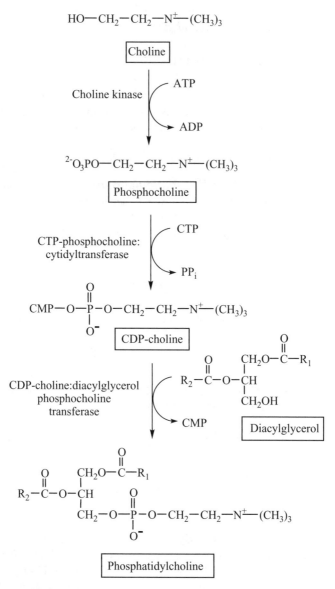

FIGURE 14-6 Synthesis of phosphatidylcholine by the salvage pathway.

include the brain and testis. The strategy of activating the diglyceride domain is also followed in the synthesis of phosphatidylglycerol and cardiolipin.

14.3.2.3 *Phosphatidylserine Synthesis.* Unlike the pathways described above that produce the other glycerophospholipids, human cells do not synthesize phosphatidylserine (PS) by condensation of the polar head group serine with the diacylglycerol backbone. Instead, PS is produced by a novel mechanism that involves polar head group exchange between phosphatidylethanolamine and free serine in a reaction catalyzed by phosphatidylserine synthase:

phosphatidylethanolamine + serine $\rightleftharpoons$ phosphatidylserine + ethanolamine

Since there is no net change in the number or types of bonds in the substrates and products, this reaction is reversible and does not require ATP or the involvement of any other high-energy compound. The ethanolamine that is released in this reaction is reincorporated into phosphatidylethanolamine by the CDP-ethanolamine pathway described above. A second isozyme of phosphatidylserine synthase preferentially utilizes phosphatidylcholine in place of phosphatidylethanolamine for synthesis of PS.

14.3.2.4 Synthesis of Sphingomyelin. Like PS, sphingomyelin is not synthesized by condensation of its polar head group (phosphocholine) with the lipid domain (ceramide). Instead, synthesis involves the transfer of the polar head group from phosphatidylcholine to ceramide:

ceramide + phosphatidylcholine $\rightarrow$ sphingomyelin + diacylglycerol

14.3.3 Phospholipases

Membrane phospholipids are dynamic molecules that undergo rapid turnover and extensive modification reactions characterized by hydrolysis and resynthesis. The enzymes that catalyze phospholipid hydrolysis are called *phospholipases* and are grouped into classes that designate which bond is cleaved (Fig. 14-7).

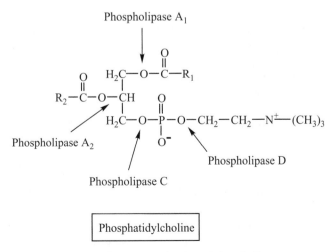

FIGURE 14-7   Sites of action of phospholipases.

14.3.3.1 *Phospholipase A₁*. Phospholipase A_1 is an enzyme that hydrolyzes the fatty acid in the 1-position of glycerophospholipids. Some lipases that act primarily as triacylglycerol lipases, such as hepatic lipase and lipoprotein lipase, also have phospholipase A_1 activity.

14.3.3.2 *Phospholipase A₂ (PLA₂)*. Phospholipases A_2 release the fatty acid in the 2-position of phospholipids. These phospholipases are much more abundant and widely distributed than phospholipases A_1, possibly because the structure of phospholipids in membranes and micelles renders the acyl bond in the 2-position more accessible to the enzyme. There are several intracellular PLA_2's, including a cytosolic phospholipase A_2 that is specific for arachidonic acid, activated by micromolar Ca^{2+} concentrations, and involved primarily in the regulation of agonist-stimulated prostaglandin and leukotriene biosynthesis. Extracellular PLA_2's include pancreatic PLA_2, which hydrolyzes phospholipids in the small intestine and the secretory PLA_2's, which are small extracellular esterases that participate in the inflammatory response.

14.3.3.3 *Phospholipase B*. Phospholipase B has lysophospholipase as well as phospholipase activity and is able to remove both fatty acids from glycerophospholipids.

14.3.3.4 *Phospholipase C*. Phospholipase C is a phosphodiesterase that cleaves phospholipids so as to separate the phosphorylated polar head group from the diacylglycerol moiety. One especially important phospholipase C is specific for phosphatidylinositol bisphosphate (PIP_2) and catalyzes the following reaction (Fig. 14-8):

phosphatidylinositol bisphosphate $+ H_2O \rightarrow$ inositol trisphosphate $+$ diacylglycerol

Although PIP_2 represents only a small fraction of the inositol phospholipids, it plays a major role in signal transduction in that hydrolysis of PIP_2 generates two products, both of which are active as intracellular second messengers. Inositol trisphosphate (IP_3) triggers the release of calcium from the endoplasmic reticulum, and diacylglycerol activates protein kinase C.

14.3.3.5 *Phospholipase D*. Phospholipase D is a phosphodiesterase that cleaves on the polar head group side of the phosphodiester bond of phospholipids; for example:

phosphatidylcholine $+ H_2O \rightarrow$ phosphatidic acid $+$ choline

In some cells (e.g., neutrophils) a phospholipase D generates phosphatidic acid during agonist-stimulated cellular activation. Phosphatidic acid then regulates secretory and degranulation responses, possibly by recruiting other proteins to the

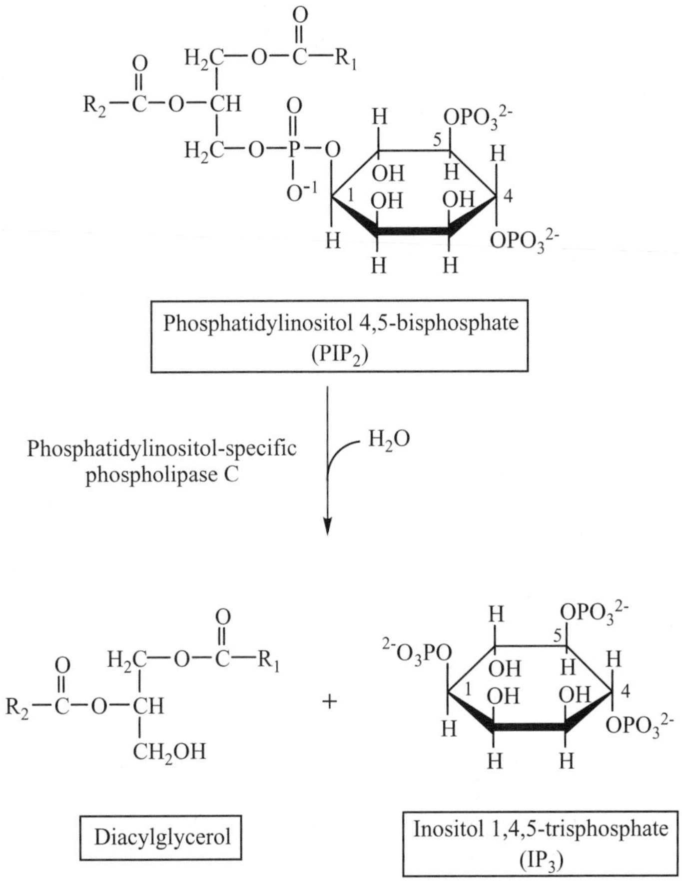

Phosphatidylinositol 4,5-bisphosphate
(PIP$_2$)

Phosphatidylinositol-specific
phospholipase C

H$_2$O

Diacylglycerol

Inositol 1,4,5-trisphosphate
(IP$_3$)

FIGURE 14-8 Phosphatidylinositol 4,5-bisphosphate as a precursor to intracellular second messengers.

membrane. Phosphatidic acid can also be hydrolyzed further to generate additional diacylglycerol for protein kinase C activation or to generate lysophosphatidic acid, which is released from the cell as an extracellular signaling molecule. Extracellular C- and D-type phospholipases cleave glycosylphosphatidylinositol anchors (see Fig. 14-2), releasing this particular class of membrane-bound proteins from the cell surface.

*14.3.3.6 **Sphingomyelinase.*** Sphingomyelinase is analogous to phospholipase C in that it catalyzes reactions that cleave sphingomyelin on the lipid side of the phosphodiester bridge:

$$\text{sphingomyelin} + \text{H}_2\text{O} \rightarrow \text{ceramide} + \text{phosphocholine}$$

Mammalian cells contain multiple sphingomyelinases that generate ceramide in response to a wide spectrum of agents, including 1,25-dihydroxyvitamin D_3, cytokines, and corticosteroids. Ceramide activates several phosphoprotein phosphatases as well as a protein kinase; increased concentrations of ceramide can lead to growth inhibition, differentiation, and in some cases, apoptosis.

14.3.4 Phospholipid Remodeling Reactions

Membrane phospholipids undergo extensive modification reactions. These reactions serve to convert the phospholipids generated by the biosynthetic pathways described above to the mixture of specific phospholipid structures required by the cell.

14.3.4.1 Deacylation/Reacylation Reactions. In many cases, the initially synthesized phospholipids do not have the appropriate fatty acids in the 1- and 2-positions of the glycerol backbone. As noted earlier, phosphatidic acid usually has palmitate or stearate in the 1-position and linoleate or oleate in the 2-position. Remodeling reactions are therefore required to introduce arachidonic acid into the 2-position of PC, PE, and PI, where it is then available for mobilization to initiate eicosanoid synthesis. In the lung, type II pneumocytes utilize remodeling reactions to generate dipalmitoylphosphatidylcholine (dipalmitoyllecithin), the major lipid component of pulmonary surfactant.

Remodeling is initiated by the action of a phospholipase A_2, which removes the fatty acid from the 2-position of the phospholipid, generating the corresponding lysophospholipid. There are two mechanisms for reacylating the lysophospholipid: direct acylation, which uses fatty acyl-CoA as the fatty acid donor, and transacylation, in which the fatty acid that acylates the lysophospholipid is taken from another phospholipid in a transesterification reaction.

Synthesis of Platelet-Activating Factor (PAF). An example of a set of deacylation/reacylation reactions is the pathway for the synthesis of PAF (Fig. 14-9). PAF is an alkyl lipid with a phosphocholine head group. Unlike typical glycerophospholipids, PAF contains an acetyl moiety in the 2-position instead of a long-chain fatty acid. PAF is synthesized by inflammatory cells, such as neutrophils and macrophages, and is a potent mediator of hypersensitivity, acute inflammatory reactions, and allergic reactions. Synthesis of PAF involves phospholipase A_2–mediated removal of the long-chain fatty acid from the 2-position of the ether-linked phospholipid followed by acetylation with acetyl-CoA. Since most of the long-chain fatty acid that is released is arachidonate, PAF production is often accompanied by eicosanoid synthesis. Inactivation of PAF involves hydrolysis to remove the acetyl moiety followed by reacylation to regenerate an ether-linked membrane phospholipid.

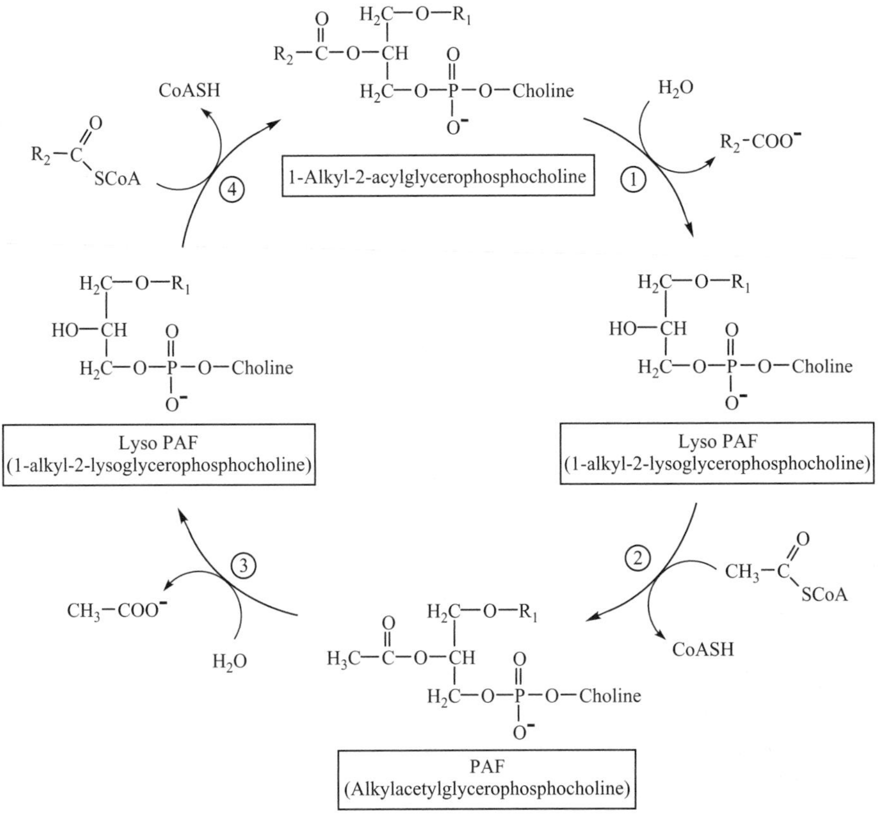

FIGURE 14-9 Role of remodeling reactions in the synthesis and inactivation of platelet activating factor (PAF), 1-alkyl-2-acetyl-glycerophosphocholine. The enzymatic steps are catalyzed by ①, phospholipase A$_2$; ② acetyl-CoA:alkyllysoglycerophosphate acetyltransferase; ③. PAF acetylhydrolase; ④. acyltransferase. Alternative transacylation mechanisms exist for the both the deacylation ① and reacylation ④ steps.

14.3.4.2 De Novo *Synthesis of Choline and Ethanolamine.* Although inositol is synthesized from glucose 6-phosphate, there is no direct pathway for synthesizing either ethanolamine or choline. Instead, the ethanolamine and choline moieties of phospholipids are synthesized from the serine moiety of phosphatidylserine (PS). The generation of PE by decarboxylation of PS occurs in the mitochondria of many cell types (Fig. 14-10):

$$phosphatidylserine \rightarrow phosphatidylethanolamine + CO_2$$

By contrast, the conversion of PE to PC occurs only in hepatocytes. The pathway, which is catalyzed by a single enzyme, phosphatidylethanolamine

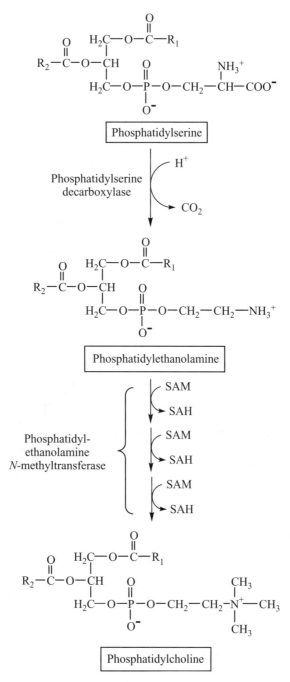

FIGURE 14-10 Synthesis of phosphatidylcholine from phosphatidylserine. SAM, *S*-adenosylmethionine; SAH, *S*-adenosylhomocysteine.

N-methyltransferase, uses three molecules of S-adenosylmethionine (SAM) to transfer three methyl groups successively to the nitrogen of phosphatidylethanolamine (Fig. 14-10):

phosphatidylethanolamine $+$ 3S-adenosylmethionine $\rightarrow$ phosphatidylcholine

$+$ 3S-adenosylhomocysteine

The subsequent metabolism of S-adenosylhomocysteine is discussed in Chapter 21.

As described above, the activation of choline to CDP-choline provides a pathway for incorporating free choline into phospholipids. The CDP-choline pathway captures and reutilizes choline released during the catabolism of PC, sphingomyelin, and the neurotransmitter acetylcholine, as well as choline derived from the diet. The pathway for incorporating choline into PC is therefore commonly referred to as a *salvage pathway*. Similarly, the CDP-ethanolamine pathway is used to salvage free ethanolamine released during PE turnover, as well as the ethanolamine released from PE during the synthesis of PS by base exchange and ethanolamine derived from the diet.

If humans can utilize the salvage pathway to incorporate dietary choline into glycerophospholipids, why is there still a need for de novo choline synthesis via phosphatidylethanolamine N-methyltransferase (PEMT)? It appears that the PEMT pathway has a special role in the secretion of VLDL by the liver, and may play a role in regulating plasma homocysteine levels.

14.3.4.3 Phosphorylation of Phosphatidylinositol. Another example of phospholipid synthesis occurring by modification of a preexisting phospholipid is the phosphorylation of the inositol moiety of phosphatidylinositol. The major products of this pathway are phosphatidylinositol 4,5-bisphosphate (PI-4,5-P$_2$), which is the substrate for phospholipase C, and phosphatidylinositol 3-phosphate (PI-3-P), which is involved in endosomal trafficking.

14.4 REGULATION OF PHOSPHOLIPID METABOLISM

The regulation of phospholipid synthesis is complex and incompletely understood. The rate-limiting step in the salvage pathway for phosphatidylcholine synthesis is the CTP : phosphocholine cytidylyltransferase (CCT) reaction. Translocation of CCT from the cytosol to the endoplasmic reticulum in response to physiological signals is associated with a large increase in CCT activity.

14.5   DISEASES INVOLVING PHOSPHOLIPIDS

14.5.1   Neonatal Respiratory Distress Syndrome

Around the twenty-sixth week of gestation, type II cells in the fetal lung begin synthesizing the components of surfactant, including dipalmitoylphosphatidylcholine.
Failure to produce adequate amounts of surfactant during the third trimester results
in *respiratory distress syndrome* (RDS) in the neonate (a.k.a., *hyaline membrane disease*), which is a major cause of infant mortality worldwide, particularly in preterm
infants. The maturity of the fetal lung can be assessed by an increased phosphatidyl-
choline/sphingomyelin (P/S) ratio in amniotic fluid. Most cases of RDS can be
prevented if mothers who go into preterm labor are given glucocorticoids, which
stimulate surfactant production in the fetus. Respiratory failure due to insufficiency
of surfactant can also occur in adults when their type II cells have been destroyed
by severe infections or as an adverse side effect of chemotherapeutic drugs (e.g.,
bleomycin).

14.5.2   Choline Deficiency

Although choline synthesis in the liver does contribute to the choline pool, it is not
always sufficient to satisfy a person's daily choline requirement. Choline is now
recognized as an essential nutrient, especially for people with low dietary intakes of
protein whose methionine pool may be inadequate. Choline deficiency compromises
hepatic VLDL synthesis and secretion and can therefore result in fatty liver. There is
evidence that choline deficiency may also compromise brain development, memory
function, and cardiovascular health. Meats, eggs, and vegetables (e.g., cauliflower,
lettuce) are good sources of choline.

14.5.3   Lupus

Systemic lupus erythematous (SLE) is an autoimmune disorder that affects the kidneys and the cardiovascular and central nervous systems. The hallmark of the disease
is the production of antibodies directed against antigens of the host's own cells and
tissues, which may be membrane phospholipids, usually phosphatidylserine, as well
as nuclear proteins (e.g., histones) or even double-stranded DNA. Phosphatidylserine
is normally concentrated on the cytosolic face of the plasma membrane; cellular
damage may result in inappropriate phosphatidylserine exposure on the cell surface,
giving rise to antibody formation. In lupus, antibodies are sometimes also directed
against cardiolipin, another acidic phospholipid, which is normally localized to the
inner mitochondrial membrane (Fig. 14-11).

14.5.4   Barth Syndrome

Barth syndrome is a genetic cardiomyopathy caused by mutations in the tafazzin gene,
which is involved in the remodeling of the fatty acids in mitochondrial cardiolipin. The

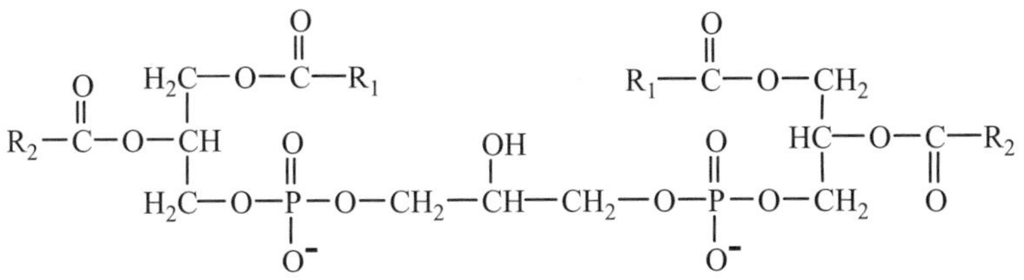

FIGURE 14-11   Structure of cardiolipin.

protein tafazzin transfers specific fatty acids from phosphatidylcholine to cardiolipin and is essential to the synthesis of the symmetrical cardiolipin molecules, such as tetralinoleoylcardiolipin, which is essential for normal mitochondrial function.

14.5.5   Niemann–Pick Disease, Types A and B

Mutations in the gene for lysosomal sphingomyelinase result in sphingomyelin accumulation in cells, particularly those of the reticuloendothelial system, and cause hepatosplenomegaly, jaundice, and neurologic disturbances. Type A Niemann–Pick disease is more severe, with early death resulting from excess deposition of sphingomyelin in the central nervous system. Patients with type B Niemann–Pick have more residual enzyme activity than those with type A Niemann–Pick disease, generally do not exhibit neurologic involvement, and usually survive into adulthood. Types A and B Niemann–Pick, should not be confused with type C_1 or C_2 Niemann–Pick Disease, in which the underlying molecular defect involves mutation in the gene *NPC1*, which is required for the transport of cholesterol and other lipids out of the lysosome.

CHAPTER 15

EICOSANOIDS

15.1 FUNCTIONS OF EICOSANOIDS

The eicosanoids are a complex family of bioactive lipid messengers generated by oxygenation of 20-carbon polyunsaturated fatty acids, primarily arachidonic acid (Fig. 15-1). Eicosanoids are local-acting autocrine and paracrine hormones that stimulate cells adjacent to their site of synthesis. In general, eicosanoids have a short half-life, usually on the order of minutes. They are not stored in cells but instead are released as soon as they are synthesized.

Eicosanoids fall into two main classes: (1) prostanoids that have a ring structure, including prostaglandins, thromboxanes, and prostacyclins, and (2) linear eicosanoids consisting of leukotrienes, lipoxins, and hydroxyeicosatetraenoic acids (HETEs).

15.1.1 Prostaglandins Are Eicosanoids with Ring Structures

The term *prostaglandin* reflects the original isolation of these molecules from seminal fluid, into which they are secreted by the seminal glands (rather than the prostate). Prostaglandins act to modulate many physiological functions, including blood pressure, uterine contraction, and the production of pain and fever. Prostaglandins are designated PGA, PGD, PGE, or PGF, based on the functional groups on the cyclopentane ring that is comprised of carbons 8 through 12 (Fig. 15-2). For example, PGE_2 contains a 9-keto and an 11-hydroxy group, while $PGF_{2\alpha}$ contains two hydroxy groups; α designates the stereochemistry of the C9 hydroxyl group. All of the

Medical Biochemistry: Human Metabolism in Health and Disease By Miriam D. Rosenthal and Robert H. Glew
Copyright © 2009 John Wiley & Sons, Inc.

oxyeicosatetraenoic acids (HETEs) are closely related to noncysteinyl
nes but lack the conjugated series of three double bonds (Fig. 15-4).
and LTB$_4$ regulate neutrophil and eosinophil function; specifically, they
hemotaxis, stimulate adenylyl cyclase activity, and induce polymorphonu-
ulocytes to degranulate and release hydrolytic enzymes from lysosomes.
xins are another class of linear eicosanoids that are derived from arachi-
Their structures are distinct from those of the leukotrienes and HETEs
contain three hydroxyl groups and a conjugated tetraene system (Fig.
and LTB$_4$ have many physiological functions, including antiangiogenic
enhanced clearance of exudates resulting from pulmonary edema, and
om reperfusion injury.

sanoid Receptors

itiate their physiologic effects by binding to G-protein-coupled recep-
sma membranes of target cells. The receptors are named IP, EP, and
nate the specific ligand they bind: EP1 and EP2 designate multiple
same prostaglandin. Various prostaglandin receptors signal through
reases in cAMP, G$_q$-mediated increases in intracellular free calcium,
decreases in cAMP.
classes of leukotriene receptors. Binding of LTB$_4$ or related hydroxy-
s) to B-LT receptors elicits chemotactic responses in leukocytes. The
enes bind to cys-LT receptors and stimulate contraction of smooth
TEs may also be incorporated into the phospholipids of membranes
ere the presence of fatty acyl chains containing a polar hydroxyl
ipid packing and thus the normal structure and function of the

ids from Dihomo-γ-Linolenic Acid and
Acid

or the synthesis of human eicosanoids is arachidonic acid, which
es of prostaglandins (two of the arachidonate double bonds are
e cyclization reaction) and the 4-series of leukotrienes. Al-
n also be synthesized from dihomo-γ-linolenic acid (20:3ω6),
sor of arachidonic acid, there is little synthesis of 1-series
eries leukotrienes in humans. PGE$_1$ has, however, been uti-
pharmacological agent; one application of its vasodilatory
iagnosis and treatment of erectile dysfunction.
d (EPA, 20:5ω3) is the other physiologically significant pre-
noids in humans. The main dietary source of EPA is fish
cold-water marine fish; some EPA is also synthesized from

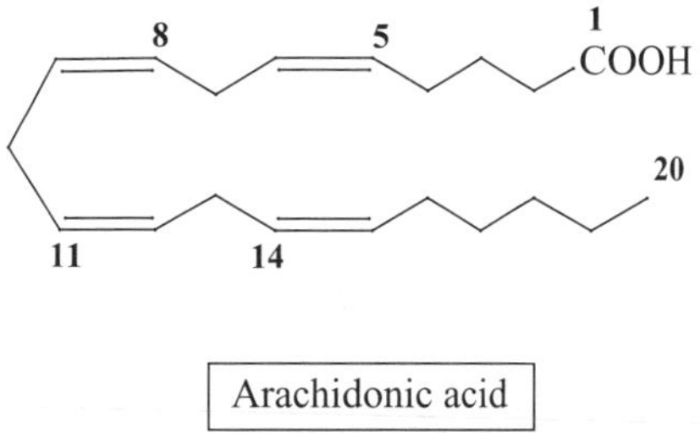

Arachidonic acid

FIGURE 15-1   Structure of arachidonic acid.

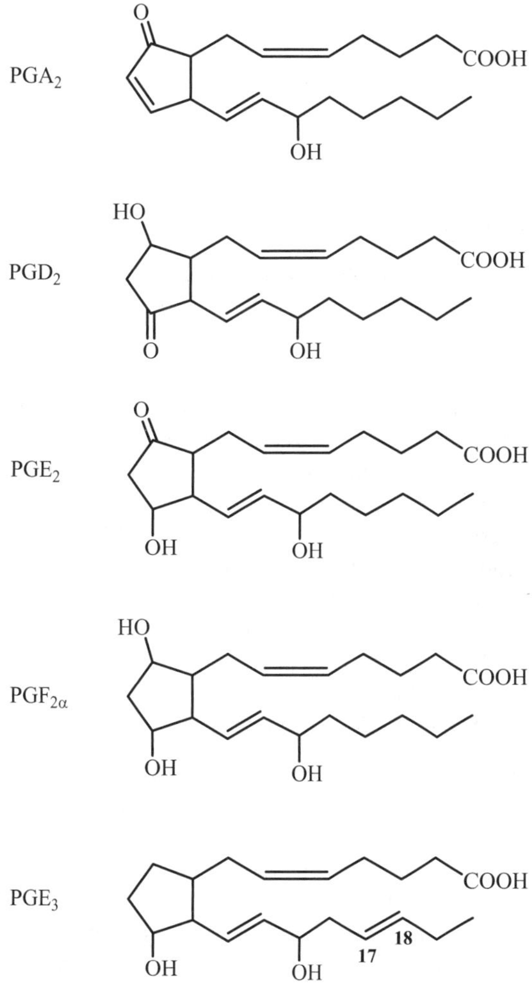

FIGURE 15-2   Structures of some common prostaglandins.

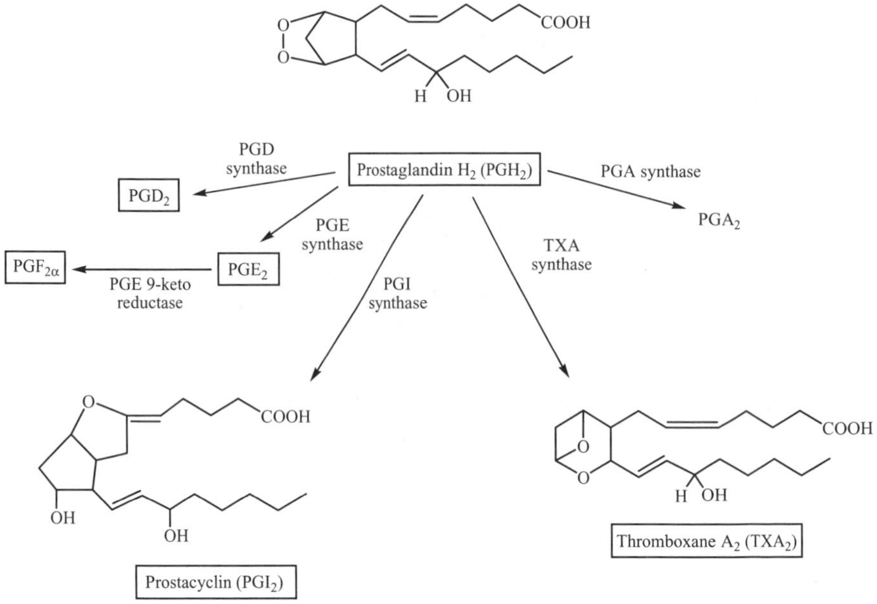

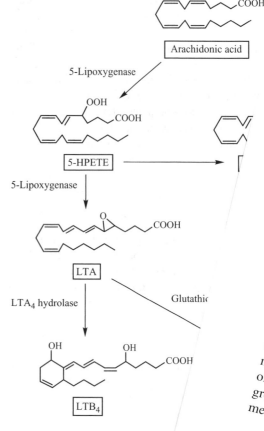

FIGURE 15-3 Synthesis of the different prostaglandins from their common precursor, prostaglandin H$_2$.

prostanoids that are derived from arachidonic acid have numeral 2 as a subscript, referring to the number of carbon–carbon double bonds in the two side chains.

Prostacyclin (PGI$_2$) and thromboxane (TX$_2$) are two prostanoids that have somewhat unusual ring structures. Relative to PGE$_2$ for example, prostacyclin has an additional oxygen-containing ring between C6 and C9. Thromboxanes have a six-membered oxygen-containing ring instead of the cyclopentane ring of classic prostaglandins (Fig. 15-3). The term *thromboxane* refers to the platelet-aggregating activity, which has thrombus-forming potential.

15.1.2 Leukotrienes and Lipoxins are Linear Eicosanoids

Unlike the prostanoids, which contain a ring element in their structure, the leukotrienes are linear molecules (Fig. 15-4). The term leukotriene derives from their cell of origin (leukocytes) and the fact that their structures contain three carbon–carbon double bonds in conjugation. The most important leukotrienes in humans are LTA$_4$ and LTB$_4$ and their cysteinyl derivatives, LTC$_4$, LTD$_4$, and LTE$_4$; all are derived from arachidonic acid and contain four double bonds. The cysteinyl-leukotrienes constitute the slow-reacting substance of anaphylaxis (SRS-A) and promote smooth muscle contraction, constriction of pulmonary airways, trachea and intestine, and increases in capillary permeability (edema).

FIGURE 15

Hydr
leukotrie
5-HETE
mediate c
clear gran

The lip
donic acid
in that the
15-5). LXA
properties,
protection fr

15.1.3 Eic

Eicosanoids ir
tors on the pla
so on, to desig
receptors for th
G$_s$-mediated in
or G$_i$-mediated

There are two
acids (i.e., HETE
cysteinyl-leukotri
muscle cells. HE
of target cells, wr
group modulates
membrane.

15.1.4 Eicosan
Eicosapentaenoi

The major precursor
gives rise to the 2-ser
removed as part of t
though eicosanoids ca
the immediate precur
prostaglandins and 3-s
lized extensively as a
effects has been in the
Eicosapentaenoic ac
cursor of human eicosa
oil, particularly oil from

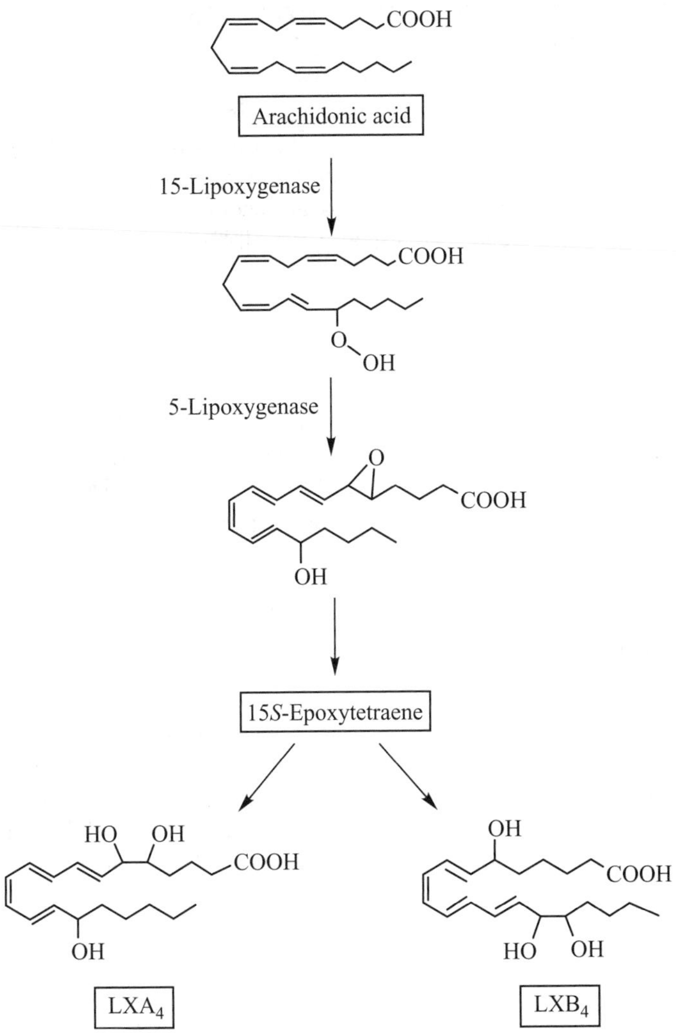

FIGURE 15-5   Structures and synthesis of lipoxins.

dietary α-linolenic acid (18:3ω3). EPA is the precursor to a family of eicosanoids, each of which has one more double bond than the corresponding eicosanoid derived from arachidonate (i.e., PGE_3, TXA_3, LTC_5) (Fig. 15-2). An increased dietary intake of fish oil can raise the ratio of membrane phospholipid EPA to arachidonate from the average of less than 0.1 to as high as 0.5. Although dietary fish oils have been shown to be cardioprotective, anti-inflammatory, and anticarcinogenic, it is still unclear how much these benefits are due to the partial replacement of arachidonate-derived eicosanoids with those synthesized from EPA.

15.2 SITES OF SYNTHESIS OF PROSTAGLANDINS AND LEUKOTRIENES

With the exception of red blood cells, prostaglandins are produced and released by nearly all human cells and act on adjacent cells of the same organ. The cyclooxygenases and related enzymes in the eicosanoid synthesis pathways are localized to the cytoplasmic surfaces of the nuclear envelope and endoplasmic reticulum. The pattern of both eicosanoid production and response are cell-type specific. The following are selected examples.

15.2.1 Gastrointestinal Tract

Prostaglandins serve a cytoprotective role in the stomach. PGE_2 is synthesized by epithelial and smooth muscle cells in the stomach, where it reduces gastric acid secretion while stimulating the production of protective mucus. For this reason, synthetic prostaglandins are helpful in promoting the healing of gastric ulcers.

15.2.2 Cardiovascular System

In blood vessels, different prostaglandins have opposing effects. For example, platelets produce thromboxane A_2 (TXA_2), which promotes platelet aggregation, whereas vascular endothelial cells produce prostacyclin (PGI_2), which inhibits platelet aggregation. Both PGE_2 and PGI_2 are vasodilators that lower systemic arterial pressure, thereby increasing local blood flow and decreasing peripheral resistance. By contrast, both TXA_2 and $PGF_{2\alpha}$ (produced by vascular smooth muscle) are vasoconstrictors.

15.2.3 Kidney

PGE_2 is the major prostaglandin in the kidney, and the collecting ducts are the main site of its production. The kidney also produces $PGF_{2\alpha}$, PGD_2, and thromboxane A_2. PGE_2 dilates renal blood vessels and increases blood flow through the kidney. PGE_2 is also an important stimulator of renin release, thus contributing to the regulation of sodium excretion and the glomerular filtration rate.

15.2.4 The Lungs

Monocytes and neutrophils in the lungs produce LTB_4, 5-HETE, and the cysteinyl-leukotrienes (LTC_4, LTD_4, and LTE_4), which are bronchoconstrictors. LTC_4 is more potent than histamine in contracting nonvascular smooth muscles of bronchi.

15.2.5 Female Reproductive Tract

PGE_2 within the ovarian follicle is essential for ovulation. During parturition, prostaglandins soften tissues in the cervix and stimulate uterine contractions to expel the fetus.

15.3 CONDITIONS WHEN EICOSANOID SYNTHESIS IS UP-REGULATED

Prostaglandin and leukotriene synthesis in human cells and tissues is often triggered by hormonal or neural excitation, or muscular activity. For example, histamine increases prostaglandin production in the gastric mucosa. Also, prostaglandins are released during labor and after cellular injury (e.g., platelets exposed to thrombin, lungs irritated by dust).

15.3.1 Inflammation

Prostaglandins, PGE_2 in particular, are mediators of the edema, erythema (redness of the skin), and the fever and pain associated with inflammation. Inflammatory reactions most often involve the joints (e.g., rheumatoid arthritis), skin (e.g., psoriasis), and eyes and are usually treated with corticosteroids that inhibit prostaglandin synthesis. PGE_2, generated in immune cells (e.g., macrophages, mast cells, B cells), evokes chemotaxis of T cells. It is thought that pyrogens (fever-inducing agents) activate the prostaglandin synthesis pathway with release of PGE_2 in the hypothalamus, where body temperature is regulated.

15.3.2 Activation of Neutrophils, Monocytes, and Macrophages

Synthesis of leukotrienes and hydroxyeicosatetraenoic acids (HETEs) is up-regulated under conditions of allergy and inflammation. Binding of IgE antibodies to membrane receptors stimulates mast cells to release HETEs, which then activate other cells. Similarly, HETEs (especially 5-HETE) and LTB_4 produced by activated leukocytes induce degranulation of neutrophils and eosinophils.

15.4 METABOLISM OF EICOSANOIDS

15.4.1 Release of Arachidonic Acid

Synthesis of eicosanoids is initiated by release of arachidonic acid, primarily by the action of phospholipase A_2's (PLA_2), which hydrolyze the fatty acid from the *sn*-2 position of membrane phospholipids:

$$\text{phospholipid} + H_2O \rightarrow \text{1-acyllysophospholipid} + \text{arachidonate}$$

Several PLA$_2$ enzymes are responsible for arachidonate mobilization. One is a cytosolic enzyme that is activated in response to signal-transduction cascades. There are also several small secretory PLA$_2$'s which are active as extracellular enzymes; they are produced in response to sepsis and inflammation and found in high concentrations in the synovial fluid of patients with arthritis.

15.4.2   Synthesis of Prostaglandins

15.4.2.1 Prostaglandin G/H Synthase. The key enzyme of prostaglandin biosynthesis is the bifunctional enzyme prostaglandin G/H synthase (PGS). PGS is a single-polypeptide-chain enzyme that has two catalytic sites. One catalytic site has cyclooxygenase activity which catalyzes the addition of two molecules of molecular oxygen to arachidonate to form the initial prostaglandin, which is PGG$_2$ (Fig. 15-6). The second catalytic site is a glutathione-dependent, heme-containing

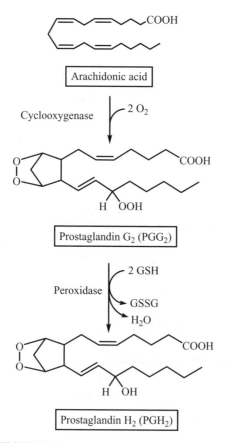

FIGURE 15-6   Synthesis of prostaglandin H$_2$.

peroxidase that converts the hydroperoxide group ($-$OOH) on carbon 15 of PGG_2 to a hydroxyl group.

There are two major isozymes of prostaglandin G/H synthase or cyclooxygenase (COX), commonly designated COX-1 and COX-2. COX-1 is primarily a constitutive enzyme of gastric mucosa, platelets, vascular endothelium, and kidney, whereas COX-2 is inducible and expressed in activated macrophages and monocytes as well as smooth muscle and epithelial cells.

Recent studies have identified a third PGS isozyme, COX-3, which is made from the COX-1 gene but retains intron 1 in its mRNA. COX-3 is expressed in the cerebral cortex and is inhibited by analgesic and antipyretic drugs, such as acetaminophen (Tylenol), that do not inhibit COX-1 and COX-2.

15.4.2.2 Synthesis of Prostaglandins. A family of prostaglandin synthases (i.e., PGD synthase, PGE synthase, PGI synthase) convert PGH_2 to the various prostaglandins (Fig. 15-3). PGH_2 is also the precursor of thromboxane A_2, in which the cyclopentane ring is replaced by a six-membered oxygen-containing ring. The one prostaglandin that is not synthesized directly from PGH_2 is $PGF_{2\alpha}$, which is synthesized from PGE_2 by PGE 9-keto reductase.

15.4.3 Synthesis and Catabolism of Linear Eicosanoids

15.4.3.1 Lipoxygenases. Lipoxygenases are dioxygenases that attach both atoms of molecular oxygen to a particular carbon atom of arachidonic acid (e.g., position 5, 12, or 15). In humans, the most important leukotrienes are the 5-lipoxygenase (5-LOX) products, which mediate inflammatory disorders; their synthesis is initiated by addition of a hydroperoxy group to arachidonic acid to produce 5-hydroperoxyeicosatetraenoic acid (5-HPETE) (Fig. 15-4). 5-LOX is activated by an accessory protein called *5-lipoxygenase-activating protein* (FLAP), an arachidonic acid transfer protein that presents the fatty acid substrate to the 5-LOX enzyme.

HPETE-hydroperoxides are highly reactive, unstable metabolic intermediates. The reduction of hydroxyeicosatetraenoic acids (HPETEs) to the hydroxyeicosatetraenoic acid (e.g., 5-HETE) occurs spontaneously or is catalyzed by peroxidases.

15.4.3.2 Leukotriene Synthesis. 5-LOX contains two enzymatic activities: the dioxygenase activity that converts arachidonic acid to 5-HPETE and a dehydrase activity that transforms 5-HPETE to the epoxide leukotriene A_4 (LTA_4) (Fig. 15-4). LTA_4 is an important branch point in the pathway of leukotriene synthesis. It can be converted to leukotriene B_4 by LTA_4 hydrolase, which opens up the epoxide ring. Alternatively, LTA_4 can be converted to LTC_4 by LTC_4 synthase, which catalyzes the conjugation of LTA_4 with glutathione. Sequential removal of glutamate and glycine residues by specific peptidases yields the leukotrienes LTD_4 and LTE_4, respectively.

15.4.3.3 Lipoxin Synthesis. Lipoxins are synthesized by multicellular processes involving the sequential actions of two lipoxygenases. One such sequence involves synthesis of LTA_4 by 5-lipoxygenase in granulocytes followed by

15-hydroxylation of LTA_4 by the 12/15-lipoxygenase in platelets and hydrolysis of the epoxide ring to generate LXA_4 (Fig. 15-5).

15.4.3.4 *Cytochrome P450 Epoxygenase Pathway for Eicosanoid Synthesis.*

In addition to the lipoxygenase family of enzymes, linear eicosanoids are also synthesized by cytochrome P450 epoxygenases. The resulting epoxy-eicosatrienoic acids (EETs) are converted into a variety of hydroxyeicosatrienoic acids and dihydroxyeicosatrienoic acids. The products differ from those of the lipoxygenase pathway in that they contain three rather than four double bonds.

15.4.4 Catabolism of Eicosanoids

Most eicosanoids are extremely unstable molecules. Thromboxane A_2, for example, has a half-life of about 30 seconds in water and is rapidly transformed into inactive thromboxane B_2 (TXB_2). Similarly, prostacyclin I_2 (PGI_2) spontaneously breaks down to 6-keto $PGF_{1\alpha}$. Within the cell, prostaglandins are often inactivated by the action of a prostaglandin 15-hydroxydehydrogenase. The 15-keto prostaglandins are then oxidized further by oxidation from the omega (ω) end as well as by β-oxidation from the carboxy end, and the resulting more polar molecules are excreted by the kidney.

15.5 REGULATION OF EICOSANOID SYNTHESIS AND ACTIVITY

15.5.1 Regulation of Arachidonate Mobilization

Activation of phospholipase A_2 is crucial for the release of arachidonic acid, which serves as a substrate for eicosanoid synthesis. Cytosolic phospholipase A_2 ($cPLA_2$) is activated in a cell-specific manner by a variety of agonists (e.g., by thrombin in platelets). Steroidal anti-inflammatory drugs such as prednisone and betamethasone block prostaglandin release in part by inducing the synthesis of members of the phospholipase A_2-inhibitory family of proteins called *lipocortins* or *annexins*. Glucocorticoids may also directly suppress transcription of the genes for $cPLA_2$ and for secretory PLA_2.

15.5.2 Regulation of Prostaglandin G/H Synthase

Prostaglandin G/H synthase represents the committed step in prostaglandin and thromboxane synthesis. Whereas COX-1 is constitutive in many cell types, synthesis of COX-2 in various cells is induced by a variety of cytokines and lipid mediators (e.g., sphingosine 1-phosphate). Although glucocorticoids had long been assumed to block prostaglandin synthesis at the level of phospholipase A_2, there is increasing evidence that they also suppress induction of COX-2 in many cell types (e.g., airway smooth muscle cells).

Prostaglandin synthase, a major target for pharmacological intervention, is inhibited by nonsteroidal anti-inflammatory drugs (NSAIDs) such as aspirin (acetylsalicylic acid), ibuprofen, indomethacin, and phenylbutazone. Most NSAIDs act as reversible inhibitors of the cyclooxygenase component of PGH_2. Aspirin acts differently in that it irreversibly inhibits PGS by acetylating the hydroxyl group of a particular serine hydroxyl at the active site of cyclooxygenase. Low-dose aspirin regimens are often used to decrease the risk of thrombosis and coronary heart disease in older persons. This therapy is effective because circulating platelets are unable to synthesize more prostaglandin synthase to replace that which has been inactivated.

Recent pharmaceutical efforts have focused on the development of selective COX-2 inhibitors such as Celebrex (celecoxib), with the goal of developing anti-inflammatory and pain-blocking drugs less likely to cause the gastric toxicity associated with chronic use of NSAIDs that block COX-1. Both Celebrex and Vioxx (now withdrawn from the market) have been associated with increased adverse cardiovascular events, possibly related to decreased production of antithrombotic PGI_2 while not inhibiting COX-1-mediated synthesis of thromboxane A_2 in platelets.

15.5.3 Regulation of Leukotriene Metabolism

Current therapies for asthma include use of 5-lipoxygenase inhibitors such as zileuton, and cysteinyl-leukotriene (cysLT) receptor antagonists such as montelukast.

15.6 DISEASES INVOLVING EICOSANOIDS

15.6.1 Aspirin-Intolerant Asthma

People with aspirin-intolerant asthma develop bronchoconstriction in response to aspirin and other NSAIDs, which, by inhibiting cyclooxygenase activity, leave more arachidonic acid available to enter the linear eicosanoid pathways. Aspirin-sensitive persons appear to have polymorphisms in one or more genes related to cysteinyl-leukotriene production, such as LTC_4 synthase itself or the EP2 receptor through which PGE_2 inhibits leukotriene synthesis.

15.6.2 Ulcers

Because prostaglandins have a protective effect on the lining of the stomach, long-term inhibition of COX-1 with NSAIDs can promote gastrointestinal bleeding.

15.6.3 Increased Tendency for Bleeding

Whereas the omega-3 fatty acids in fish oils have beneficial effects on cardiovascular health because they modify the ratio of prothrombotic and antithrombotic prostanoid activities, high doses (greater than 3 g/day) have been associated with increased bleeding times, cardiovascular disease, and stroke.

15.6.4 Cancer

COX-2 expression is high in many cancers, including those of the prostate, breast, and colon. High concentrations of PGE_2 promote survival of tumor cells by inhibiting apoptosis, stimulating cell proliferation, and promoting angiogenesis.

15.6.5 Therapeutic Uses of Eicosanoids

Exogenous prostaglandins have a number of therapeutic uses. For example, in the fetus, PGE_2 maintains the patency of the ductus arteriosus prior to birth. In infants born with congenital abnormalities that can be corrected surgically, infusion of PGE_2 will maintain blood flow through the ductus until surgery is performed. Conversely, if the ductus remains open after birth in an otherwise normal infant, closure can be hastened by the cyclooxygenase inhibitor indomethacin. PGE_2 has also been used to induce cervical ripening and uterine contractions, leading to parturition.

Misoprostol is a synthetic PGE_1 analog used to prevent NSAID-induced ulcers. In many countries, misoprostol is also used in combination with the synthetic steroid RU486 to block the action of progesterone and induce medical (as opposed to surgical) abortions.

CHAPTER 16

GLYCOLIPIDS AND GLYCOPROTEINS

16.1 FUNCTIONS OF GLYCOCONJUGATES

Glycoconjugates are a diverse group of molecules that contain one or more sugars covalently attached to protein or lipid. The carbohydrate domains of glycoconjugates often contain modified amino or acidic sugar derivatives that contribute to their strongly hydrophilic nature and structural specificity (Fig. 16-1). The glycoconjugates can be grouped into three major classes, described below.

16.1.1 Glycolipids

Glycolipids are membrane sphingolipids that contain one or more sugar moieties attached to the hydrophobic ceramide domain. The major glycolipids are cerebrosides, globosides, and gangliosides. Glycosphingolipids in the plasma membrane tend to concentrate along with sphingomyelin, cholesterol, and phosphatidylinositol-anchored proteins in lipid domains called *rafts*, which are involved in receptor-mediated plasma-membrane signal transduction and cell–cell adhesion.

16.1.1.1 Cerebrosides. Cerebrosides are ceramide monohexosides, the most common of which is *galactocerebroside*, also called *galactolipid* (Fig. 16-2). Most galactocerebroside is found in the brain, where it is a major component of the myelin sheath.

Medical Biochemistry: Human Metabolism in Health and Disease   By Miriam D. Rosenthal and Robert H. Glew
Copyright © 2009 John Wiley & Sons, Inc.

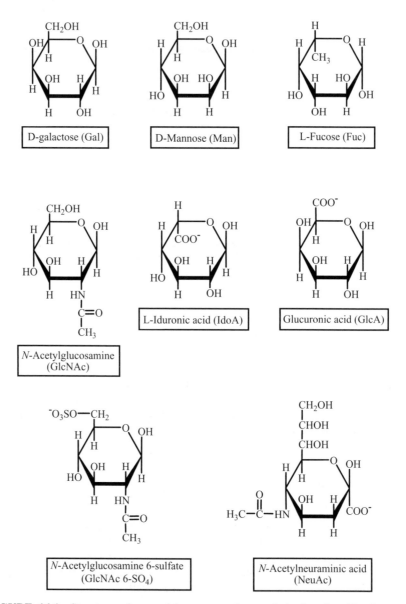

FIGURE 16-1   Structures of some of the sugars and sugar derivatives found in glycoconjugates.

16.1.1.2 Globosides.

Globosides are sphingolipids that contain two or more sugar residues, usually a combination of galactose, glucose, and N-acetylgalactosamine. The oligosaccharides of globosides are uncharged and contain no free amino groups. Prominent globosides include lactosylceramide [ceramide-β-glc(4-1)-β-gal], which is present in the erythrocyte membrane, and ceramide

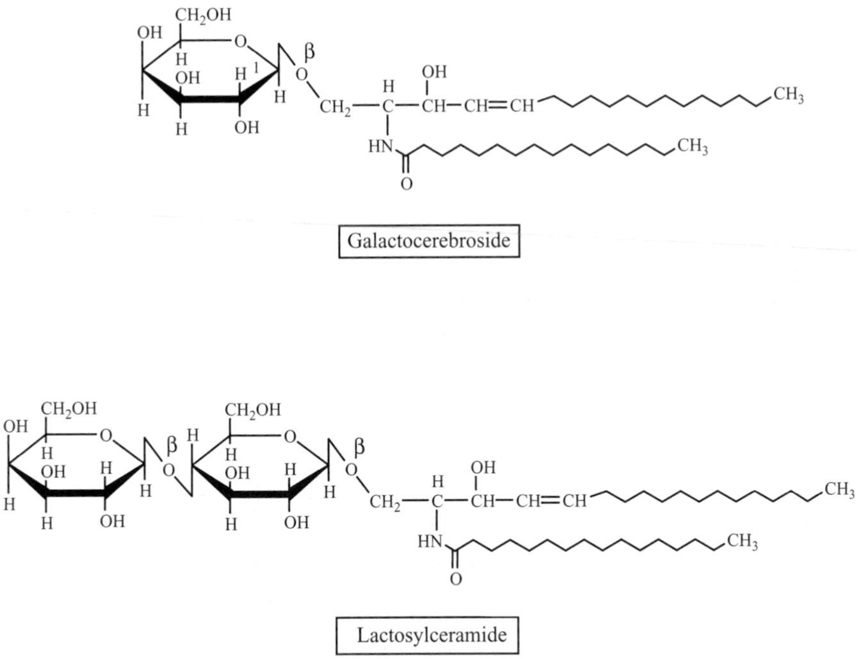

Galactocerebroside

Lactosylceramide

FIGURE 16-2   Structures of a cerebroside (galactocerebroside) and a globoside (lactosyl-ceramide).

galactosyllactoside [ceramide-β-glc(4-1)-β-gal-(4-1)-α-gal], which is prominent in membranes of the nervous system (Fig. 16-2). Still other globosides contain carbohydrate domains that provide the ABO antigenic determinants (Fig. 16-3) on the surfaces of cells, particularly erythrocytes. ABO carbohydrate antigens are also found on membrane glycolipids. In some persons, the ABO carbohydrates are also found in plasma in soluble form associated with secreted glycoproteins or on free oligosaccharides.

16.1.1.3   Gangliosides. Gangliosides are sialic acid–containing glycosphingolipids which are highly concentrated in ganglion cells of the central nervous system, particularly in the nerve endings (Fig. 16-4). Lesser amounts of gangliosides are present in the plasma membrane of cells of most extraneural tissues. Sialic acid, which is found in glycoproteins and mucins as well as gangliosides, is the *N*-acetyl derivative of the nine-carbon amino sugar neuraminic acid (Neu) (Fig. 16-1). The carbohydrate domains of the various gangliosides serve as receptors for many different classes of ligands, including cytokines, microbial toxins (e.g., *Vibrio cholera* toxin), microbes, viruses, and hormones. Gangliosides play roles in diverse cellular processes, such as cell–cell recognition, cell homing and adhesion, and growth regulation and differentiation.

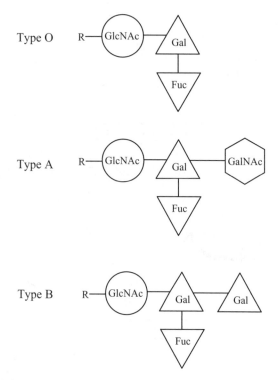

FIGURE 16-3 ABO antigens. R, ceramide or glycoprotein.

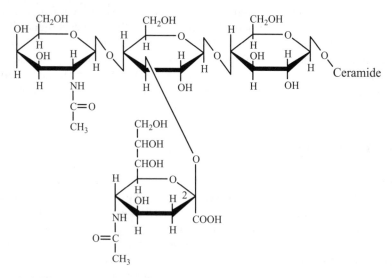

FIGURE 16-4 Ganglioside GM$_2$, which accumulates in Tay–Sachs disease.

16.1.2 Glycoproteins

Proteins with covalently attached carbohydrate chains (oligosaccharides) are found on the outer surface of the plasma membrane of cells, as part of the extracellular matrix, and in the blood; indeed, except for albumin, nearly all of the proteins in plasma are glycoproteins. Cell-associated glycoproteins include receptors (e.g., the LDL receptor) and ion exchangers in membranes (e.g., the chloride–bicarbonate exchanger known as band 3 of erythrocytes). Glycoproteins also play important roles in cell growth and development, and in the communication that occurs between cells.

A particular glycoprotein can have one or as many as 30 oligosaccharide chains, with the carbohydrate accounting for as little as 1% to as much as 70% of the mass of the glycoprotein. The oligosaccharide chains of a particular glycoprotein usually influence one or more of its biological properties, including intracellular transport, solubility, viscosity, susceptibility to inactivation (by heat, extremes of pH, and proteolysis), and the tendency to aggregate.

The carbohydrate chains of glycoproteins are grouped into two classes, depending on how they are linked to the protein. O-linked oligosaccharides are linked to protein through a glycosidic bond to the hydroxyl group of a threonine or serine residue, whereas N-linked oligosaccharides are attached to the amide nitrogen of an asparagine residue in the protein. Many proteins, including the LDL receptor, contain both N- and O-linked oligosaccharide domains. Figure 16-5 illustrates two types of N-asparagine-linked oligosaccharide chains commonly found in mammalian glycoproteins: the high-mannose type and the complex type.

16.1.2.1 *Mucins.* Mucins are a subclass of glycoproteins that are abundantly glycosylated. They are high molecular weight (200 to 10,000 kDa) glycoproteins that contain dozens to several hundred oligosaccharides O-linked to serine or threonine residues. Oligosaccharides usually account for 50 to 90% of the total mass of the mucin molecule. The sugars that comprise a particular oligosaccharide chain of a mucin may include N-acetylgalactosamine (GalNAc), galactose (Gal), N-acetylglucosamine (GlcNAc), fucose, and sialic acid.

Secreted mucins (such as salivary mucin) aggregate into oligomeric gels which form a protective layer over the digestive, respiratory, and reproductive tracts and provide lubrication as well as a barrier against pathogens and toxins. Other mucins are integral membrane proteins and remain cell-associated. Cancer cells often synthesize abnormal mucins, whose structures can perturb the normal function of a cell, including its immunologic and adhesive properties and its potential to invade and metastasize.

16.1.3 Proteoglycans

Proteoglycans (formerly called *mucopolysaccharides*) are a class of highly acidic molecules that function as lubricants and structural components of connective tissue. They also mediate adhesion of cells to the extracellular matrix. Proteoglycans are characterized by long glycosaminoglycan (GAG) chains that are attached covalently to a core protein. Most of the structure of the GAG chains of proteoglycans

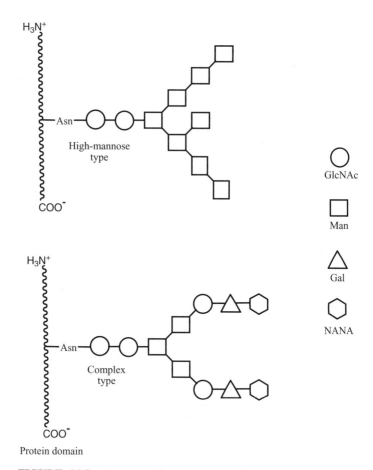

FIGURE 16-5   Structures of some common N-linked oligosaccharides.

is made up of disaccharide repeat units (Fig. 16-6). For example, in heparan sulfate the disaccharide repeat unit glucuronic acid–N-acetylglucosamine. In general, one of the two sugars of the disaccharide repeating unit of a GAG is an amino sugar (N-acetylglucosamine or N-acetylgalactosamine); the other is an acidic sugar (glucuronic acid or iduronic acid). Many of the sugars of GAGs contain sulfate in either O- or N-sulfate linkage.

The core protein of a proteoglycan can contain as many as several hundred GAG chains per molecule. The high density of negative charges due to the carboxyl-containing sugars (e.g., glucuronic acid or iduronic acid) and the sulfate residues in the GAG chains cause proteoglycans to form an extended linear structure that resembles a bottle brush.

Hyaluronic acid is a glycosaminoglycan that is not a proteoglycan. Instead, the extremely long GAG chain of hyaluronic acid is secreted into the extracellular matrix as a polysaccharide that is not attached covalently to a core protein. Unlike the proteoglycans, the sugar residues of hyaluronic acid are not sulfated.

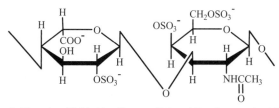

D-Glucuronic acid (GlcA) sulfate   N-Acetyl-D-galactosamine (GalNAc) sulfate

Chondroitin 4-sulfate

L-Iduronic acid (IdoA) sulfate   N-Acetyl-D-galactosamine (GalNAc) sulfate

Dermatan sulfate

D-Glucuronic acid (GlcA) sulfate   D-Glucosamine (GlcNH$_2$) sulfate

Heparan sulfate

L-Iduronic acid (IdoA) sulfate   D-Glucosamine (GlcNH$_2$) sulfate

Heparin

FIGURE 16-6   Structures of repeating disaccharide units in proteoglycans.

16.2   WHERE ARE GLYCOCONJUGATES FOUND IN THE BODY?

16.2.1   Glycocalyx

Carbohydrates are a major component of the external surface of cell membranes. Both glycolipids and glycoproteins are integral components of the asymmetrical plasma membrane, with the carbohydrate moieties facing outward. Other glycoproteins and proteoglycans are adsorbed onto the extracellular surface of the membrane. One such proteoglycan is heparan sulfate (Fig. 16-6), which is the receptor for various growth factors, including vascular endothelial growth factor (VEGF). Heparan sulfate proteoglycan also serves as the ligand that binds lipoprotein lipase to the luminal surface of the capillary endothelium. Heparin, an anticoagulant, is more heavily sulfated than is heparan sulfate.

16.2.2   Extracellular Matrix

Proteoglycans are major components of the extracellular matrix where different classes of proteoglycans confer specific chemical and physical properties on the tissue. For example, chondrocytes (cartilage cells) secrete a variety of proteoglycans, of which a major component is a particular chondroitin sulfate named *aggrecan*. Following its secretion, aggrecan molecules aggregate spontaneously to form a supramolecular structure known as *hyaluronan* which endows cartilage with its load-bearing properties. Hyaluronic acid is a component of the extracellular matrix of skin and connective tissue where its viscous and elastic nature allows it to function as a lubricant and shock absorber in the synovial fluid of joints.

16.2.3   Plasma Proteins

As noted above, nearly all plasma proteins are glycoproteins, including many with enzymatic activity, such as lecithin : cholesterol acyltransferase (LCAT) and ceruloplasmin.

16.2.4   Lysosomal Proteins

Although most glycosylated proteins are secreted from cells, others are targeted to lysosomes. These include the lysosomal-associated membrane glycoproteins as well as lysosomal acid hydrolases.

16.3   METABOLISM OF GLYCOCONJUGATES

16.3.1   Synthesis of Sugar Residues

Glycoconjugates contain several sugar residues and sugar derivatives that have not been described in previous chapters (Fig. 16-1). These include mannose, L-fucose, glucuronic acid, iduronic acid, and sialic acid, as well as amino sugars such as glucosamine. As is the case with the synthesis of glycogen, synthesis of glycocon-jugates requires activated sugars, usually in the form of their UDP derivatives (e.g., UDP-glucose, UDP-xylose; UDP-N-acetylgalactosamine). Exceptions include man-nose and L-fucose, which are activated as GDP sugars, and sialic acid, which is activated as the CMP derivative.

Mannose 6-phosphate is formed by aldose–ketose isomerization of fructose 6-phosphate and then activated by GTP to GDP-mannose. GDP-fucose, the acti-vated form of the L-deoxysugar, is synthesized by NADH-dependent reduction of GDP-mannose. UDP-glucuronate is a sugar acid formed by oxidization of UDP-glucose. Following its incorporation into glycosaminoglycans, glucuronate may be isomerized to iduronate. UDP-glucuronate can also be converted to UDP-xylose.

Synthesis of amino sugars is initiated by glutamine : fructose 6-phosphate amido-transferase, which catalyzes the synthesis of glucosamine 6-phosphate. The amino groups of amino sugars are generally acetylated using acetyl-CoA. Sialic acid (N-acetylneuraminic acid) is synthesized by condensation of the carbon backbone of phosphoenolpyruvate with N-acetylmannosamine 6-phosphate.

16.3.2   Synthesis of Sphingolipids

16.3.2.1   Cerebrosides. Galactocerebroside and glucocerebroside are usually synthesized from ceramide and UDP-galactose or UDP-glucose, respectively, by galactosyl and glucosyl transferases that are associated with the endoplasmic reticulum:

$$\text{ceramide} + \text{UDP-galactose} \rightarrow \text{galactocerebroside} + \text{UDP}$$

Alternatively, in some tissues, synthesis of glucocerebroside proceeds by gluco-sylation of sphingosine by glucosyltransferase:

$$\text{sphingosine} + \text{UDP-glucose} \rightarrow \text{glucosylsphingosine} + \text{UDP}$$

followed by fatty acylation:

$$\text{glucosylsphingosine} + \text{acyl-CoA} \rightarrow \text{glucocerebroside} + \text{CoASH}$$

Unlike sphingomyelin, whose N-acyl group is usually stearate, glycolipids contain behenic acid, a saturated 22-carbon fatty acid or some other very long-chain saturated fatty acid. Galactocerebroside is the major cerebroside found in membranes; by contrast, glucocerebroside is primarily an intermediate in the synthesis of more complex sphingolipids.

16.3.2.2   Sulfatides. Galactocerebroside 3-sulfate, a sulfuric acid ester of galac-tocerebroside, is the major sulfolipid of brain, accounting for about 15% of the lipids of white matter. It is synthesized from galactocerebroside and 3′-phosphoadenosine

5′-phosphosulfate (PAPS; see Chapter 21) in a reaction is catalyzed by sulfotrans-ferase:

$$\text{galactocerebroside} + \text{PAPS} \rightarrow$$

$$\text{galactocerebroside 3-sulfate} + \text{phosphoadenosinephosphate}$$

16.3.2.3   *Globosides.*

Globosides and gangliosides are synthesized in the Golgi apparatus by enzymes that transfer sugars sequentially onto a cerebroside. One such example is the synthesis of the glycosphingolipids containing A, B, and O blood group antigens (Fig. 16-3). The core structure of the O (or H) oligosaccharide is formed by sequential addition of *N*-acetylglucosamine, galactose, and fucose to galactocerebroside. Persons with a gene for the type A transferase are able to transfer *N*-acetylgalactosamine to the core structure to synthesize the A antigen from the O core structure, while those with the type B transferase transfer galactose to synthesize the B antigen. Some people have genes for both the A and B transferases and therefore synthesize both A and B antigens. The "O gene" is actually a mutation which results in premature termination of translation so that no active transferase A or B is formed; persons homozygous for the gene for the O blood group therefore synthesize only the core O antigen.

16.3.2.4   *Gangliosides.*

Most gangliosides are built on lactosylceramide (Gal-β1,4-Glc-β1,1′-Cer), which is formed by the transfer of galactose from UDP-galactose to glucosylceramide (glucocerebroside). Additional sugars, including sialic acid, are then added stepwise to the growing glycan chain. Humans contain at least five different sialyltransferases, each with a different specificity with regard to the acceptor.

16.3.3   Synthesis of Glycoproteins

The oligosaccharide chains of glycoproteins can either be O-linked (to serine or threonine residues of the protein) or N-linked (to the amide nitrogen in the side chain of asparagine). O-linked glycosylation of proteins, like that of sphingolipids, occurs by stepwise glycosylation in the Golgi complex. By contrast, the synthesis of N-linked oligosaccharide chains involves a more complex sequence of reactions which is initiated in the endoplasmic reticulum and continues in the Golgi complex.

The first phase of N-linked oligosaccharide synthesis takes place on an isoprene lipid, dolichol pyrophosphate (Fig. 16-7). The core oligosaccharide, containing two *N*-acetylglucosamine, nine mannose, and three glucose residues, is assembled on dolichol pyrophosphate by the successive transfer of glycosyl residues from their respective nucleoside diphosphate sugar donors (e.g., GDP-mannose, UDP-glucose). The transfer of the two *N*-acetylglucosamine and five of the mannose residues occurs in the rough endoplasmic reticulum (ER). The dolichol moiety with its attached

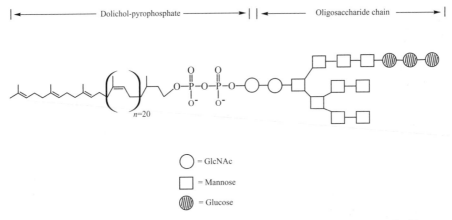

FIGURE 16-7   Structure of dolichol attached to a fully assembled oligosaccharide.

oligosaccharide chain is then flipped across the ER membrane into the lumen of the ER, where the four additional mannose residues plus the three glucose residues are then added. Once the target polypeptide is transported through a channel in the ER membrane and its signal sequence has been cleaved, the core oligosaccharide is then transferred en bloc from the dolichylpyrophosphate to the polypeptide. The asparagine that accepts the core oligosaccharide always occurs in the sequence Asn-X-Thr/Ser, where X is any amino acid except proline.

The next phase of N-linked oligosaccharide-chain processing involves α-glucosidase-catalyzed stepwise removal of the three glucose residues and one mannose from the core oligosaccharide. Two ER proteins, calnexin and calreticulin, assure the correct folding of the glycoprotein. The properly folded glycoprotein then moves by vesicular transport to the Golgi complex, where it undergoes a variety of additional posttranslational modifications, including the removal of additional mannose residues and the sequential addition of single residues each of N-acetylglucosamine, galactose, fucose, and sialic acid.

16.3.4 Synthesis of Proteoglycans

The initial synthesis of proteoglycans resembles that of O-linked glycoproteins in that sugars are added to the protein one at a time in the lumen of the endoplasmic reticulum. A xylose residue is attached to the hydroxyl group of a serine, followed by addition of two galactose residues to form what is called the *link trisaccharide*. Two glycosyltransferases enzymes then alternate adding sugar residues to generate the repeating disaccharide units (i.e., glucuronic acid and glucosamine) when heparin is being synthesized. Sulfation of sugar residues occurs after the sugars have been attached to the growing oligosaccharide chain, and as with the sulfation of galactosylceramide, PAPS is the donor of the activated sulfate moiety.

16.3.5   Catabolism of Glycoconjugates

Phagocytic cells, particularly the macrophages of the reticuloendothelial system located primarily in liver, spleen, and bone marrow, are especially active in glycosphingolipid catabolism. The glycosphingolipids are catabolized in lysosomes by sequential, irreversible removal of the carbohydrate residues—one at a time—followed by the hydrolysis of ceramide to sphingosine and a free fatty acid. This pathway requires enzymes that cleave specific bonds, including α- and β-galactosidases, a β-glucosidase, a neuraminidase, a hexosaminidase, a sphingomyelinase, a sulfatase, and a ceramide-specific amidase (ceramidase).

Lysosomes are also responsible for the stepwise degradation of the glycosaminoglycan (GAG) chains of proteoglycans. The lysosomal enzymes are all acid hydrolases with pH optima in the range 3.5 to 5.5. The enzymes that hydrolyze glycosphingolipids often require sphingolipid activator proteins, which promote interaction between these enzymes and their water-insoluble lipid substrates.

16.4   DISEASES OF GLYCOCONJUGATE METABOLISM

16.4.1   Sphingolipidoses

There are about one dozen different life-threatening human disorders that result from genetically-based deficiencies of lysosomal hydrolases which degrade glycolipids; collectively, they are referred to as the *sphingolipidoses*. Sphingolipid catabolism normally functions smoothly, all of the glycosphingolipids and sphingomyelin being degraded to their constituents. However, when the activity of one enzyme in the pathway is markedly reduced due to a genetic error, the substrate for that defective enzyme accumulates within the lysosomes of the tissue in which catabolism of that sphingolipid normally occurs. Examples of sphingolipidoses are described below.

16.4.1.1   Gaucher Disease.   Gaucher disease is caused by a genetic deficiency of lysosomal glucocerebrosidase. The accumulation of glucocerebroside, primarily in macrophages of the reticuloendothelial system, results in hepatomegaly, splenomegaly, anemia, and bone pain. Gaucher disease is now treated effectively by enzyme replacement therapy using recombinant glucocerebrosidase, which is produced in human cells so as to obtain appropriate glycosylation of the enzyme with oligosaccharide chains terminating in mannose residues. Mannose receptors on the surface of macrophages bind the mannose-terminated enzymes and through a process of endocytosis deliver them into lysosomes, where they degrade the accumulated lipid, glucocerebroside.

16.4.1.2   Fabry Disease.   Deficiency of lysosomal α-galactosidase A results in Fabry disease and accumulation of globotriaosylceramide (Cer → β-Glu → β-Glu → α-Gal) in tissues, mainly the walls of blood vessels. Unlike the other sphingolipidoses, which are autosomal recessive diseases, Fabry disease is X-linked. Enzyme replacement therapy is now also available for Fabry disease.

16.4.1.3 Tay–Sachs Disease. Tay–Sachs disease is a gangliosidosis caused by the absence of β-hexosaminidase A and results in neural accumulation of the ganglioside G_{M2} (Fig. 16-2). The disease is characterized by mental retardation, a cherry-red spot on the macula which reflects ganglioside accumulation in retinal ganglia, blindness, and for the most severe, infantile form, death before age 3. Because of the primary involvement of ganglion cells of the central nervous system, effective enzyme replacement therapy has not proven feasible.

16.4.2 Mucopolysaccharidoses

Deficiencies in lysosomal enzymes that are normally responsible for degrading the glycosaminoglycan (GAG) chains of proteoglycans result in accumulation of undegraded GAGs. Since there is some digestion of the oligosaccharide chains by lysosomal endoglycosidases, urinary excretion of shorter oligosaccharides is often diagnostic. The mucopolysaccharidoses are classified according to the substrate that accumulates (Table 16-1). Treatment of patients with MPS I (Hurler disease) with recombinant human α-L-iduronidase appears to reduce lysosomal storage in the liver and ameliorate some clinical manifestations of the disease.

16.4.3 Oligosaccharidoses

The genetic diseases that result from defects in the lysosomal pathway that degrades the oligosaccharide chains of glycoproteins are called *oligosaccharidoses*. They can result from deficiencies in any one of a number of enzymes, including α-mannosidase, β-mannosidase, α-fucosidase, and α-sialidase. Oligosaccharidoses are usually named for the deficient enzyme; for example, a deficiency in α-mannosidase is called α-mannosidosis.

TABLE 16-1 Selected Mucopolysaccharidoses and the Associated Enzyme Defect

Disease	Enzyme Defect	Accumulated Substance
Hurler, Scheie (MPS I)	α-Iduronidase	Dermatan sulfate, heparan sulfate
Hunter (MPS II)	Iduronate sulfatase	Dermatan sulfate, heparan sulfate
Sanfilippo A (MPS IIIA)	Heparan *N*-sulfatase	Heparan sulfate
Sanfilippo B (MPS-B)	*N*-Acetylglucosaminidase	Heparan sulfate
Morquio A (MPS IVA)	*N*-Acetylgalactosamine-6-sulfatase	Keratin sulfate
Maroteaux–Lamy (MPS VI)	*N*-Acetylgalactosamine-4-sulfatase	Dermatan sulfate

MPS, mucopolysaccharadosis.

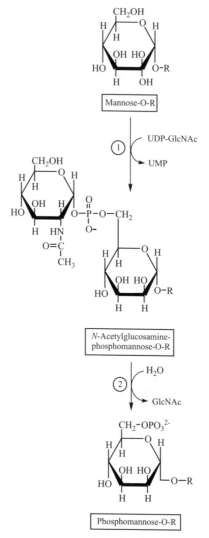

FIGURE 16-8 Generation of mannose 6-phosphate tag on the oligosaccharide chains of lysosomal enzymes: ① UDP-*N*-acetylglucosamine-1-phosphotransferase; ② *N*-acetylglucosamine-1-phosphodiester-α-*N*-acetylglucosaminidase. R, lysosomal enzyme.

16.4.4 I-Cell Disease

I-cell disease (mucolipidosis II) is a rare autosomal lysosomal storage disorder characterized by the accumulation of mucopolysaccharides, sphingolipids, and glycolipids inside lysosomes of visceral and mesenchymal cells. Patients with I-cell disease secrete large amounts of many different lysosomal enzymes into body fluids but have deficient enzyme activity within the lysosomes.

The trafficking of lysosomal enzymes, which contain N-asparagine-linked oligosaccharides, to the lysosome requires the addition of a mannose 6-phosphate recognition marker. Two enzymes are required to attach a phosphate group to a mannose moiety of oligosaccharide chains of lysosomal enzymes (Fig. 16-8). The first reaction is catalyzed by UDP-N-acetylglucosamine-1-phosphotransferase, commonly called *phosphotransferase*. The second enzyme in the pathway, N-acetylglucosamine-1-phosphodiester-α-N-acetylglucosaminidase, removes the terminal α-N-acetylglucosamine residue, leaving the phosphate group attached to the underlying mannose residue.

Lysosomal enzymes bearing the mannose 6-phosphate marker bind to the mannose 6-phosphate receptor in the trans Golgi, are packaged into clathrin-coated vesicles, and are transported to late endosomes, where the low pH causes the lysosomal enzymes to dissociate from the receptors. Patients with I-cell disease have a deficiency of the phosphotransferase, which impairs targeting of enzymes to the lysozyme.

16.4.5   Abnormal Glycosylation

Congenital disorders of glycosylation (CDGs) include abnormalities in either the glycosyltransferases, which elongate, or the glycosidases, which process the oligosaccharide chains of N-linked glycoproteins, O-linked glycoproteins, or both. The CDG I family comprises defects in the assembly of the dolichol-linked glycan and its transfer to the protein, whereas the CDG II family includes defects in the processing of the protein-bound glycans. CDG disorders affect many different glycoconjugates, including clotting factors, collagen, red cell membrane glycophorin, and α_1-antitrypsin.

16.4.6   Cholera

A ganglioside on intestinal mucosal cells binds cholera toxin, an 84-kDa protein secreted by the pathogen *Vibrio cholerae*. The toxin consists of one A subunit and five B subunits. The *choleragenoid domain*, as the B subunits are called, binds to the ganglioside G_{M1}. The A subunit then enters the cell and acts as an ADP-ribosyltransferase that transfers ADP-ribose of NAD onto the $G_{\alpha s}$ subunit of a G protein. This leads to activation of adenylyl cyclase, which stimulates secretion of chloride ion and produces diarrhea. Gangliosides may also bind other toxins (e.g., tetanus toxin) and viruses, such as the influenza viruses.

CHAPTER 17

CHOLESTEROL SYNTHESIS AND TRANSPORT

17.1 FUNCTIONS OF CHOLESTEROL

Cholesterol is the major sterol in humans. The sterol structure of cholesterol (Fig. 17-1) consists of four fused rings, three six-carbon and one five-carbon, designated A to D. Cholesterol has a hydroxyl group at C3, a C5–C6 carbon–carbon double bond, and two methyl groups, attached at positions C10 and C13 of the sterol ring. In addition, cholesterol has a branched eight-carbon hydrocarbon chain attached to the D ring at C17.

17.1.1 Cholesterol Is a Structural Component of Cellular Membranes

Cholesterol is a ubiquitous and essential component of mammalian cell membranes. It is also present in small amounts in the outer membrane of mitochondria. Cholesterol is especially abundant in myelinated structures of the central nervous system, with 25% of the body's cholesterol located in the brain. In contrast to plasma, where most of the circulating cholesterol exists esterified to a fatty acid, most cholesterol in cellular membranes is present in the free (unesterified) form. The fluidity of membranes is regulated in part by changing their cholesterol content.

17.1.2 Cholesterol Is a Major Component of Bile

Cholesterol is abundant in bile (normal concentration is about 15 mg per 100 mL, only 4% of which is esterified). The solubilization of free cholesterol in bile is achieved

Medical Biochemistry: Human Metabolism in Health and Disease   By Miriam D. Rosenthal and Robert H. Glew
Copyright © 2009 John Wiley & Sons, Inc.

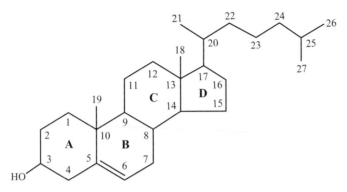

FIGURE 17-1 Structure of cholesterol.

in part by the detergent property of phosphatidylcholine, which is produced in liver and secreted into bile. Bile acids, which are metabolites of cholesterol, also aid in solubilizing cholesterol in bile. Increased biliary secretion of cholesterol or decreased secretion of phospholipids or bile acids into bile may lead to deposition of cholesterol-rich gallstones. Indeed, the name *cholesterol* was derived some 200 years ago from the Greek words *chole* (bile) + *stereos* (solid). Cholesterol and phospholipids in bile protect gallbladder membranes from potentially irritating or harmful effects of bile salts. In the absence of dietary intake of cholesterol (i.e., vegan or low-fat diets), the cholesterol in bile also provides enterocytes with a source of cholesterol for chylomicron synthesis.

17.1.3 The Cholesterol Synthesis Pathway Provides a Mechanism for Increasing the Hydrophobicity of Proteins

The activated prenyl groups farnesyl pyrophosphate (15 carbons) (Fig. 17-2) and geranylgeranyl pyrophosphate (20 carbons) can donate their hydrophobic domains to many proteins involved in cell signaling, including the γ-subunits of trimeric G proteins, Ras, and the nuclear lamins A and B. Posttranslational prenylation of these proteins increases their tendency to associate with membranes, and prenylation is usually required for their full activity. Covalent attachment of cholesterol to protein is required for the membrane tethering that is necessary for the activation of the Hedgehog family of proteins, which are essential to embryonic patterning; the cholesterol moiety is attached via an ester bond to the C-terminal glycine during autocatalytic cleavage and maturation of the initially soluble protein.

17.1.4 The Cholesterol Synthesis Pathway Provides Key Metabolic Intermediates

Cholesterol is the precursor of the various steroid hormones and bile acids. 7-Dehydrocholesterol, the immediate precursor of cholesterol, is converted to vitamin D_3 when skin is exposed to ultraviolet light. Isoprene units, generated as

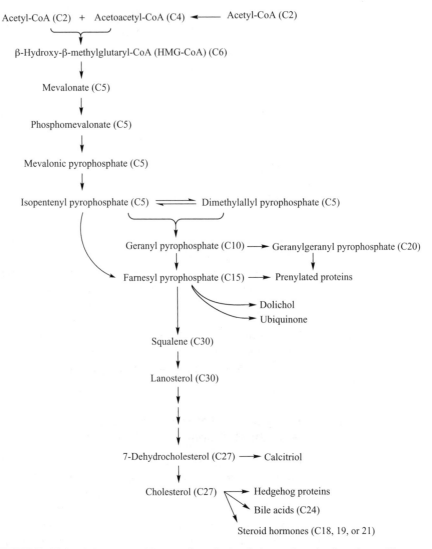

FIGURE 17-2 Substrates and intermediates in the cholesterol synthesis pathway. The numbers in parentheses indicate the number of carbon atoms in each molecule.

intermediates in the pathway of cholesterol synthesis, are also precursors in the synthesis of larger molecules, including heme A, dolichol pyrophosphate, and coenzyme Q (ubiquinone).

17.2 CHOLESTEROL HOMEOSTASIS

Cholesterol is both derived from the diet and synthesized de novo in the body. Humans cannot metabolize cholesterol to CO_2 and water. Excretion of cholesterol and of bile acids synthesized from cholesterol occurs by way of the liver, gallbladder, and intestine.

17.2.1 Synthesis of Cholesterol

Although synthesis of cholesterol occurs to some extent in virtually all cells (except red blood cells), in adults this capacity is greatest in liver, intestine, adrenal cortex, and reproductive tissues, including ovaries, testes, and placenta. The brain is the most cholesterol-rich organ of the body; however, since plasma lipoproteins that transport cholesterol in the circulation do not cross the blood–brain barrier, all cholesterol in the brain must be synthesized within the central nervous system. Cholesterol synthesis in the brain occurs at a high rate during the period of active myelination and declines substantially thereafter. Within the cell, cholesterol synthesis takes place in the cytosol and endoplasmic reticulum.

17.2.2 Intestinal Absorption and Excretion of Cholesterol

The average adult who consumes a Western diet takes in about 500 mg of dietary cholesterol. Cholesterol also enters the small intestine as a component of bile (Fig. 17-3). Although there is wide individual variation, on average about half of the cholesterol that enters the small intestine is absorbed into the body. Animal products contain both cholesterol and cholesteryl esters; the latter are hydrolyzed in the small intestine by pancreatic cholesteryl esterase:

$$\text{cholesteryl ester} + H_2O \rightarrow \text{cholesterol} + \text{fatty acid}$$

Cholesterol is absorbed by the cells of the intestinal mucosa and incorporated into the surface of chylomicrons. Cholesterol in excess of that required for the chylomicron surface is esterified to cholesteryl esters and incorporated into the triacylglycerol-rich chylomicron core. Both free and esterified cholesterol are delivered to the liver as components of chylomicron remnants.

Cholesterol that is not secreted from the intestinal mucosa in chylomicrons is returned to the intestinal lumen as a component of sloughed mucosal cells and excreted. The fecal sterols are a mixture of cholesterol and cholesterol metabolites, such as cholestanol and coprostanol, generated by intestinal bacteria (Fig. 17-3).

17.3 PATHWAY OF CHOLESTEROL SYNTHESIS

The pathway of cholesterol synthesis is summarized in Figure 17-2. All of the carbon atoms of cholesterol are derived from acetyl-CoA, which can be obtained from several

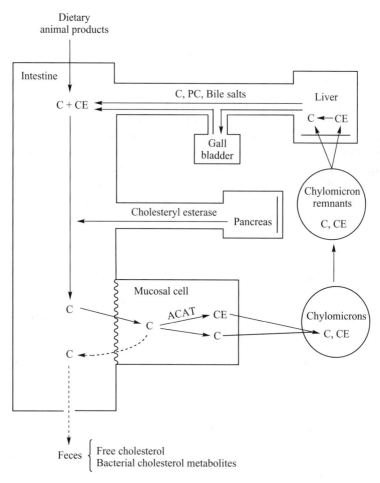

FIGURE 17-3 Intestinal absorption and excretion of cholesterol. ACAT; acyl-CoA–cholesterol acyltransferase; C, cholesterol; CE, cholesteryl ester; PC, phosphatidylcholine.

sources, including the pyruvate dehydrogenase reaction, β-oxidation of fatty acids, oxidation of amino acids, and ethanol. As with fatty acid synthesis, nearly all of the acetyl-CoA used for cholesterogenesis is generated in the mitochondrion; the acetyl moieties are transported to the cytosol in the form of citrate. Reducing power in the form of NADPH is required for cholesterogenesis and it is provided mainly by glucose 6-phosphate dehydrogenase and 6-phosphogluconate dehydrogenase of the hexose monophosphate pathway and by the malic enzyme. Cholesterol synthesis is driven largely by hydrolysis of the high-energy thioester bonds of acetyl-CoA and the phosphoanhydride bonds of ATP. The synthesis of one molecule of cholesterol consumes 18 molecules of acetyl-CoA, 16 molecules of NADPH, and 36 molecules of ATP.

17.3.1   Synthesis of β-Hydroxy-β-methylglutaryl-CoA

As is the case in the pathway that produces ketone bodies, synthesis of cholesterol involves β-hydroxy-β-methylglutaryl-CoA as an intermediate. Synthesis starts with the condensation of two molecules of acetyl-CoA to form acetoacetyl-CoA in a reaction catalyzed by acetoacetyl-CoA thiolase (acetyl-CoA:acetyl-CoA acetyltransferase) (Fig. 17-4):

$$2\,\text{acetyl-CoA} \rightarrow \text{acetoacetyl-CoA} + \text{CoASH}$$

A third molecule of acetyl-CoA is then used to form the branched-chain compound β-hydroxy-β-methylglutaryl-CoA (HMG-CoA). This reaction is catalyzed by HMG-CoA synthase (Fig. 17-4):

$$\text{acetyl-CoA} + \text{acetoacetyl-CoA} \rightarrow \text{β-hydroxy-β-methylglutaryl-CoA} + \text{CoASH}$$

Hepatocytes contain two distinct HMG-CoA synthases: a cytosolic enzyme that is involved in cholesterol synthesis, and a mitochondrial enzyme which functions in the pathway of ketone body synthesis.

17.3.2   Synthesis of Mevalonic Acid

The first compound unique to cholesterol synthesis is mevalonic acid. Mevalonic acid is formed from HMG-CoA by the enzyme HMG-CoA reductase, which is an intrinsic protein of the endoplasmic reticulum whose catalytic C-terminal domain extends into the cytosol. The NADPH-dependent reaction catalyzed by HMG-CoA reductase is irreversible and represents the rate-limiting step in cholesterol biosynthesis. This reaction consumes two molecules of NADPH, results in hydrolysis of the thioester bond of HMG-CoA, and generates the primary alcohol group of mevalonate (Fig. 17-4):

$$\text{HMG-CoA} + 2\text{NADPH} \rightarrow \text{mevalonate} + \text{CoASH} + 2\text{NADP}^+ + 2\text{H}^+$$

17.3.3   Synthesis of Isoprene Pyrophosphates

Mevalonate is converted to 5-pyrophosphomevalonate by the stepwise transfer of the terminal phosphate group of two molecules of ATP. The two reactions are catalyzed by mevalonate kinase and phosphomevalonate kinase, respectively (Fig. 17-5):

$$\text{mevalonate} + \text{ATP} \rightarrow \text{5-phosphomevalonate} + \text{ADP}$$

$$\text{5-phosphomevalonate} + \text{ATP} \rightarrow \text{5-pyrophosphomevalonate} + \text{ADP}$$

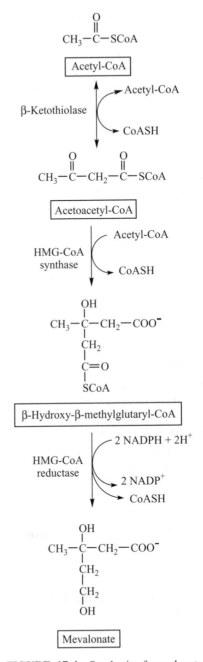

FIGURE 17-4 Synthesis of mevalonate.

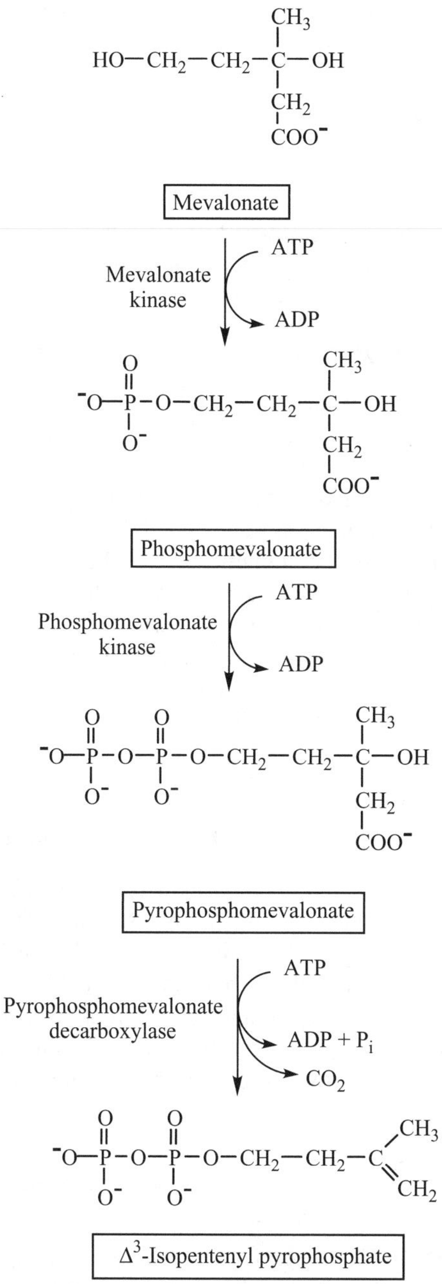

$$
\begin{array}{c}
CH_3 \\
| \\
HO-CH_2-CH_2-\overset{\displaystyle |}{C}-OH \\
| \\
CH_2 \\
| \\
COO^-
\end{array}
$$

| Mevalonate |

Mevalonate
kinase

ATP

ADP

$$
\begin{array}{cc}
\overset{\displaystyle O}{\underset{\displaystyle O^-}{\overset{\displaystyle \|}{^-O-P}}}-O-CH_2-CH_2-\overset{\displaystyle CH_3}{\underset{\displaystyle \underset{\displaystyle COO^-}{CH_2}}{\overset{\displaystyle |}{C}}}-OH
\end{array}
$$

| Phosphomevalonate |

Phosphomevalonate
kinase

ATP

ADP

$$
^-O-\overset{\displaystyle O}{\underset{\displaystyle O^-}{\overset{\displaystyle \|}{P}}}-O-\overset{\displaystyle O}{\underset{\displaystyle O^-}{\overset{\displaystyle \|}{P}}}-O-CH_2-CH_2-\overset{\displaystyle CH_3}{\underset{\displaystyle \underset{\displaystyle COO^-}{CH_2}}{\overset{\displaystyle |}{C}}}-OH
$$

| Pyrophosphomevalonate |

Pyrophosphomevalonate
decarboxylase

ATP

ADP + P_i

CO_2

$$
^-O-\overset{\displaystyle O}{\underset{\displaystyle O^-}{\overset{\displaystyle \|}{P}}}-O-\overset{\displaystyle O}{\underset{\displaystyle O^-}{\overset{\displaystyle \|}{P}}}-O-CH_2-CH_2-C\overset{\displaystyle CH_3}{\underset{\displaystyle CH_2}{}}
$$

| Δ^3-Isopentenyl pyrophosphate |

FIGURE 17-5 Synthesis of isopentenyl pyrophosphate.

5-Pyrophosphomevalonate is then decarboxylated by pyrophosphomevalonate decarboxylase to generate Δ^3-sopentenyl pyrophosphate.

$$5\text{-prophosphomevalonate} + ATP \rightarrow \Delta^3\text{-isopentenyl pyrophosphate} + CO_2$$
$$+ ADP + P_i$$

Isopentenyl pyrophosphate is converted to its allylic isomer 3,3-dimethylallyl pyrophosphate by isopentenyl pyrophosphate isomerase.

17.3.4 Condensation of Isoprene Pyrophosphates

Next, six isoprene pyrophosphate molecules condense to form the 30-carbon polyprene molecule squalene, which is devoid of oxygen atoms (Fig. 17-6). First, prenyltransferase condenses two isoprene pyrophosphate molecules, 3,3-dimethylallyl pyrophosphate and 3-isopentenyl pyrophosphate (in a head-to-tail manner), to release one molecule of pyrophosphate (PP_i) and form the 10-carbon molecule geranyl pyrophosphate. A second prenyltransferase then adds another 3-isopentenyl pyrophosphate unit to form the 15-carbon intermediate, farnesyl pyrophosphate, with the release of a second molecule of pyrophosphate. The last condensation step in cholesterol synthesis involves head-to-head fusion of two molecules of farnesyl pyrophosphate to form squalene. The reaction, catalyzed by squalene synthase, occurs on the endoplasmic reticulum, requires NADPH, and releases two pyrophosphate groups.

17.3.5 Synthesis of Lanosterol and Its Conversion to Cholesterol

As seen in Figure 17-7, rotation about carbon–carbon bonds permits squalene to assume an overall shape resembling that of cholesterol. Cholesterol synthesis from squalene proceeds through the intermediate lanosterol, which contains the fused tetracyclic ring system and an eight-carbon side chain. The endoplasmic reticulum enzyme that catalyzes this cyclization reaction is bifunctional and has both squalene epoxidase and lanosterol cyclase activity. Cyclization is initiated by epoxide formation between the future C2 and C3 of cholesterol, the epoxide being formed at the expense of NADPH:

$$\text{squalene} + O_2 + NADPH + H^+ \rightarrow \text{squalene 2,3-epoxide} + H_2O + NADP^+$$

The hydroxylation at C3 triggers the subsequent cyclization of squalene to lanosterol, with many carbon–carbon bonds being formed in a concerted fashion (Fig. 17-7). This reaction sequence requires the addition of an OH group to C3, the shifting of two methyl groups, and the elimination of a proton. The OH group of lanosterol projects above the plane of the A ring (i.e., in the β-orientation).

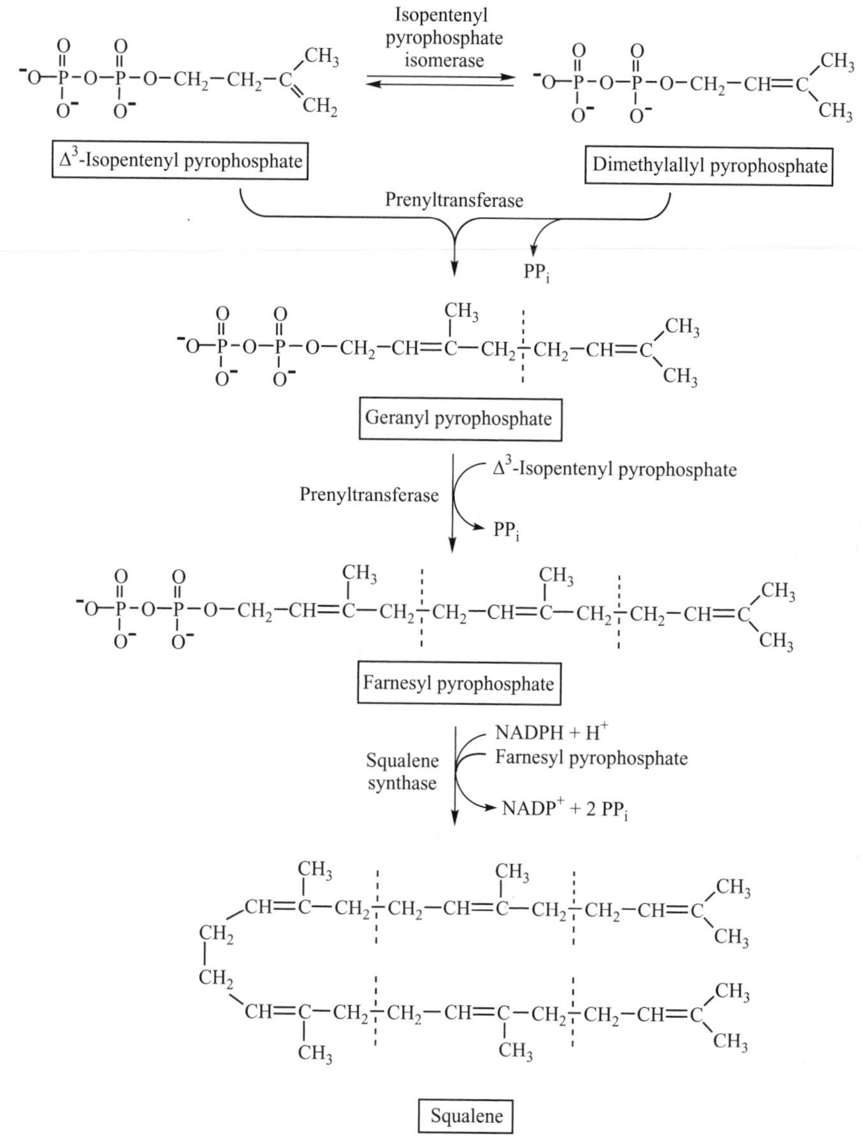

FIGURE 17-6   Condensation of isoprene pyrophosphate units generates squalene.

Transformation of the 30-carbon lanosterol to the 27-carbon cholesterol molecule requires at least eight different enzymes which catalyze removal of the methyl group at C14, removal of two methyl groups at C4, migration of the double bond from C8 to C5, and reduction of the double bond between C24 and C25 in the side chain (Fig. 17-7).

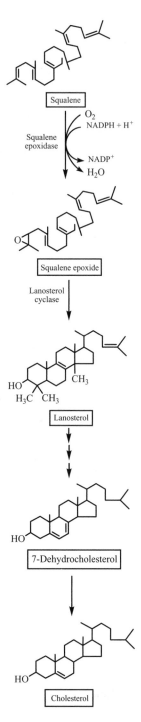

FIGURE 17-7   Synthesis of cholesterol from squalene.

17.3.6   Esterification of Cholesterol

Two key enzymes can esterify cholesterol: One is an intracellular enzyme, acyl-CoA: cholesterol acyltransferase (ACAT), which catalyzes the reaction

$$\text{fatty acyl-CoA} + \text{cholesterol} \rightarrow \text{cholesteryl ester} + \text{CoASH}$$

The other is an extracellular enzyme called lecithin:cholesterol acyltransferase (LCAT), which esterifies cholesterol molecules associated with the lipoprotein HDL. In this reaction, the source of the fatty acid is the phospholipid phosphatidylcholine:

$$\text{phosphatidylcholine} + \text{cholesterol} \rightarrow \text{cholesteryl ester} + \text{2-lysophosphatidylcholine}$$

17.3.7   Hydrolysis of Cholesteryl Esters

There are two distinct intracellular cholesteryl hydrolases. One is a lysosomal enzyme that hydrolyzes cholesteryl esters internalized through receptor-mediated endocytosis via the LDL receptor (see below), releasing free cholesterol for use by the cell. The other is a cytosolic enzyme that acts on cholesteryl esters in lipid droplets present in steroidogenic cells and which releases free cholesterol for synthesis of steroid hormones.

17.4   TRANSPORT OF CHOLESTEROL

17.4.1   Lipoproteins Transport Cholesterol in the Plasma

Cholesterol has very low solubility in water; at $30°C$, the limit of solubility is approximately 0.01 mg per 100 mL. The total fasting cholesterol concentration in plasma of healthy people is usually 150 to 200 mg per 100 mL, which is about twice the normal plasma glucose concentration. Such a high concentration of cholesterol in plasma is possible due to cholesterol-rich plasma lipoproteins that solubilize and disperse cholesterol (see Table 12-1). Only about 30% of the total plasma cholesterol is free (unesterified); the rest is esterified with a long-chain fatty acid, usually linoleic acid, which increases the hydrophobicity of cholesterol. Along with proteins and phospholipids, free cholesterol is a key component of the surface coat of the plasma lipoproteins, whereas cholesteryl esters and TAGs are located within the core of the lipoproteins (see Fig. 12-3).

17.4.1.1   Chylomicrons. Chylomicrons are the main vehicle for transporting dietary-derived lipids, including cholesterol and cholesteryl esters in the circulation (Fig. 17-8). Like other lipoproteins, cholesterol is part of the surface coat of chylomicrons. Dietary cholesterol is also esterified within the intestinal mucosal cells and incorporated into the core of the developing chylomicrons. As the chylomicrons

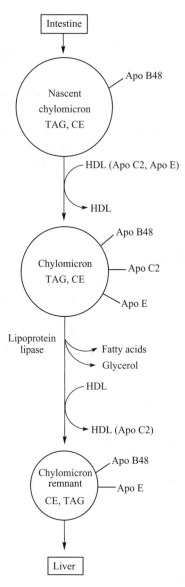

FIGURE 17-8 Simplified pathway for the transport of exogenous cholesterol in the blood. C, cholesterol; CE, cholesteryl ester, TAG, triacylglycerol.

circulate in the plasma, they lose TAG through hydrolysis by lipoprotein lipase (LpL), and shrink to become smaller particles called *chylomicron remnants*. Apoprotein E (apo E) on the surface of the chylomicron remnants facilitates binding of the remnants to the LDL receptor-related protein (LRP) on hepatocytes. The chylomicron remnants then enter the liver cells by receptor-mediated endocytosis. This process delivers exogenous cholesterol and other lipophilic molecules (e.g., fat-soluble vitamins) to the liver.

17.4.1.2 VLDL. Very low-density lipoproteins (VLDL) are synthesized and secreted by the liver. Although the main function of VLDL is to export endogenous TAG that are made in the liver, VLDL is also involved in the transport of both free cholesterol and cholesteryl esters between tissues. Very low-density lipoproteins are the main vehicle for exporting both dietary-derived and endogenously synthesized cholesterol from hepatocytes into the plasma. Like chylomicrons, VLDL changes its composition and size as it circulates; it loses TAG through hydrolysis by LpL and by acquiring additional cholesteryl esters from HDL. As their triacylglycerol component undergoes hydrolysis, VLDL particles become remnants of various sizes (sometimes called IDL, or intermediate-density lipoproteins). The VLDL remnants contain apo E and enter hepatocytes by receptor-mediated endocytosis. Approximately two-thirds of the VLDL remnants are removed from the circulation by the liver; the remaining VLDL is converted in the circulation to LDL as a result of the actions of LpL and hepatic lipase (Fig. 17-9).

17.4.1.3 LDL. In humans, low-density lipoprotein (LDL) is the major lipoprotein that transports cholesterol in blood. Unlike chylomicrons and VLDL that are rich in TAG, the core of LDL contains primarily cholesteryl esters. The surface of each LDL particle contains one molecule of apo B100. Since its concentration in plasma is positively correlated with cardiovascular disease (stroke, myocardial infarction, blood clots), LDL-cholesterol is popularly termed the "bad" cholesterol.

LDL functions primarily to deliver cholesterol and cholesteryl esters to peripheral tissues such as the adrenal glands, testes, and ovaries. Since LDL acquires cholesteryl esters from HDL, it also contributes to *reverse cholesterol transport*, whereby cholesteryl esters are transported from peripheral tissues to the liver for excretion as cholesterol or as bile salts. The half-life of plasma LDL is about 3 days.

Both hepatocytes and peripheral cells express LDL receptors which recognize apo B100 and internalize LDL via receptor-mediated endocytosis (Fig. 17-9). The receptors are then recycled back to the cell surface, and the LDL is transported to lysosomes, where hydrolysis of cholesteryl esters generates free cholesterol.

17.4.1.4 HDL. The main role of high-density lipoproteins (HDL) is reverse cholesterol transport whereby HDL extracts cholesterol from peripheral tissues and transports that cholesterol to the liver for excretion (Fig. 17-10). Circulating HDL can also donate cholesteryl esters to other lipoproteins such as VLDL and IDL. HDL also plays a central role in lipoprotein metabolism by donating proteins such as apo C2 and apo E to chylomicrons and VLDL. Since the concentration of HDL is inversely correlated with cardiovascular disease, HDL-cholesterol is described as the "good" cholesterol. The plasma concentration of HDL can be increased by physical activity and by therapeutic agents such as statins, niacin, and cholestyramine.

The primary apoprotein in HDL is apoprotein A (apo A). The major form of apo A is apo A1, which is synthesized by both the liver and the intestine. Apo A1

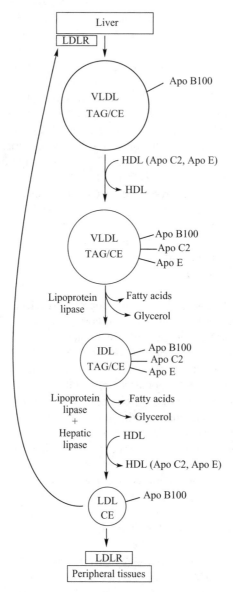

FIGURE 17-9 Simplified pathway for the transport of endogenous cholesterol. C, cholesterol; CE, cholesteryl ester; IDL, intermediate-density lipoprotein.

has a relatively long half-life (approximately 5 days), is recycled many times over, and acquires most of its lipid components in the circulation. There are three major structural forms of apo A1 circulating in the plasma: (1) amorphous or lipid-free HDL (apoA1), which does contain some phospholipid; (2) nascent or discoidal, lipid-poor HDL; and (3) mature, spherical HDL (HDL2, HDL3), which is rich in cholesteryl

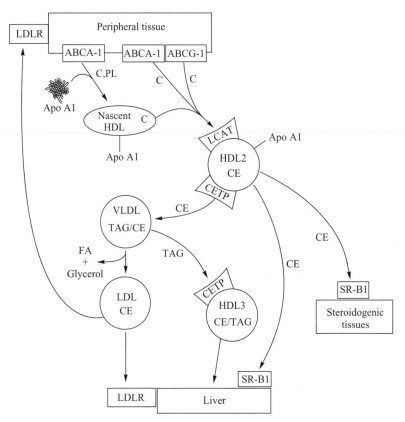

FIGURE 17-10 Simplified pathway for the formation of HDL and its role in the reverse transport of cholesterol. ABCA-1 and ABCG-1 are members of the ABC superfamily of ATP-binding casette transporters. Although not shown in the figure, HDL3 can also be converted back to HDL2 by the action of hepatic lipase. CETP, cholesteryl ester transfer protein; C, cholesterol; CE, cholesteryl ester; FA, fatty acid; G, glycerol; LDLR, LDL receptor; SR-B1, scavenger receptor B1.

esters. Lipid-free apo A1 is secreted by the liver and acquires free cholesterol and phospholipids in the plasma, thereby becoming discoidal or nascent HDL. As nascent HDL acquires additional free cholesterol, the cholesterol is esterified by the action of lecithin : cholesterol acyltransferase (LCAT) to generate cholesteryl esters, which account for the bulk of the core of the mature HDL, which is designated HDL2.

There are two mechanisms by which circulating HDL gives up some of its cholesteryl esters. One is the exchange of cholesteryl esters in HDL for TAG in VLDL, or, to a lesser extent, in LDL. The exchange process is mediated by cholesteryl ester transfer protein (CETP). Cholesteryl esters may also be removed by selective uptake by the liver, which is mediated by a scavenger receptor class BI (SR-BI) and occurs without the intracellular uptake of HDL proteins. The resulting HDL3 particles are smaller and have a higher ratio of TAG to cholesteryl esters than HDL2 has, and can be cleared by the liver or the kidney. In the process of HDL3 formation from HDL2 some of the excess apo A1 and phospholipid is released from the surface of the particles, thus regenerating lipid-poor HDL. The lipid-poor HDL can then acquire and esterify additional cholesterol, as described above. Although not shown in Figure 17-10, HDL3 can also be converted back to HDL2 by the action of hepatic lipase, which hydrolyzes TAG, releasing free fatty acids and glycerol.

17.4.2 Lipoproteins Are Remodeled While They Circulate

Plasma lipoproteins are dynamic aggregates whose compositions change as they move through the blood. One example of lipoprotein remodeling is the maturation of HDL in the circulation. Another is the gradual hydrolysis—by lipoprotein lipase—of TAG in circulating chylomicrons and VLDL (see Chapter 12). Many other proteins contribute to the remodeling of plasma lipoproteins, as described below.

17.4.2.1 *ATP-Binding Cassette Transporters.* The plasma membrane of peripheral cells contains two large proteins, ABCA-1 and ABCG-1, which mediate the release of cholesterol from those cells (Fig. 17-10). ABCA-1 and ABCG-1 are members of the ABC superfamily of ATP-binding cassette transporters that are responsible for the translocation across membranes of certain drugs, peptides, and lipids, including cholesterol. ABCA-1 converts lipid-free apo A1 into nascent HDL. When apo A1 docks with ABCA-1 on the plasma membrane, several molecules of phospholipid and free cholesterol are transferred to apoA1 in an energy (ATP)-dependent process (Fig. 17-10). Both ABCA-1 and ABCG-1 can transfer additional free cholesterol to the nascent HDL.

17.4.2.2 *Lecithin : Cholesterol Acyltransferase (LCAT).* Lecithin : cholesterol acyltransferase is a soluble enzyme that is secreted by the liver. In the circulation, LCAT binds to the surface of HDL particles, where it is activated by apo A1. As HDL acquires free cholesterol from the plasma membrane of a cell in the periphery, LCAT transfers the fatty acid from the 2-position of phosphatidylcholine (lecithin) to free cholesterol to form cholesteryl ester:

phosphatidylcholine $+$ cholesterol $\rightarrow$ cholesteryl ester $+$ lysophosphatidylcholine

LCAT prefers to use linoleic acid (18:2n-6) in this *trans*-esterification reaction. By contrast, acyl-CoA:cholesterol acyltransferase, the enzyme that generates intracellular cholesteryl esters, preferentially uses oleoyl-CoA.

17.4.2.3 *Cholesterol Ester Transfer Protein (CETP).* The transfer of cholesteryl esters between HDL and other lipoproteins is facilitated by a plasma protein called CETP, which is synthesized in and secreted by hepatocytes and adipocytes. CETP occurs in the circulation bound to the various subspecies of HDL. The function

of CETP is to catalyze the transfer of cholesteryl esters from HDL to VLDL and LDL in exchange for the transfer of TAG to HDL (Fig. 17-10).

17.4.2.4 Phospholipid Transfer Protein (PLTP). PLTP catalyzes the transfer of lipids, particularly phosphatidylcholine, between lipoproteins. As triacylglycerol hydrolysis occurs and both chylomicrons and VLDL particles shrink in size, PLTP serves to remove excess phospholipid from the surface of these lipoproteins and transfer it to HDL, providing substrate for the LCAT reaction.

17.4.3 Plasma Lipoproteins Exchange Proteins in the Circulation

When VLDL is secreted from the liver, the major lipid component is triacylglycerol and the major protein constituents are apo B100 and apo A1. Apo A1 then dissociates from the VLDL, and as it acquires cholesterol and phospholipids, it becomes HDL. Circulating VLDL acquires apo C2 and apo E from HDL. As the VLDL-associated triacylglycerol is hydrolyzed and VLDL becomes IDL and eventually LDL, VLDL particles also donate apo C2 and apo E back to HDL, so that the mature LDL particle contains only apo B100. HDL thus serves as a reservoir for apo C2 and apo E in the circulation (Figs. 17-8 and 17-9).

17.4.4 Lipoprotein Receptors

Receptor-mediated endocytosis of lipoproteins provides a mechanism both for their clearance from the circulation and for the delivery of key lipid components to target cells. Targeting of lipoproteins to sites of metabolism and removal is mediated primarily by the apoproteins on their surfaces.

17.4.4.1 LDL Receptor. The LDL receptor (LDLR) is a transmembrane glycoprotein with an apo B-100–binding domain. LDL receptors are expressed on liver cells and extrahepatic tissues and they recognize apo B100 but not the smaller apo B48 molecule present on chylomicrons and chylomicron remnants. Once the LDL receptor is occupied by LDL, the LDL:LDLR complex clusters in coated pits, which are then internalized by receptor-mediated endocytosis. Intracellularly, as the clathrin-coated vesicles lose their clathrin the LDL receptors are recycled back to the plasma membrane. The LDL-containing endosomes then fuse with lysosomes to form endolysosomes whose internal milieu is relatively acidic (approximately pH 5). Within the endolysosomes, the cholesteryl esters are hydrolyzed by "acid lipase" to free cholesterol and fatty acids, while the apo B-100 is hydrolyzed to amino acids.

17.4.4.2 LDL Receptor-Related Protein (LRP). LRP is expressed on the surface of hepatocytes but not peripheral cells. Its function is to bind and clear chylomicron remnants.

17.4.4.3 Scavenger Receptors A (SR-A's). Scavenger receptors A are a family of molecules that are expressed on tissue macrophages, Kupffer cells, and various extrahepatic endothelial cells. SR-A's take up oxidized LDL particles, which having been oxidized by free radicals are no longer recognized by the LDL receptor. Unlike LDL receptors, the scavenger receptors are not down-regulated by intracellular cholesterol. The persistence of significant amounts of oxidized LDL in the circulation can lead to excessive accumulation of oxidized LDL in the macrophages, transforming the latter into "foam cells," which eventually form atherosclerotic plaques.

17.4.4.4 Scavenger Receptors B1 (SR-B1's). A different scavenger receptor, SR-B1, mediates uptake of HDL cholesteryl esters by the liver. In contrast to the lipoprotein receptors described above, SR-B1 is not internalized by receptor-mediated endocytosis. Instead, SR-B1 permits hepatocytes selectively to remove and internalize HDL-associated cholesteryl esters. The adrenal glands, ovaries, and other steroidogenic tissues use the same SR-B1 receptor-dependent mechanism to extract cholesteryl esters from circulating HDL particles to provide cholesterol substrate for steroid hormone synthesis (Fig. 17-10).

17.4.5 Transport of Cholesterol in the Brain

As noted earlier, cholesterol does not cross the blood–brain barrier and must therefore be synthesized within the central nervous system (CNS), primarily by Schwann cells and oligodendrocytes. The CNS also has a separate lipoprotein transport system by which cholesterol is exchanged among various cells via HDL-like lipoproteins. The most abundant apolipoprotein in the CNS is apo E, which is synthesized by glial cells. The primarily mechanism for export of excess cholesterol from the brain is as the oxysterol, 24S-hydroxycholesterol (Chapter 18), rather than as cholesterol itself.

17.5 REGULATION OF CHOLESTEROL METABOLISM

17.5.1 Regulation of HMG-CoA Reductase

HMG-CoA reductase is the rate-limiting step of cholesterol synthesis, and its activity is under strict metabolic control. The simultaneous regulation of HMG-CoA reductase synthesis and turnover can alter steady-state levels of the enzyme 200-fold. The central role of HMG-CoA reductase in cholesterol homeostasis is evidenced by the effectiveness of a family of drugs called *statins* that are used to lower plasma cholesterol levels. Statins (e.g., lovastatin, pravastatin, fluvastatin, cerivastatin, atorvastatin) inhibit HMG-CoA reductase activity, particularly in liver, and decrease a person's total plasma cholesterol concentration by as much as 50%.

17.5.1.1 Transcriptional Regulation. A regulatory protein called *sterol regulatory element binding protein* (SREBP) plays a central role in regulating the expression of HMG-CoA reductase levels. Scap, the SREBP-escort protein, transports SREBP into the nucleus when SREBP binds to the sterol regulatory element (SRE) and up-regulates HMG-CoA expression. Several other proteins are involved in this regulatory process, including COPII and Insig. Cholesterol and oxysterols (e.g., 25-hydroxycholesterol) bind to COPII and Insig, respectively, and block the proteolytic activation and transport of SREBP, thereby preventing the up-regulation of HMG-CoA reductase.

The synthesis of mRNAs that encode at least three other enzymes in the cholesterol biosynthetic pathway—HMG-CoA synthase, farnesyl pyrophosphate synthase, and squalene synthase—is regulated in parallel with HMG-CoA reductase. Transcription of the LDL receptor is also regulated by the intracellular cholesterol concentration. Each of these three genes has a similar SRE in its promoter sequence that is recognized by SREBP.

There are multiple isoforms of SREBP. One of these, SREBP-2, selectively activates transcription of cholesterol biosynthetic genes and the LDL receptor gene. By contrast, SREBP-1 also activates transcription of acetyl-CoA carboxylase, fatty acid synthase, stearoyl-CoA desaturase 2, and glycerol 3-phosphate acyltransferase. SREBP-1 thus controls not only cholesterol synthesis but also the synthesis of fatty acids, TAG, and phospholipids.

17.5.1.2 *Proteolysis of HMG-CoA Reductase.* In sterol-depleted cells, HMG-CoA reductase is slowly degraded with a half-life greater than 12 hours. Insig, one of the proteins involved in the regulation of transcription of HMG-CoA, modulates proteolytic degradation of the enzyme. Binding of sterols, particularly lanosterol, and oxysterols to Insig accelerates the degradation of existing HMG-CoA reductase.

17.5.1.3 *Enzyme Phosphorylation.* HMG-CoA reductase is inhibited when it is phosphorylated by AMP-activated kinase. This kinase, which also acts to phosphorylate acetyl-CoA carboxylase, plays an important role in inhibiting multiple biosynthetic pathways when cellular energy stores are low.

17.5.2 Regulatory Role of the LDL Receptor

LDL is the major transporter of plasma cholesterol and the LDLR is the major mechanism by which cholesterol is removed from the blood. As noted above, as the intracellular concentration of free cholesterol increases, LDLR transcription is suppressed. As a result, receptor-mediated endocytosis of LDL is reduced and the plasma LDL concentration increases. There are two pharmacological mechanisms for increasing the expression of LDLRs on hepatocytes and thus decreasing circulating LDL cholesterol: statins and bile salt sequestrants.

17.5.2.1 *Statins Act to Up-regulate the LDL Receptor.* As described above, statin drugs decrease cholesterol synthesis by inhibiting HMG-CoA activity. This, in turn, results in up-regulation of the LDLR, particularly in hepatocytes.

17.5.2.2 *Bile Acid Sequestrants Up-regulate the LDL Receptor.* On a mass basis, the major metabolic products of cholesterol metabolism are the bile salts. As described in Chapter 2, bile salts are normally reabsorbed in the distal ileum.

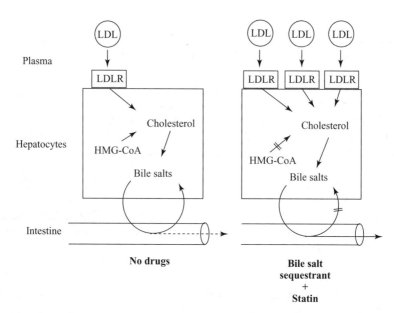

FIGURE 17-11 Synergistic effects of bile salt sequestrants and statins to up-regulate hepatic LDL-receptor expression and decrease plasma LDL-cholesterol.

Mucilaginous or soluble dietary fiber, such as that found in fruits and oat bran, binds bile salts and decrease their absorption. This, in turn, increases synthesis of bile salts in the liver, consuming cholesterol in the process and decreasing the concentration of cholesterol in hepatocytes. Bile-acid sequestrants such as cholestyramine and colestipol are anionic, insoluble polymers that bind bile acids strongly inside the resin matrix; the bile acid–sequestrant complex is excreted in the feces. Depletion of the body's bile acid pool results in up-regulation of hepatic 7α-hydroxylase, which increases the conversion of cholesterol to bile acids. The depletion of hepatic cholesterol increases the expression of LDL-receptor protein and lowers the plasma LDL-cholesterol concentration. However, since the decreased intrahepatic concentration of cholesterol also stimulates synthesis of HMG-CoA reductase, statins may be prescribed in conjunction with bile-acid sequestrants. Such a combination therapy usually results in a greater up-regulation of LDLR expression and a greater decrease in the plasma LDL-cholesterol level than is obtained from use of either a statin or a bile salt sequestrant alone (Fig. 17-11).

17.5.3 Regulation of Cholesterol Absorption

As noted earlier, much of the cholesterol that enters the small intestine is absorbed. By contrast, plant sterols (e.g., sitosterol) and related stanols (which are derivatives of sitosterol) are not well absorbed, and are selectively secreted from the enterocytes back into the intestinal lumen. Importantly, plant sterols and stanols interfere with the absorption of dietary cholesterol. For this reason, increased intake of vegetables and therapeutic doses of plant sterols and stanols have been used to lower the plasma cholesterol level. Cholesterol absorption can also be reduced by Ezetimibe (Zetia), which inhibits the transporter (NPC1L1) that moves cholesterol from the intestinal lumen into enterocytes.

17.6 ABNORMALITIES IN CHOLESTEROL METABOLISM

17.6.1 Hypercholesterolemia

Hypercholesterolemia refers to plasma levels of cholesterol that exceed the normal range. The risk of coronary heart disease (CHD) is correlated with both the total cholesterol and LDL-cholesterol (LDL-C) levels, while a high fasting HDL-C level is a negative risk factor for CHD. Since the risk of developing CHD is related mainly to the LDL-component, treatment is aimed at decreasing the level of LDL-C. The current desirable fasting plasma LDL-C level is < 100 mg/dL, 100 to 129 is considered nearly optimal, 130 to 159 is borderline high, 160 to 189 is high, and ≥ 190 is very high.

Treatment of persons with moderate hypercholesterolemia usually begins with dietary and other behavioral changes (e.g., exercise). Diets low in total fat, saturated fat, and cholesterol and relatively high in oleic acid (as in olive oil) tend to lower both the total cholesterol level and the LDL-C level. If additional reductions in cholesterol are called for, various medications are available. We have already discussed the use of sequestrants (e.g., cholestyramine, colestipol), which promote fecal excretion of bile acids; statins, which reduce endogenous cholesterol synthesis; and Ezetimibe, which blocks cholesterol absorption from the intestine. Other medications, such as niacin and gemfibrozil, can also be used to decrease plasma TAG levels and increase HDL-C levels.

17.6.2 Atherogenic Dyslipidemia

The dyslipidemic or atherogenic profile is a combination of three abnormalities in plasma lipoprotein levels: high VLDL triacylglycerol (≥150 mg/dL), low HDL-C (<40 mg/dL for men and <50 mg/dL for women), and the presence of relatively small, dense LDL particles. Independent of the concentration of either total plasma cholesterol or LDL-C, the dyslipidemic profile is a major risk factor for coronary artery disease. The dyslipidemic profile is commonly associated with insulin resistance, type II diabetes, central obesity, and hypertension, which are included in a constellation of findings that have been termed the *metabolic syndrome*. Although the etiology of the metabolic syndrome is not fully understood, the condition is often responsive to a combination of weight loss and exercise, as well as to a myriad of pharmacological agents.

17.6.3 Familial Hypercholesterolemia

A deficiency of LDLR is the most common monogenetic cause of FH. In the United States, the incidence of heterozygous FH is approximately 1 in 500 persons and that of homozygous FH is 1 in 1 million persons. FH heterozygotes usually exhibit an elevated fasting plasma LDL-C concentration (250 to 500 mg/dL; optimum, <100 mg/dL) and a normal triglyceride level. They have an increased risk of CHD, with onset in the fourth or fifth decade. Patients with heterozygous FH are generally responsive to treatment with HMG-CoA reductase inhibitors, which up-regulate the expression of the functional LDLR gene. In some patients, the addition of a bile acid sequestrant or nicotinic acid is required to achieve the desired LDL-C level.

Persons who are homozygous for FH usually have plasma LDL-C levels in the range 500 to 1200 mg/dL, even in early childhood. These patients invariably have cutaneous deposits of cholesterol called *xanthomas* on the hands, wrists, elbows, and/or knees. Coronary heart disease usually manifests within the first two decades of life. The disease is more severe in LDLR-negative patients (<2% normal LDLR activity) than in those whose cultured skin fibroblasts exhibit 2 to 20% residual LDLR activity in vitro. Due to the lack of functional receptors, patients with homozygous FH are largely unresponsive to drug therapies that up-regulate LDLR expression. Currently, the preferred treatment is LDL apheresis, which selectively removes the apoB-containing lipoproteins (VLDL, IDL, LDL) from the plasma.

17.6.4   Familial Dysbetalipoproteinemia

Familial dysbetalipoproteinemia or apo E defect is characterized by increased plasma concentrations of both cholesterol and triglycerides as a consequence of impaired clearance of chylomicron remnants and VLDL remnants by the liver. Normally, two major lipoprotein bands are visualized after electrophoresis of a fasting plasma sample from a normal person: an alpha band composed of HDL and a beta band composed of LDL; VLDL forms a small pre-beta band which runs ahead of the LDL band. Plasma from persons with familial dysbetalipoproteinemia, however, exhibits one broad band that stretches across both the beta and pre-beta positions and contains VLDL remnants of various sizes. Patients with this disease usually present in adulthood with premature coronary and/or peripheral vascular disease, xanthomas in the creases of the palms, and on the elbows and knees, as well as other locations.

There are three major allelic isoforms of apo E in the human population, designated E2, E3, and E4. Most patients with familial dysbetalipoproteinemia are homozygous for the E2 isoform (designated apo E2/E2). The E2 isoform binds less readily than other isoforms to hepatic lipoprotein receptors and delays clearance of both chylomicron remnants and VLDL from the plasma. Interestingly, although 80 to 90% of persons with familial dysbetalipoproteinemia are homozygous apo E2/E2, only 1 to 4% of apo E2/E2 homozygotes manifest this disorder, indicating that additional genetic and/or environmental factors are involved. Hypothyroidism, diabetes mellitus, obesity, and estrogen deficiency are among the factors that precipitate manifestation of familial dysbetalipoproteinemia in persons with the apo E2/E2 phenotype. Patients with this disorder usually respond well to low-cholesterol and low-fat diets, weight reduction, and drug therapy.

17.6.5 Alzheimer's Disease

There is evidence that abnormal cholesterol metabolism may be a factor in Alzheimer's disease (AD). Apo E, which is synthesized by astrocytes and microglia in the brain, is thought to play a central role in regulating cholesterol homeostasis in the central nervous system. Apo E is also associated with the β-amyloid deposits that are a neuropathologic hallmark of AD. Although the mechanisms by which apo E influences the onset and progression of AD are obscure, people who carry one E4 allele have about a twofold higher lifetime risk of developing AD than do those with other apo E genotypes, and the onset of disease occurs earlier.

17.6.6 Tangier Disease

Tangier disease is a rare disease that results from a genetic defect in reverse cholesterol transport in which there is a virtual absence of plasma HDL-C and very low levels of apo A1. Patients may present with enlarged yellow-orange tonsils loaded with cholesteryl esters and other lipophilic compounds, including retinyl esters and carotenoids. Cholesteryl esters also accumulate in reticuloendothelial cells of the thymus, spleen, liver, bone marrow, and intestinal mucosa. The disease is caused by a loss-of-function mutation in the ABC1 transporter, so that it cannot facilitate the transfer of free cholesterol from cells to apo A1 to form discoidal, nascent HDL; instead, the lipid-free apo A1 is cleared rapidly from the circulation by the kidneys. In the absence of ABC1 activity, cholesteryl esters accumulate in peripheral cells, particularly macrophages and cells of the reticuloendothelial system. The fasting plasma concentration of VLDL triacylglycerol is elevated, apparently from the loss of HDL as a reservoir for apo C2, thereby resulting in a deficiency of apo C2 that is required to activate lipoprotein lipase. Despite their lack of HDL, patients with Tangier disease do not have a markedly increased risk for premature CAD, apparently because of their low (40% of normal) LDL-C levels. The low LDL-C levels in the plasma of these patients may reflect both impaired generation of LDL from VLDL and up-regulation of hepatic LDL receptors in response to the lack of HDL-mediated return of cholesterol to the liver.

17.6.7 Familial HDL Deficiency

Familiar HDL deficiency or primary hypoalphalipoproteinemia is caused either by a mutation in the apo A1 gene or by heterozygosity for ABC1 transporter deficiency (i.e., the carrier state for Tangier disease) and is characterized by low levels of circulating HDL-C. This disorder occurs in 1% of the general population and is a common genetic cause of increased triglyceride levels in plasma and CHD. By contrast, although both complete and partial deficiencies of LCAT also result in markedly reduced HDL-C and circulating apo A1 levels, these conditions rarely lead to premature atherosclerotic disease.

17.6.8 Niemann–Pick C Disease

Niemann–Pick type C is an autosomal recessive disease that causes progressive neurological degeneration and the lysosomal accumulation of cholesterol and other lipids (e.g., gangliosides and bis-monoacylglycerol) in the central nervous system and reticulendothelial cells. The two forms of the disease are the result of mutations in two genes, *NPC1* and *NPC2*, which are required for the egress of cholesterol from late endosomes and lysosomes. In the absence of functional *NPC1* and *NPC2*, cholesterol delivered to the lysosomes via receptor-mediated endocytosis of LDL cannot be transported to the endoplasmic reticulum, thereby causing cholesterol to accumulate in lysosomes and eventually depleting the plasma membrane of cholesterol. As cells such as hepatocytes respond to the low concentration of membrane cholesterol by up-regulating expression of LDL receptors, increased internalization of LDL only exacerbates the problem. The severe neurological problems of patients with Niemann–Pick C disease may also be due, in part, to other, as yet not fully characterized functions of *NPC1* in neurons.

17.6.9 Smith–Lemli–Opitz Syndrome

Smith–Lemli–Opitz syndrome (SLOS) is the most common genetic defect of cholesterol biosynthesis, occurring in 1 in 30,000 births. Patients with SLOS are born with a variable spectrum of morphogenic anomalies, including microcephaly, dysmorphic craniofacial features, polydactyly, and congenital heart defects. SLOS was the first multiple malformation syndrome to be attributed to a deficiency of a single enzyme, in this case 3β-hydroxysterol Δ^7-reductase, which reduces the double bond of 7-dehydrocholesterol (7-DHC) in the last step of cholesterol synthesis. As a result, these patients have markedly increased amounts of 7-DHC and decreased levels of cholesterol in plasma and cultured skin fibroblasts. The multiple congenital malformations in SLOS are attributed to a lack of cholesterol for posttranslational modification of the Hedgehog proteins, which have been implicated as signaling molecules in embryonic patterning processes. Alternatively, the accumulation of 7-dehydocholesterol may alter formation of lipid rafts within the plasma membrane, and thus impair interaction of Hedgehog proteins with their membrane-bound receptors.

CHAPTER 18

STEROIDS AND BILE ACIDS

18.1 FUNCTIONS OF OXYGENATED DERIVATIVES OF CHOLESTEROL

Oxygenated derivatives of cholesterol play many roles in the body. Cholesterol is the precursor of two important classes of molecules, bile acids and steroid hormones. In addition, 7-dehydrocholesterol, the immediate precursor of cholesterol, can be converted to cholecalciferol (vitamin D_3), which ultimately produces 1,25-dihydroxycholecalciferol [1,25-$(OH)_2D_3$, calcitriol], the active hormone that regulates calcium metabolism.

18.1.1 Bile Acids

Cholesterol is a 27-carbon lipid containing a fused four-ring structure and a hydrocarbon chain (Fig. 18-1). Except for the one hydroxyl group at C3, cholesterol is completely nonpolar. By contrast, bile acids contain 24-carbon atoms and are more polar than cholesterol: The steroid ring of bile acids contains one or more additional hydroxyl groups and the shorter hydrocarbon side chain terminates in a carboxyl group (Fig. 18-2). In addition, the stereochemistry of the steroid nucleus is modified, resulting in a planar structure in which all of the hydroxyl groups are situated on the same side of the plane of the molecule (see Fig. 3-6).

The so-called "bile salts" are actually bile acids that contain an amino acid which is conjugated in amide linkage to the side chain of the carboxyl group of the bile acid.

Medical Biochemistry: Human Metabolism in Health and Disease   By Miriam D. Rosenthal and Robert H. Glew
Copyright © 2009 John Wiley & Sons, Inc.

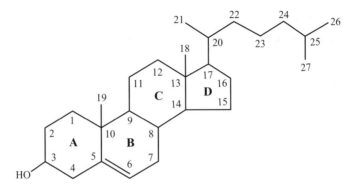

FIGURE 18-1   Numbering the carbon atoms and rings of cholesterol.

The two amino acids used most commonly by the liver to conjugate human bile acids are glycine, which is predominant in adults, and the sulfur amino acid taurine, which is predominant in infants (see Fig. 3-5). Conjugated bile acids are more ionized at the slightly acidic pH of the intestinal lumen than their nonconjugated counterparts and are therefore better emulsifying agents.

Bile salts play a major role in the digestion and absorption of triacylglycerols and cholesteryl esters. Bile salts emulsify dietary lipids in the gastrointestinal tract and stabilize the resulting mixed micelles. Along with phosphatidylcholine, bile salts solubilize the cholesterol and bile pigments present in bile, preventing formation of precipitates (stones) of cholesterol or bilirubin in the gallbladder and bile ducts. In addition, formation of bile salts represents the only significant metabolic mechanism for eliminating excess cholesterol from the body.

18.1.2   Steroid Hormones

Cholesterol serves as the precursor for the synthesis of all five classes of steroid hormones (Fig. 18-3). Three of these classes are 21-carbon structures: glucocorticoids, mineralocorticoids, and progestogens. The *glucocorticoids* (e.g., hydrocortisone,

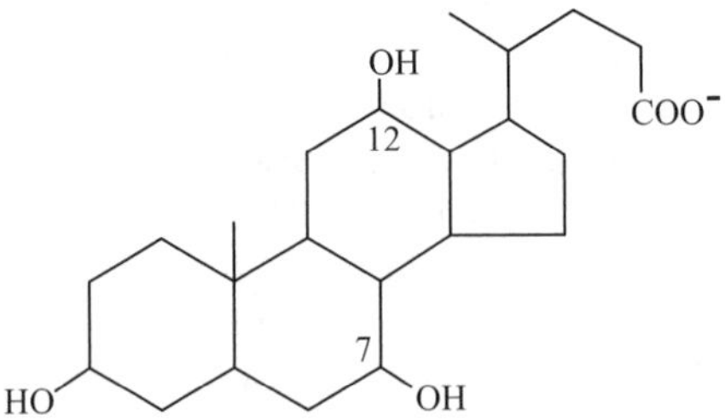

FIGURE 18-2   Cholic acid is the major bile acid produced in adults. Chenodeoxycholic acid (not shown) lacks the hydroxyl group at C12.

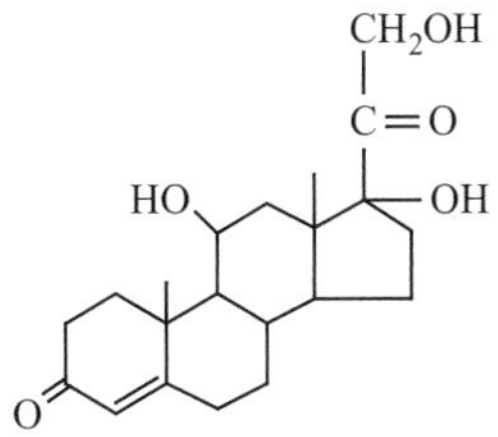

Hydrocortisone

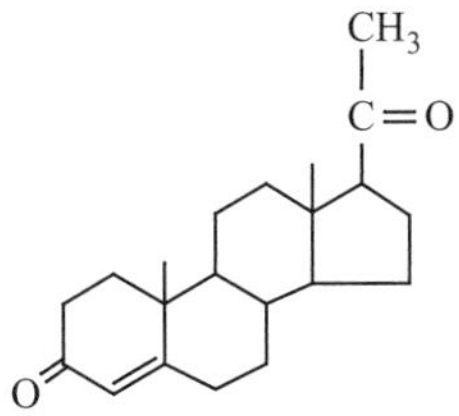

Progesterone

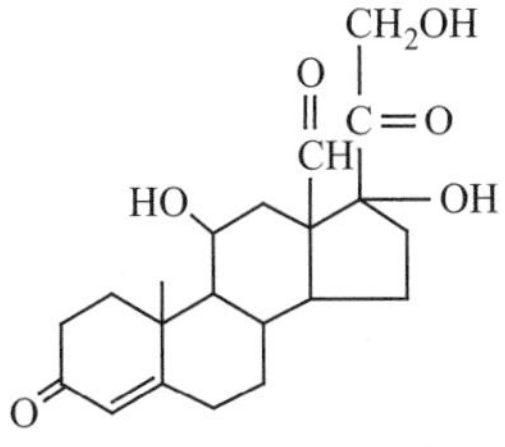

Aldosterone

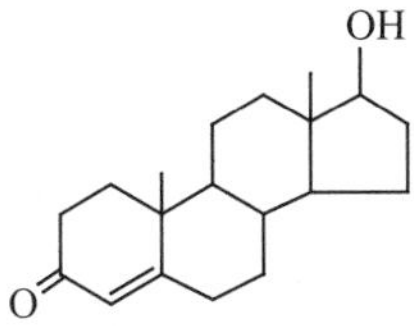

Testosterone

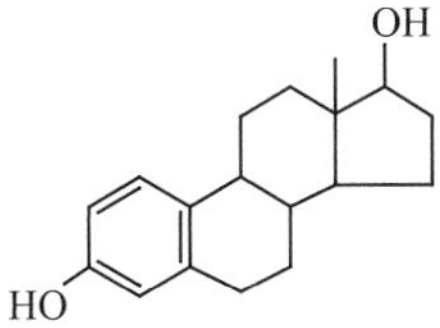

Estradiol

FIGURE 18-3 Representative members of each of the five classes of steroid hormones.

a.k.a. cortisol) mediate a person's response to physiological stress and are involved in coordinating carbohydrate, fat, and protein metabolism. *Mineralocorticoids* (e.g., aldosterone) regulate the body's electrolyte and water balance via their action on the renal tubules. Progesterone, a *progestogen*, regulates menses and breast development and is required for implantation of the fertilized ovum and maintenance of pregnancy. Testosterone and other androgens contain 19 carbons and are responsible for masculine characteristics, whereas estrogens, such as estradiol, have 18 carbons and are the female steroid sex hormones. As can be seen from the structures of the representative steroid hormones shown in Figure 18-3, all of these steroids have been modified from the original 27-carbon cholesterol molecule by shortening or elimination of the hydrocarbon side chain and by the addition of polar hydroxyl or keto substituents on the steroid ring structure. The basic steroid structure has been further modified in estrogens by the elimination of a methyl group and desaturation to generate an aromatic A ring.

Because of their lipid nature, steroid hormones pass readily through the plasma membrane of all target cells. Once inside its target cell, the steroid hormone binds to its specific receptor protein. The steroid receptors act in the nucleus, where the hormone-receptor complex binds to specific DNA sequences, resulting in increased transcription of those particular genes and ultimately an increase in the level of the protein encoded by those genes.

18.1.3   1,25-Dihydroxycholecalciferol (Calcitriol)

1,25-Dihydroxycholecalciferol [1,25-$(OH)_2D_3$] is a 27-carbon molecule that shares many similarities with the steroid hormones described above. It is actually derived from 7-dehydrocholesterol, the immediate precursor of cholesterol, rather than from cholesterol and differs from the classic steroid hormones in that the B ring of choles- terol has been opened (Fig. 18-4). 1,25-$(OH)_2D_3$ plays an important role in Ca^{2+} homeostasis: It increases the plasma calcium concentration both by promoting in- testinal absorption of dietary calcium and by stimulating Ca^{2+} release from bone when dietary calcium is insufficient. However, the major function of 1,25-$(OH)_2D_3$ in bone is to promote mineralization. 1,25-Dihydroxycholecalciferol also has impor- tant effects on differentiation in many cell types other than bone and may be protective against colorectal cancers and other cancers.

18.2   LOCALIZATION OF THE SYNTHESIS OF OXYGENATED DERIVATIVES OF CHOLESTEROL

18.2.1   Bile Acids

The synthesis of the primary bile acids, cholic acid and chenodeoxycholic acid, and their conjugation with taurine or glycine to form bile salts occurs exclusively in the liver. As described in Chapter 3, the bile salts are secreted from the liver into the biliary tract and enter the duodenum, where they facilitate digestion and absorption of lipids. The conjugated bile acids are reabsorbed in the distal ilium and transported through the blood back to the liver. Some of the secondary bile acids, formed when bacterial enzymes in the intestine reduce (dehydroxylate) primary bile acids, are also reabsorbed and recycled.

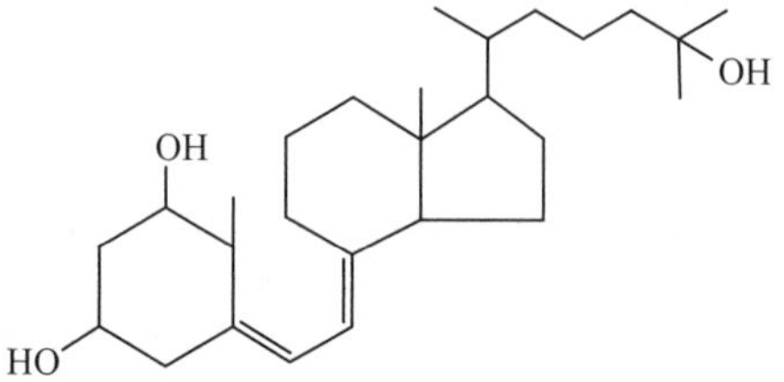

FIGURE 18-4   Structure of 1,25-dihydrocycholecalciferol [1,25-$(OH)_2D_3$].

18.2.2 Steroid Hormones

Estrogen is synthesized in the ovarian follicle, whereas the corpus luteum which develops following ovulation produces progesterone and some estrogen. The major site of androgen synthesis in men is the testis. In women, ovarian cells synthesize androgens as well as estrogens; indeed, androstenedione synthesis in the theca cells of the ovary is essential to the production of estradiol by the follicle. The adrenal cortex synthesizes glucocorticoids, mineralocorticoids, and the androgens androstenedione and dehydroepiandrosterone (DHEA).

18.2.3 1,25-Dihydroxycholecalciferol

The precursor for [1,25-$(OH)_2$-D_3] synthesis is cholecalciferol, which is either formed nonenzymatically in the skin from 7-dehydrocholesterol (vitamin D_3) or obtained in the diet. Conversion of vitamin D_3 to the active hormone 1,25-$(OH)_2$-D_3, requires two successive hydroxylation reactions in the liver and kidney, respectively.

18.3 PATHWAYS OF CHOLESTEROL METABOLISM

18.3.1 Monooxygenases

Many of the enzymes that catalyze oxygenation of cholesterol and cholesterol derivatives are monooxygenases or mixed-function oxidases. Mixed-function oxidases catalyze reactions in which one atom of molecular oxygen oxidizes an organic substrate, while the other atom of molecular oxygen oxidizes NADPH or NADH:

$$R-H + NADPH + H^+ + O_2 \rightarrow R-OH + NADP^+ + H_2O$$

The monooxygenases that modify cholesterol are members of the cytochrome P450 (CYP) superfamily of enzymes and usually utilize NADPH as a cofactor or second substrate. These monooxygenases are all membrane-bound and localized to either the endoplasmic reticulum or the inner mitochondrial membrane. The cytochrome P450 enzymes usually have two names, one identifying them as a member of the cytochrome P450 family (e.g., CYP7A1) and the other identifying the cholesterol carbon atom that is modified in the reaction (e.g., 7α-hydroxylase).

18.3.2 Synthesis of Bile Acids

The classical or neutral pathway of bile acid synthesis involves modification of the cholesterol ring structure before the hydrocarbon side chain is cleaved (Fig. 18-5). However, there is also an alternative or acidic pathway of bile acid synthesis in which the side-chain modification reactions occur first, followed by modification of the sterol nucleus. In adult humans, 75 to 95% of bile salts are synthesized via the classical pathway. The acidic pathway is more prevalent in the fetus and neonate.

The intracellular trafficking of intermediates in bile acid synthesis is complex, with reactions in the pathway occurring in mitochondria, peroxisomes, the endoplasmic reticulum, and cytosol.

18.3.2.1 Classical Pathway of Bile Acid Synthesis. The initial—and regulated—step of bile acid synthesis is hydroxylation of C7 on the B ring of cholesterol by 7α-hydroxylase (CYP7A1) (Fig. 18-5):

$$\text{cholesterol} + \text{NADPH} + \text{H}^+ + \text{O}_2 \rightarrow \text{7α-hydroxycholesterol} + \text{NADP}^+ + \text{H}_2\text{O}$$

This initial 7α-hydroxylation step is followed by a sequence of 14 reactions that:

- Epimerize the 3β group to form a 3α-hydroxyl group
- Hydrogenate the C5–C6 double bond to form a saturated bond

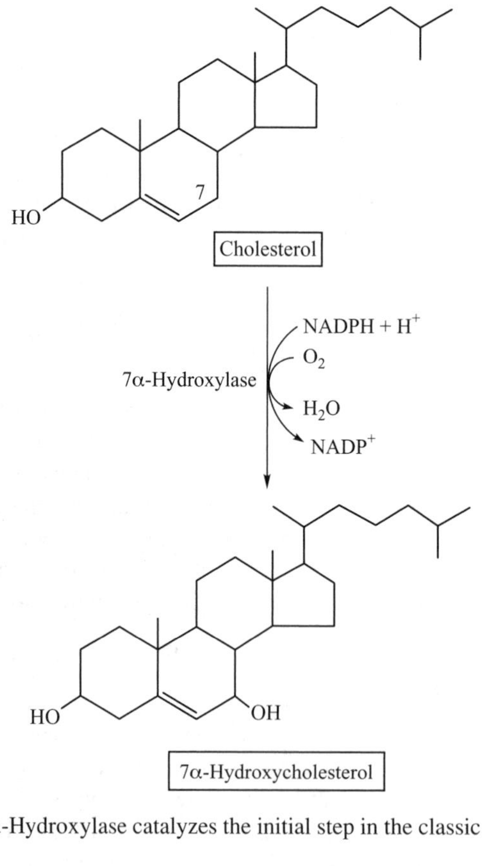

FIGURE 18-5 7α-Hydroxylase catalyzes the initial step in the classic pathway of bile acid synthesis.

- Introduce a hydroxyl group on the C12 carbon
- Cleave three carbons from the hydrocarbon side chain
- Oxidize the terminal carbon of the side chain to generate choloyl-CoA

Choloyl-CoA is then conjugated with either glycine or taurine to form glycocholate or taurocholate, respectively.

Synthesis of chenodeoxycholoyl-CoA follows the same pathway as that for choloyl-CoA but omits the hydroxylation of C12. The activity of the enzyme sterol 12α-hydroxylase thereby controls the resulting ratio of cholic acid to chenodeoxycholic acid and thus the overall detergent potency of the bile acid pool.

18.3.2.2 Alternate Pathway of Bile Acid Synthesis. The first step in the alternate pathway for bile acid synthesis is catalyzed by several different sterol hydroxylases which generate oxysterols by hydroxylating cholesterol at carbon 24, 25, or 27 (Fig. 18-6). The oxysterols are then 7α-hydroxylated by oxysterol hydroxylases CYP7B1 or CYP39A1 rather than by CYP7A1. Subsequent steps in the alternate pathway for bile acid synthesis utilize the same reactions as those of the classical pathway. The major product of the classical pathway is cholic acid whereas the alternate pathway generates relatively more chenodeoxycholic acid than cholic acid.

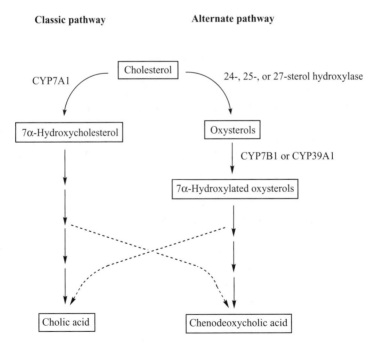

FIGURE 18-6 Classic and alternate pathways for bile acid synthesis. The dashed lines indicate minor pathways.

Although bile acid synthesis occurs only in the liver, the initial steps of the acidic pathway (generation and hydroxylation of oxysterols) can occur in a variety of extrahepatic tissues. For example, the brain, which cannot export cholesterol, expresses 24-sterol hydroxylase and exports the resulting oxysterol into the blood. The 27-hydroxylation of cholesterol by peripheral cells may also contribute to reverse transport of cholesterol by converting cholesterol into a more water-soluble product.

18.3.3   Synthesis of Steroid Hormones

The cholesterol utilized by the adrenal cortex and the gonads for steroid hormone synthesis is derived principally from LDL-cholesterol in blood. LDL binds to LDL receptors on steroidogenic tissues and is internalized by receptor-mediated endocytosis. The LDL-cholesteryl esters are then hydrolyzed by lysosomal cholesteryl esterase.

$$\text{cholesteryl ester} + H_2O \rightarrow \text{cholesterol} + \text{fatty acid}$$

The resulting cholesterol is then reesterified by acyl-CoA cholesterol acyl transferase (ACAT) and stored as cholesteryl esters in lipid droplets in the cytosol. Initiation of steroid hormone synthesis thus requires a second hydrolysis of cholesteryl esters by a cytosolic cholesteryl esterase. Cholesterol can also be synthesized de novo within the adrenal cortex and gonads.

18.3.3.1   Synthesis of Pregnenolone.   The 21-carbon steroid pregnenolone is the common precursor for all of the steroid hormones. It is synthesized from cholesterol by the cytochrome P450 cholesterol side-chain cleavage enzyme (CYP11A, cholesterol desmolase) (Fig. 18-7). The removal of six carbons from the side chain as isocaproate and the introduction of a keto group on C20 is accomplished by three successive monooxygenase reactions, which result in hydroxylation of C20 and C22, followed by cleavage of the C20–C22 carbon–carbon bond. Three molecules of O_2 and three molecules of NADPH are consumed in the process of converting cholesterol to pregnenolone.

The side-chain cleavage enzyme complex is located on the inner mitochondrial membrane. Synthesis of pregnenolone therefore begins with the transport of cholesterol from the cytosol into the mitochondrion. In most steroidogenic cells except the placenta, the transport of cholesterol into the mitochondrion is mediated by a carrier called *steroidogenic acute regulatory protein* (StAR); the placenta lacks StAR but expresses another protein that has similar functional properties. Pregnenolone is transported back into the cytosol, where the subsequent steps of steroid synthesis occur.

18.3.3.2   Synthesis of Progesterone.   Pregnenolone is converted to progesterone by 3β-hydroxysteroid dehydrogenase-isomerase (3β-HSD) (Fig. 18-8).

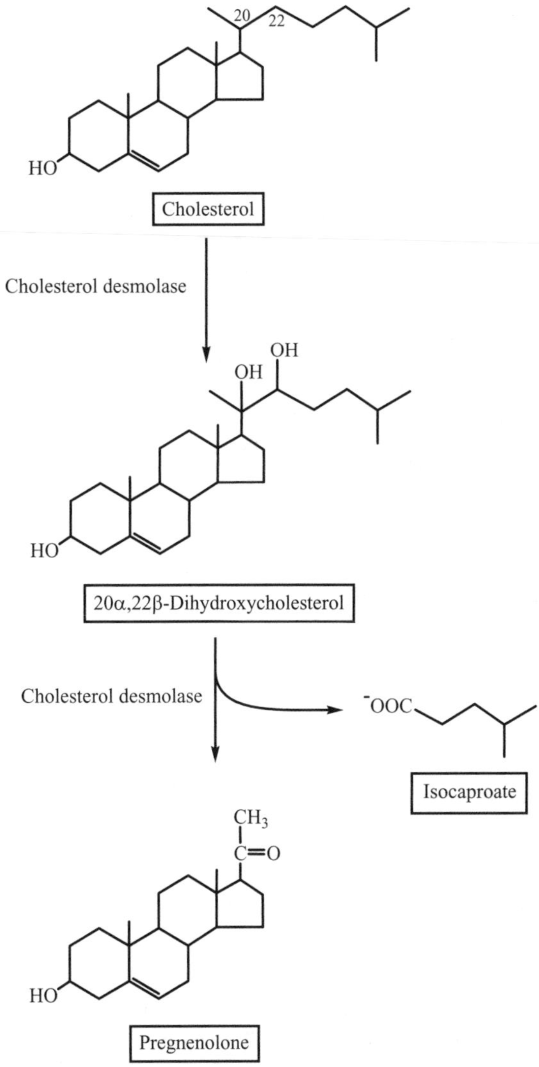

FIGURE 18-7 Synthesis of pregnenolone from cholesterol.

18.3.3.3 Synthesis of Hydrocortisone (Cortisol).

Pregnenolone is converted by 17α-hydroxylase to 17-hydroxypregnenolone, which is converted by 3β-HSD to 17-hydroxyprogesterone. 17-Hydroxyprogesterone is then hydroxylated sequentially at C21 and then at C11 to form hydrocortisone (Fig. 18-9). Cortisone administered therapeutically is inactive, and is activated in the liver by 11β-dihydroxysteroid dehydrogenase, which generates hydrocortisone by reduction of the keto group on C11 of cortisone.

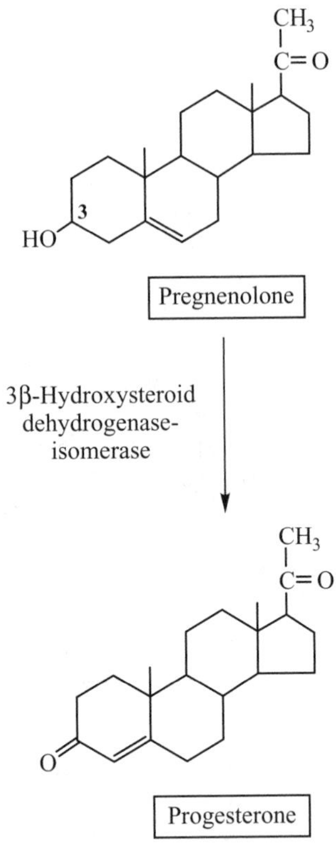

FIGURE 18-8 Conversion of pregnenolone to progesterone.

18.3.3.4 Synthesis of Aldosterone. Aldosterone synthesis is initiated by 21-hydroxylation of progesterone to form 21-hydroxyprogesterone, also known as deoxycorticosterone (Fig. 18-10). The subsequent steps in aldosterone synthesis involve two hydroxylation reactions at C11 and C18, respectively, followed by oxidation of the C18 hydroxyl group to a keto group. These three steps are all catalyzed by the same enzyme, CYP11B2.

18.3.3.5 Synthesis of Androgens. The pathway for the synthesis of testosterone in the testis is shown in Figure 18-11. Pregnenolone is first hydroxylated by CYP17 to 17-hydroxpregnenolone. CYP17 also has 17,20-lyase activity, thus allowing for the synthesis of dehydroepiandrosterone (DHEA). Synthesis of testosterone from DHEA involves successive hydroxylations by 3β-hydroxysteroid dehydrogenase (generating androstenedione) and then 17β-hydroxysteroid dehydrogenase. Both DHEA and androstenedione have weak androgenic activity and are synthesized in the adrenals as well as the gonads.

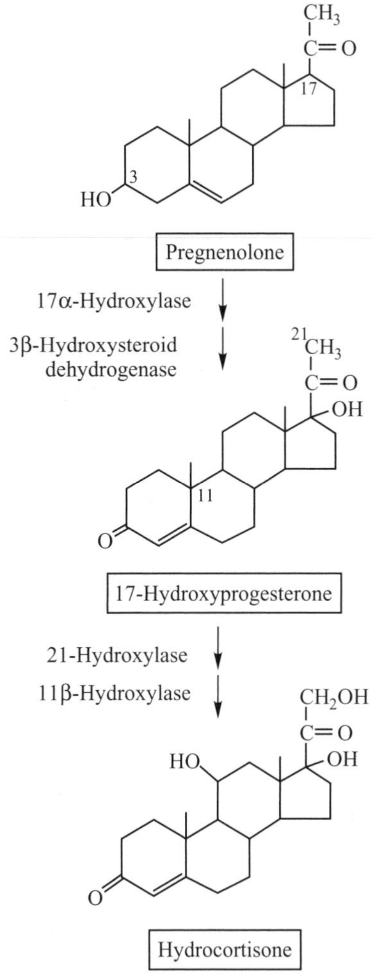

FIGURE 18-9   Synthesis of hydrocortisone from pregnenolone.

18.3.3.6 Synthesis of Estrogens. Aromatase (CYP19A1), the key enzyme complex that synthesizes estrogens from androgens, catalyzes a series of reactions that start with two successive hydroxylations of C19 of testosterone (Fig. 18-12). Subsequent cleavage of C19 as formate generates an 18-carbon steroid with an aromatic A ring. In women, aromatase is present in both the ovaries and peripheral tissues and acts on both testosterone and androstenedione, producing estradiol and estrone, respectively. Aromatase inhibitors have become an important treatment in women with breast cancer.

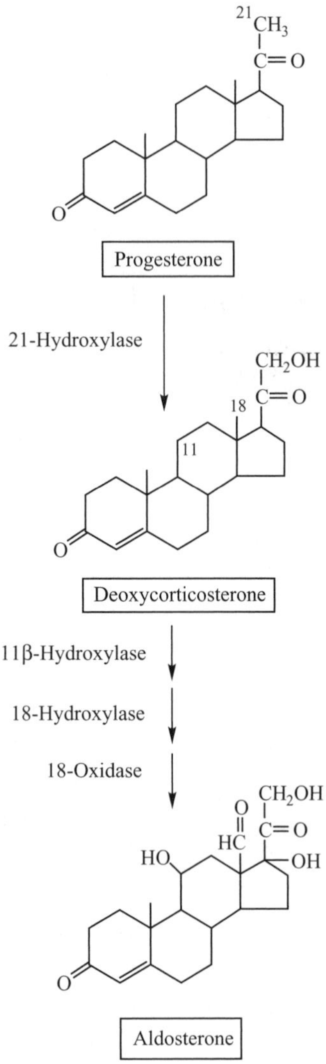

FIGURE 18-10 Synthesis of aldosterone from progesterone.

18.3.4 Synthesis of 1,25-Dihydroxycholecalciferol

Cholecalciferol (vitamin D_3) can either be obtained from the diet or formed in the skin by the nonenzymatic action of ultraviolet (UV) light on 7-dehydrocholesterol (Fig. 18-13). Ergocalciferol (vitamin D_2) is formed by the action of UV light on a structurally similar plant sterol (ergosterol). Vitamin D_3 is preferred for vitamin D supplementation.

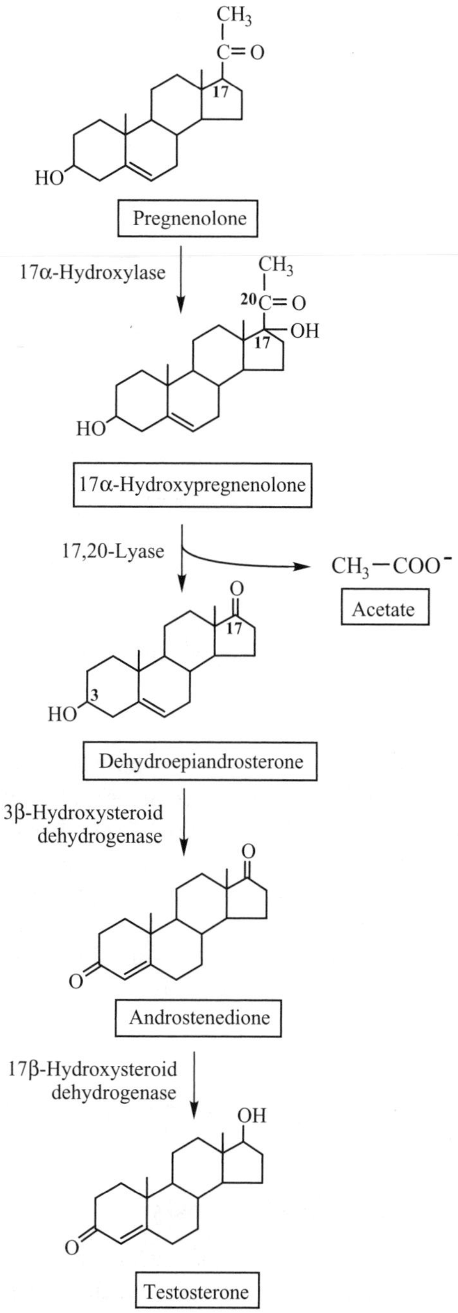

FIGURE 18-11 Synthesis of androgens.

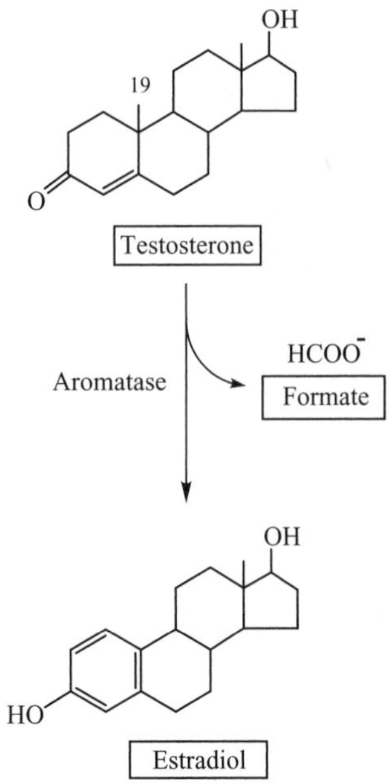

FIGURE 18-12 Conversion of testosterone to estradiol.

Conversion of cholecalciferol to its active hormonal form is a multiorgan process. First, cholecalciferol is hydroxylated in the liver by 25-hydroxylase to form 25-hydroxycholecalciferol (Fig. 18-13). A second hydroxylation step, catalyzed by 1α-hydroxylase in the kidney, generates the active hormone 1,25-dihydroxycholecalciferol. 1α-Hydroxylase is also expressed in placenta and placenta. Kidney, bone, cartilage, and intestine contain a 24-hydroxylase that converts 25-dihydroxycholecalciferol to the inactive 24,25-dihydroxycholecalciferol, thus preventing formation of excess active 1,25-$(OH)_2D_3$.

18.4 REGULATION OF THE SYNTHESIS OF METABOLITES OF CHOLESTEROL

18.4.1 Regulation of Bile Acid Synthesis

On a mass basis, bile acids are the main products of cholesterol metabolism. Under normal conditions, only 5 to 10% of the bile salts that enter the gut are actually excreted in the feces; their replacement requires de novo synthesis of approximately 400 mg/day of bile salts. Certain dietary fibers and pharmaceutical *bile acid sequestrants* such as cholestyramine decrease the reabsorption of bile acids and, as a result, increase hepatic synthesis of bile acids from cholesterol. Cholestyramine is therefore used to reduce the plasma concentration of cholesterol (see Chapter 17).

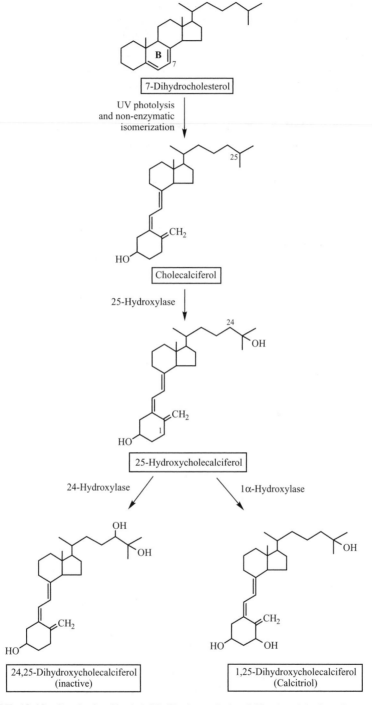

FIGURE 18-13 Synthesis of both 1,25-dihydroxycholecalciferol and the inactive molecule 24,25-dihydroxycholecalciferol from 7-dehydrocholesterol (vitamin D_3).

The regulated step in the classic pathway of bile salt synthesis is the initial hydroxylation of cholesterol by 7α-hydroxylase, an enzyme with a relatively short half-life. Synthesis of new 7α-hydroxylase protein is inhibited at the transcriptional level by primary bile acids. Cholestyramine and related resins that bind bile salts and decrease the efficiency of their reuptake by the ileum up-regulate 7α-hydroxylase and thus increase the metabolism of hepatic cholesterol to bile salts.

18.4.2 Regulation of Steroid Synthesis

Synthesis of steroid hormones by steroidogenic glands is stimulated by corresponding trophic hormones synthesized by the pituitary. For example, adrenocorticotrophic hormone (ACTH) stimulates the synthesis of steroids in the adrenal cortex, while the gonadotrophins LH (luteinizing hormone) and FSH (follicle-stimulating hormone) regulate steroid hormone synthesis in the gonads.

Binding of the trophic hormones to their corresponding plasma membrane receptors initiates a signal transduction cascade that results in increased concentrations of cAMP, which stimulates protein kinase A (PKA) (Fig. 18-14). PKA, in turn, phosphorylates and activates cholesteryl ester hydrolase, which makes cholesterol available

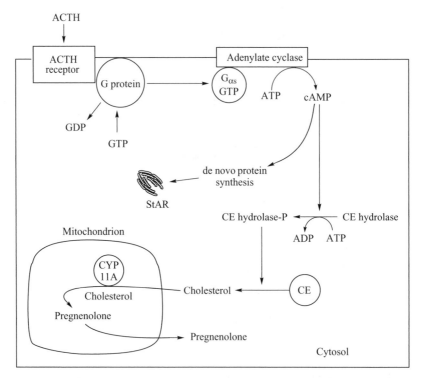

FIGURE 18-14 ACTH stimulates hydrolysis of cholesteryl esters and transport of free cholesterol into the mitochondrion. CE, cholesteryl ester; CYP 11A, cholesterol side-chain cleavage enzyme; PKA, protein kinase A; StAR, steroidogenic acute regulatory protein.

for steroid hormone synthesis. The action of PKA also induces de novo synthesis of StAR and thus stimulates transport of cholesterol to the inner mitochondrial membrane where the cholesterol P450 side-chain cleavage enzyme is located.

What determines which steroid hormones are produced by a particular cell? The common initial step in steroid hormone synthesis is the one that generates pregnenolone. The metabolic fate of pregnenolone is dependent primarily on the particular enzymes present within that cell. For example, cells of the zona fasciculata in the adrenal cortex contain 17α-hydroxylase, which converts pregnenolone to 17α-hydroxypregnenolone, which is subsequently metabolized to hydrocortisone. By contrast, cells of the zona glomerulosa in the adrenal cortex convert pregnenolone to aldosterone. Since several of the enzymes of steroid hormone pathways contribute to the generation of multiple hormones in the same gland, impairment of an enzyme activity resulting in decreased synthesis of some steroids may increase the availability of precursor molecules for other steroidogenic pathways.

18.4.3 Regulation of 1,25-Dihydroxycholecalciferol Synthesis

Formation of $1,25\text{-}(OH)_2D_3$ is regulated primarily by the level of 1α-hydroxylase. Synthesis of 1α-hydroxylase in the proximal renal tubule is stimulated by parathyroid hormone and is inhibited by its product, $1,25\text{-}(OH)_2D_3$, and by a high concentration of extracellular calcium.

18.5 ABNORMAL FUNCTION

18.5.1 Genetic Defects in Bile Acid Synthesis

The pathways that convert cholesterol to cholate and deoxycholate are affected by several inborn errors of metabolism, all of which result in progressive liver disease typically manifested in the neonatal period. Affected persons have cholestasis (blockage of bile flow) accompanied by hyperbilirubinemia, steatorrhea, and deficiencies of the fat-soluble vitamins. The disorders are diagnosed by the finding of unusual bile acids (e.g., 7α-hydroxy-3-oxo-4-cholenoic acid) in the urine. Treatment with oral chenodeoxycholic acid is usually effective in restoring the normal secretion of bile, improving the digestion and absorption of dietary lipids, and inhibiting the production of abnormal bile acids.

18.5.2 Congenital Adrenal Hyperplasia

The most common cause of congenital adrenal hyperplasia (CAH) is a genetic deficiency of the activity of the steroidogenic enzyme 21-hydroxylase (CYP21). This enzyme defect results in decreased synthesis of hydrocortisone. The resulting increase in pituitary ACTH secretion in turn produces adrenal hyperplasia and increases androgen production. Affected persons exhibit hydrocortisone insufficiency (hypoglycemia, hypotension) and are dependent on exogenous hydrocortisone. In severe cases, aldosterone deficiency also occurs, resulting in dehydration, hyponatremia due to salt wasting, and hyperkalemia.

Congenital adrenal hyperplasia due to 21-hydroxylase deficiency is also the most common cause of genital ambiguity in newborn females. Lack of hydrocortisone results in increased ACTH secretion, which overstimulates 17-hydroxyprogesterone synthesis, resulting in increased synthesis of DHEA and androstenedione. Exposure of the female fetus to androgens in utero results in prenatal virilization (masculinization of the external genitalia). In affected newborn males, there are no abnormalities of the external genitalia. However, if a male infant with CAH is not treated promptly, excess androgen synthesis results in precocious development of secondary sex characteristics and increased stature. Hydrocortisone replacement therapy is effective in decreasing ACTH levels and suppressing excess androgen production. Hydrocortisone also restores normal glucocorticoid function; however, the more severely affected patients also require mineralocorticoid replacement therapy. In at-risk pregnancies, early maternal treatment with dexamethasone may prevent or significantly reduce virilization of the genitalia of female neonates.

18.5.3   Congenital Lipoid Adrenal Hyperplasia

A rare type of adrenal hyperplasia results from a genetic deficiency of StAR, the steroidogenic acute regulatory protein. Loss of the ability to transport cholesterol into mitochondria results in a loss of all steroidogenesis (defective glucocorticoid, mineralocorticoid, and androgen production) and accumulation of cholesteryl esters in the cytosol of the adrenal cortex. The lack of hydrocortisone production leads to increased ACTH stimulation, resulting in unusually large adrenal glands filled with cholesteryl esters. The condition is lethal unless appropriate mineralocorticoid- and glucocorticoid-replacement therapy is provided. Steroid biosynthesis is impaired in the gonads as well. Affected newborn males will show incomplete masculinization of the external genitalia. Affected females usually do not manifest symptoms of ovarian failure until puberty, which is the expected time of estrogen synthesis.

18.5.4   Rickets and Osteomalacia

Vitamin D deficiency results in rickets in children and osteomalacia in adults. Both diseases reflect impaired mineralization of newly synthesized organic bone matrix. In adults, the undermineralization occurs after linear growth has stopped and involves bone only. However, in children whose bones are still growing, decreased mineralization of osteoid occurs in growing cartilage in the growth plate as well as in bone, and bony deformities, including the characteristic bowed long bones of the legs, may occur.

Vitamin D deficiency usually results from a combination of inadequate exposure to sunlight and decreased dietary intake of vitamin D. Breast-fed infants who do not receive vitamin D supplements are at particular risk because human milk usually contains insufficient vitamin D. Also at high risk for lack of endogenously synthesized vitamin D are dark-skinned people living at latitudes where there is less sunlight, or persons who do not expose their skin to sunlight. Dietary vitamin D is present mainly in foods of animal origin, especially liver, eggs, and vitamin D–fortified milk and dairy products. A decreased serum level of 25-hydroxycholecalciferol is diagnostic of vitamin D deficiency. Severe kidney or liver disease reduces conversion of vitamin D to 25-hydroxycholecalciferol and/or the hormonally active $1,25\text{-}(OH)_2D_3$, resulting in rickets and osteomalacia even in the presence of adequate sunlight and/or dietary intakes of vitamin D.

In many parts of the world, rickets is caused by a dietary deficiency of calcium rather than vitamin D. Rickets may also result from an inherited deficit in renal 1α-hydroxylase or from genetic disorders that cause hypophosphatemia due to inhibition of phosphate reabsorption in the renal tubule.

CHAPTER 19

NITROGEN HOMEOSTASIS

19.1 FUNCTIONS OF NITROGEN METABOLISM

Each day, all humans turn over 1 to 2% of their total body protein, most of which is muscle protein. About three-fourths of the amino acids liberated during this process are reutilized for the synthesis of new proteins; the remaining 20 to 25% is catabolized. To maintain health, dietary protein must replace the degraded amino acids and provide amino acids that are precursors for the synthesis of proteins and other critical nitrogen-containing compounds, such as heme and nucleic acids.

There is no distinct storage form for amino acids in the body as there is for glucose in the form of glycogen; all body proteins serve one or another important function. The turnover of some proteins, particularly those in muscle, is increased under conditions of fasting and starvation when amino acids are needed to provide substrates for gluconeogenesis.

In general, catabolism of amino acids released during protein turnover and dietary amino acids in excess of those needed for protein synthesis involves (1) removal of the α-amino group and (2) oxidation of the carbon skeletons to CO_2 and H_2O, generating ATP in the process.

19.1.1 Urea

Humans excrete excess dietary nitrogen primarily as urea in the urine (Fig. 19-1). At plasma concentrations greater than 50 μM, ammonia ($NH_3 + NH_4^+$) is toxic to the

Medical Biochemistry: Human Metabolism in Health and Disease   By Miriam D. Rosenthal and Robert H. Glew
Copyright © 2009 John Wiley & Sons, Inc.

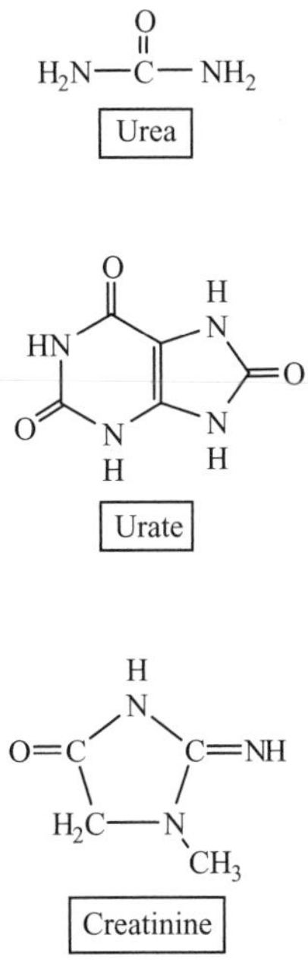

FIGURE 19-1   Major nitrogenous compounds excreted in the urine.

central nervous system. Urea, one of the simplest of organic compounds, is formed from CO_2, ammonium ion, and the α-amino group of aspartate by means of the urea cycle. Synthesis of urea provides a mechanism for detoxifying ammonia.

19.1.2   Urinary Excretion of Other Nitrogenous Compounds

Urine contains nitrogenous compounds other than urea, including uric acid, creatinine, and ammonia, the excretion of which serves other, distinct functions or represents breakdown products of certain metabolites (Fig. 19-1). For example, creatinine is a breakdown product of muscle creatine phosphate and, as such, provides the clinician with a convenient measure of muscle mass. For this reason, it is sometimes useful to quantify the urinary content of other metabolites by normalizing their values to urinary creatinine. Uric acid is the end product of purine catabolism. The kidney also excretes some nitrogen directly in the form of ammonium ions, which serve to buffer acidic, anionic waste products such as β-hydroxybutyrate, acetoacetate, and sulfate. Excretion of ammonium ions is thus increased during ketoacidosis and other metabolic conditions where excess organic acids are produced.

19.1.3   Other Losses of Nitrogen from the Body

Although the major route of nitrogen excretion is urea, there is also a loss of approximately 1.6 g of nitrogen per day (equivalent to 10 g of protein) in the feces.

Fecal nitrogen arises both from sloughing of cells into the gut and from incompletely absorbed dietary protein. In addition, there are minor losses of nitrogen-containing substances from the skin (both as sweat and shed skin cells) and from hair loss, nasal secretions, and menstrual fluid.

19.1.4 Nitrogen Balance

To maintain nitrogen homeostasis, the quantity of nitrogen excreted must be balanced by dietary nitrogen intake. A healthy adult who consumes 100 g of protein (16 g of nitrogen) per day will excrete 16 g of nitrogen per day, of which approximately 15 g will be in the form of urea. If the same person were to increase his or her protein intake further, the renal excretion of urea would be increased to eliminate the excess nitrogen.

19.2 LOCALIZATION OF NITROGEN METABOLISM

The synthesis of urea occurs in the liver, which is the only tissue in humans that contains the enzyme arginase. The liver is also a major site of aminotransferase activity. The resulting urea is released into the blood and excreted by the kidney. Under conditions of vigorous exercise, some urea is also eliminated from the body in sweat.

The major source of amino acid nitrogen that is excreted is from turnover of muscle proteins. A large fraction of the excess nitrogen from amino acid catabolism in muscle is released into the blood as alanine and glutamine. Most of the alanine secreted by muscle is utilized by the liver, with the carbon skeleton serving as a precursor for gluconeogenesis, while the amino acid nitrogen group is incorporated into urea. The glutamine secreted by muscle is used mainly by the kidney. The ammonium ions generated during the catabolism of glutamine in the kidney serve to buffer organic acids and other anions (e.g., sulfate) in the urine, while the carbon skeleton (α-ketoglutarate) becomes a substrate for renal gluconeogenesis.

19.3 PHYSIOLOGICAL STATES IN WHICH THE NITROGEN DETOXIFICATION AND EXCRETION PATHWAYS ARE ESPECIALLY ACTIVE

Since the synthesis of urea increases when there is a need to eliminate excess nitrogen from the body, urea synthesis is higher when a person is consuming a high-protein diet as opposed to a diet low in protein. By contrast, the excretion of other nitrogenous compounds (e.g., creatinine, uric acid, ammonium ion) is relatively unaffected by the amount of nitrogen in the diet.

Urea synthesis is also increased in the fasted state. This may seem paradoxical: If urea synthesis decreases as the protein intake decreases, one might expect urea synthesis to be lowest when the protein intake is zero. However, as discussed in Chapter 9, carbon skeletons of a number of amino acids play an important role as substrates for gluconeogenesis. A person consuming carbohydrate without any dietary protein will carry out minimal urea synthesis and excrete only about 2 g of nitrogen per day as urea. By contrast, the same person who is fasted and therefore actively breaking down muscle proteins to support gluconeogenesis may excrete as much as 10 to 12 g of nitrogen per day as urea.

19.4 REACTIONS OF UREA SYNTHESIS

19.4.1 Overall Pathway

Each urea molecule contains two nitrogen atoms and one carbon atom (Fig. 19-1). The two nitrogen atoms of urea originate directly from ammonium ion and aspartate, although both nitrogens are ultimately derived from the amino groups of all 20 common amino acids and the non-α-amino group nitrogens of asparagine, glutamine, and histamine. The single carbon atom of urea is derived from bicarbonate. The urea molecule is assembled on ornithine, which is an α-amino acid but not one used for protein synthesis. The assembly of urea on ornithine generates the amino acid arginine; subsequent hydrolysis of arginine by arginase releases urea and regenerates ornithine:

$$\text{arginine} \rightarrow \text{ornithine} + \text{urea}$$

Indeed, excess dietary arginine is itself an immediate precursor of urea. Since arginase is expressed only in liver, that organ is the sole site of urea synthesis.

19.4.2 Intracellular Localization of the Urea Cycle

Synthesis of urea involves both mitochondrial and cytosolic enzymes (Fig. 19-2). The first two steps of the urea cycle that generate the intermediate citrulline occur in the mitochondrion. Citrulline is then transported out of the mitochondrion into the cytosol, where it is converted to arginosuccinate and then to arginine. After arginase releases a molecule of urea from arginine, ornithine, the other product, is transported back into the mitochondrion for another round of urea synthesis.

19.4.3 Aminotransferase Reactions Generate Glutamate

Aminotransferases catalyzes reversible transfer of an α-amino group from an amino acid to an acceptor α-ketoacid, producing a new amino acid and an α-ketoacid derived from the donor amino acid (Fig. 19-3). The most common acceptor utilized by aminotransferases is α-ketoglutarate. Thus, the process of transamination results ultimately in the collection of amino acid nitrogen in glutamate.

Pyridoxal phosphate (PLP), the active form of vitamin B_6, is a cofactor in transamination reactions (Fig. 19-4). The aldehyde group of PLP can accept the α-amino

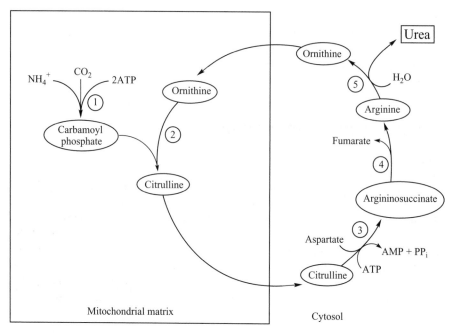

FIGURE 19-2 Localization of the reactions of the urea cycle: ①, carbamoyl phosphate synthase I; ②, ornithine transcarbamoylase; ③, arginosuccinate synthase; ④ arginosuccinate lyase; ⑤ arginase.

group from an amino acid, generating pyridoxamine phosphate, which in turn donates that amino group to an α-ketoacid, regenerating PLP. The activated intermediate in this process is a Schiff base. PLP-containing enzymes also catalyze many other reactions involving amino acids, including the decarboxylation reactions involved in the synthesis of the neurotransmitters dopamine and serotonin.

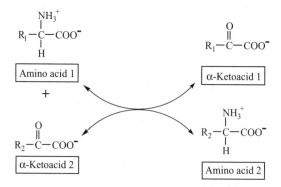

FIGURE 19-3 Transamination of amino acids.

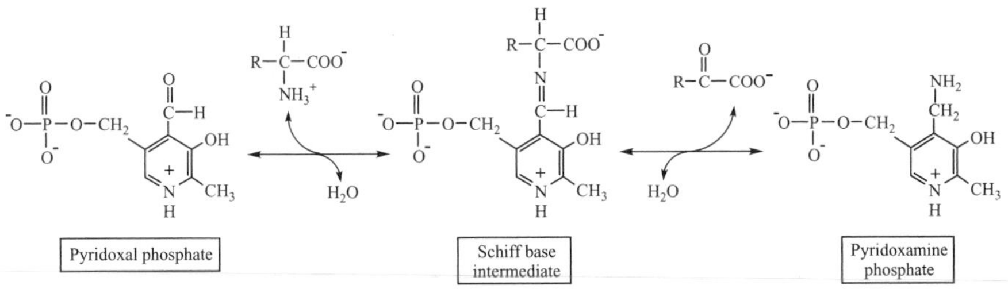

FIGURE 19-4   Role of pyridoxal phosphate in aminotransferase reactions.

19.4.3.1 *Liver Is the Major Site of Aminotransferase Activity.* The two principal liver transaminases are alanine aminotransferase (ALT), which catalyzes the reaction

$$\text{alanine} + \alpha\text{-ketoglutarate} \rightleftharpoons \text{pyruvate} + \text{glutamate}$$

and aspartate aminotransferase (AST), which catalyzes the reaction

$$\text{aspartate} + \alpha\text{-ketoglutarate} \rightleftharpoons \text{oxaloacetate} + \text{glutamate}$$

Most of the NH_4^+ generated when muscle proteins are broken down during a fast is exported in the form of alanine. In the liver, ALT catalyzes the transamination of alanine, generating pyruvate which can be utilized for gluconeogenesis, and glutamate which provides nitrogen atoms for urea synthesis. Some of the glutamate nitrogen is released as ammonium ions by the enzyme glutamate dehydrogenase (see below). Concurrently, AST utilizes some of the glutamate to generate aspartate by transfer of the amino group from glutamate to oxaloacetate. The NH_4^+ from the glutamate dehydrogenase reaction and the aspartate from the AST reaction provide the two nitrogens for urea synthesis.

19.4.3.2 *Increased Plasma Levels of ALT and AST Are Indicative of Liver Damage.* Aminotransferases are intracellular enzymes that have both cytosolic and mitochondrial isoforms. When there is liver damage, as occurs with cirrhosis or viral hepatitis, these aminotransferase are released from the hepatocytes. Increased plasma levels of ALT and AST are thus markers of liver damage. In the older clinical literature, these enzymes are sometimes referred to as SGPT (serum glutamate:pyruvate transaminase) and SGOT (serum glutamate:oxaloacetate transaminase), respectively.

19.4.4 Generation of Ammonium Ions from Amino Acids

Two steps are required to generate ammonium ions from most of the 20 common amino acids. The first is an aminotransferase reaction in which the α-amino group of

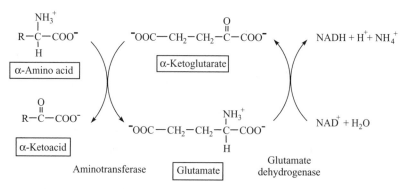

FIGURE 19-5 Central role of glutamate in the release of amino acid nitrogen as ammonium ions.

an amino acid is transferred to α-ketoglutarate, which converts the latter into gluta-mate. The second step is the NAD$^+$-dependent oxidative deamination of glutamate, which releases a free ammonium ion, regenerating α-ketoglutarate in the process (Fig. 19-5). Alternate catabolic pathways exist for those amino acids (e.g., serine, histidine, proline) that do not undergo transamination. Catabolism of amino acids and other nitrogenous molecules by bacteria in the gut is also a significant source of NH$_4^+$ in the body.

19.4.4.1 Glutamate Dehydrogenase.

Glutamate dehydrogenase catalyzes an oxidative deamination reaction in which the amino group of glutamate is released as NH$_4^+$. This reaction occurs in the mitochondrial matrix and is readily reversible:

$$\text{glutamate} + \text{NAD}^+ \rightleftharpoons \alpha\text{-ketoglutarate} + \text{NH}_4^+ + \text{NADH} + \text{H}^+$$

The sequential activities of an aminotransferase that generates glutamate and gluta-mate dehydrogenase, which generates NH$_4^+$, thus provide a pathway for the generation of ammonium ions from the α-amino groups of most of the amino acids (Fig. 19-5).

19.4.4.2 Deamination of Other Amino Acids.

Although all 20 common amino acids, except threonine, lysine, and proline, can be transaminated, many of the amino acids can also be catabolized by pathways that do not involve transamination. One such example is the dehydratase-catalyzed reactions that remove water from both serine and threonine and generate intermediates that contain unstable imino groups (Fig. 19-6). Spontaneous addition of water to these intermediates results in release of NH$_4^+$ and the generation of pyruvate and α-ketobutyrate, respectively.

Histidine is deaminated in one step by histidine ammonia lyase or histidase:

$$\text{histidine} \rightarrow \text{urocanate} + \text{NH}_4^+$$

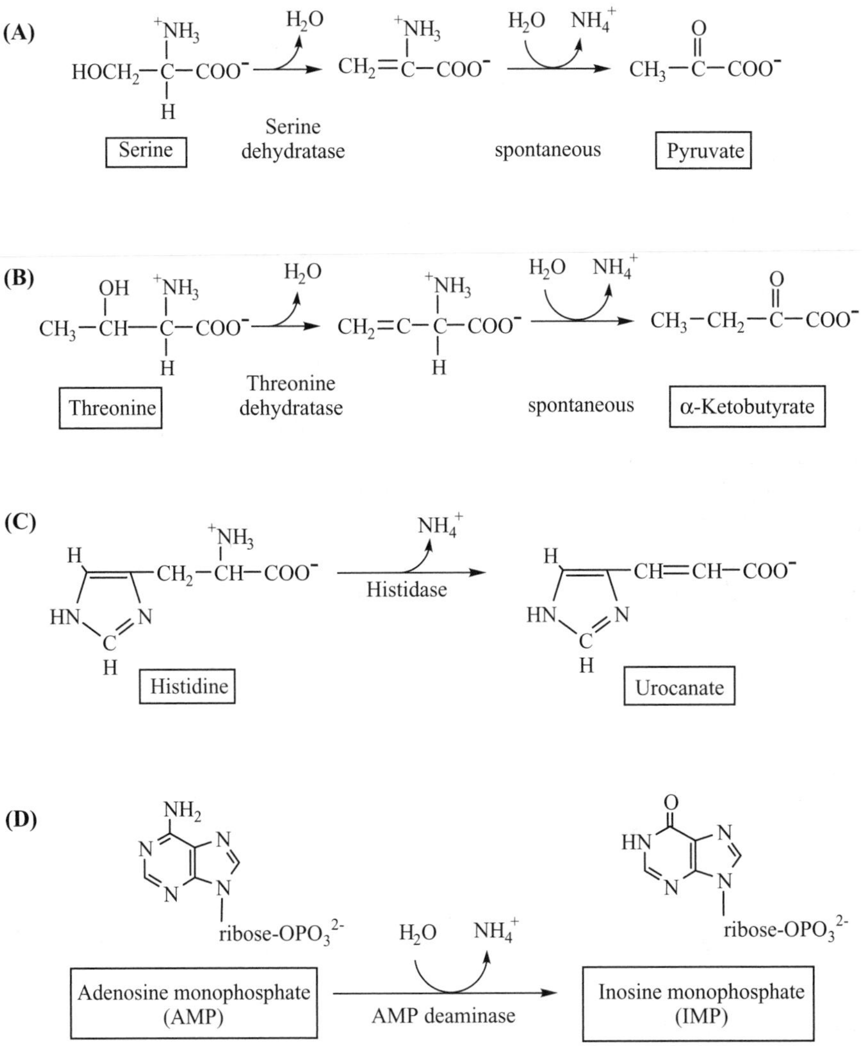

FIGURE 19-6 Alternate pathways for generation of ammonium ions: (A) serine dehydratase; (B) threonine dehydratase; (C) histidase; (D) AMP deaminase.

Subsequent metabolism of urocanate generates glutamate and NH_4^+ and results in the donation of one carbon from urocanate to the tetrahydrofolate pool in the form of N^5,N^{10}-methylenetetrahydrofolate.

The nitrogen group of proline forms part of its five-membered ring structure. Catabolism of proline involves oxidation and opening of the ring to generate glutamate semialdehyde, which is then converted to glutamate (see Chapter 20).

19.4.4.3 Purine Nucleotide cycle. Ammonium ion is also generated through the deamination of AMP to IMP (inosine monophosphate) by adenosine monophosphate deaminase (Fig. 19-6):

$$AMP + H_2O \rightarrow IMP + NH_4^+$$

This reaction is especially active in exercising muscle which generates AMP. When ATP levels are low, muscle can generate additional ATP directly from ADP by means of the myokinase (adenylate kinase) reaction:

$$2ADP \rightleftharpoons ATP + AMP$$

Removal of the resulting AMP is necessary if the reaction is to continue to the right. The pathway by which AMP is deaminated to IMP and IMP is subsequently utilized for resynthesis of AMP is referred to as the *purine nucleotide cycle* and is discussed further in Chapter 23.

19.4.4.4 Deamination of Glutamine and Asparagine. The amide nitrogens of glutamine and asparagine are removed hydrolytically by glutaminase:

$$glutamine + H_2O \rightarrow glutamate + NH_4^+$$

and asparaginase:

$$asparagine + H_2O \rightarrow aspartate + NH_4^+$$

Glutaminase is particularly important in the kidney, where NH_4^+ is used to buffer organic acids (e.g., ketones) and sulfate ions that are excreted in the urine.

19.4.5   The Urea Cycle

The reactions of the urea cycle are illustrated in Figure 19-7.

19.4.5.1 Synthesis of Carbamoyl Phosphate. The initial step in the detoxification of NH_4^+ is the synthesis of the nitrogenous organic compound carbamoyl phosphate:

$$HCO_3^- + NH_4^+ + 2ATP \rightarrow carbamoyl\,phosphate + 2ADP + P_i$$

The enzyme that catalyzes this reaction is carbamoyl phosphate synthetase I (CPS I), and it is localized to mitochondria. Carbamoyl phosphate synthetase II (CPS II) is a cytosolic enzyme that generates carbamoyl phosphate for pyrimidine synthesis.

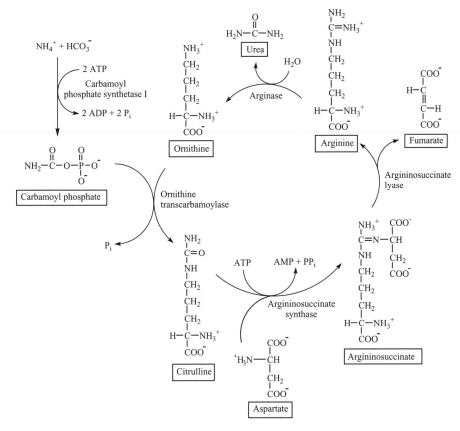

FIGURE 19-7   The urea cycle.

19.4.5.2 *Ornithine Transcarbamoylase.*
Ornithine transcarbamoylase (OTC) catalyzes the transfer of the carbamyl group from carbamoyl phosphate to the amino group in the side chain of the amino acid ornithine, generating citrulline:

$$\text{ornithine} + \text{carbamoyl phosphate} \rightarrow \text{citrulline} + P_i$$

Although both ornithine and citrulline are α-amino acids, they are not utilized for protein synthesis. Citrulline is transported out of the mitochondrion in exchange for ornithine.

19.4.5.3 *Argininosuccinate Synthase and Argininosuccinate Lyase.*
The effect of the next two steps in the urea cycle is to add an amino group to citrulline to generate arginine. This amino group is provided by the amino acid aspartate in the

reaction catalyzed by argininosuccinate synthase:

$$\text{citrulline} + \text{aspartate} + \text{ATP} \rightarrow \text{argininosuccinate} + \text{AMP} + \text{PP}_i$$

Subsequently, argininosuccinate is cleaved, retaining the amino group from aspartate and generating fumarate:

$$\text{argininosuccinate} \rightarrow \text{arginine} + \text{fumarate}$$

Fumarate is hydrated by cytosolic fumarase to form malate, which is oxidized to oxaloacetate, and then transaminated to regenerate aspartate.

19.4.5.4   Arginase. The urea cycle is completed by the hydrolysis of arginine, which generates urea plus ornithine:

$$\text{arginine} + \text{H}_2\text{O} \rightarrow \text{urea} + \text{ornithine}$$

Urea is secreted from hepatocytes and transported to the kidney for excretion in the urine. Ornithine generated by arginase is transported back into mitochondria to continue the cyclic process of urea synthesis.

19.4.5.5   De Novo Synthesis of Ornithine and Arginine. Some of the ornithine utilized in the urea cycle is derived through the action of arginase on dietary arginine. Ornithine can also be synthesized from glutamate by way of a pathway that involves reduction of glutamate by NADPH to produce glutamate semialdehyde, which contains one aldehyde and one carboxyl group (Fig. 19-8). Ornithine transaminase then acts on glutamate semialdehyde to generate ornithine:

$$\text{glutamate semialdehyde} + \text{glutamate} \rightleftharpoons \text{ornithine} + \alpha\text{-ketoglutarate}$$

Synthesis of Arginine from Ornithine Is Not Limited to the Liver. As described above, the enzymes of the urea cycle convert ornithine into arginine. Some of the

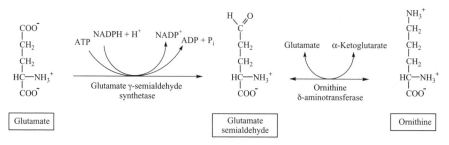

FIGURE 19-8   Synthesis of ornithine.

enzymes of the urea cycle are present in tissues other than the liver. For example, the intestinal mucosa can convert ornithine to citrulline, and the kidney can convert the resultant citrulline to arginine. However, since the kidney and intestine lack arginase, they cannot synthesize urea.

Arginine Is the Precursor of Nitric Oxide. Nitric oxide (NO) is an oxygen-containing free radical which functions as a neurotransmitter and vasodilatory auto-coid or local hormone. At high concentrations, NO combines with molecular oxygen or superoxide to form other reactive oxygen species which can contribute to chronic inflammation and to neurodegenerative diseases. NO is synthesized from arginine in a reaction catalyzed by NO synthase:

$$\text{arginine} + O_2 + \text{NADPH} + H^+ \rightarrow \text{NO} + \text{citrulline} + \text{NADP}^+$$

There are three isozymes of NO synthase. The neural (nNOS) and endothelial (eNOS) forms are regulated by the intracellular calcium concentration and produce NO for its neurotransmitter and autocoid roles. By contrast, the isozyme induced in activated macrophages (iNOS) produces NO that contributes to the bacteriocidal response. The other product of the NO synthase reaction is citrulline, which is converted back to arginine through the sequential actions of argininosuccinate synthase and argininosuccinate lyase. Both argininosuccinate synthase and argininosuccinate lyase are induced under conditions that increase expression of iNOS and allow for regeneration of arginine in the absence of the full urea cycle.

19.5   REGULATION OF UREA SYNTHESIS

19.5.1   Allosteric Regulation of Carbamoyl Phosphate Synthetase I by *N*-Acetylglutamate

Carbamoyl phosphate synthetase I catalyzes the regulated step of urea synthesis. This enzyme is activated by *N*-acetylglutamate. Like fructose 2,6-bisphosphate which regulates glycolysis and gluconeogenesis, *N*-acetylglutamate is not an intermediate in the metabolic pathway that it regulates, but instead, is synthesized for the sole purpose of providing allosteric modulation of a key regulatory enzyme.

N-Acetylglutamate synthase catalyzes the following reaction:

$$\text{glutamate} + \text{acetyl-CoA} \rightarrow N\text{-acetylglutamate} + \text{CoASH}$$

N-Acetylglutamate synthase is allosterically activated by arginine (Fig. 19-9). Thus, *N*-acetylglutamate synthesis occurs only when the supply of both glutamate and ornithine is adequate.

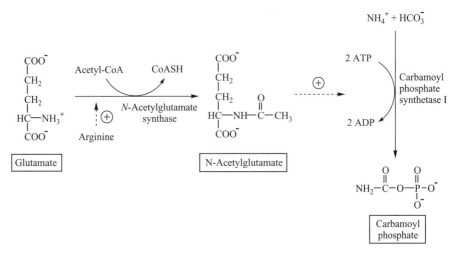

FIGURE 19-9   Regulation of carbamoyl phosphate synthetase I. The dashed arrows indicate the roles of allosteric regulators.

19.5.2   Induction of Enzymes

The levels of the urea cycle enzymes in the liver increase 10- to 20-fold in response to increased concentrations of amino acids and ammonia. Induction of urea cycle enzymes thus occurs under conditions of starvation and stress (e.g., sepsis, burns, trauma) when there is accelerated breakdown of body proteins to provide substrates for gluconeogenesis and the synthesis of acute-phase proteins, whose plasma concentrations increase during inflammatory states. The enzymes of the urea cycle are also induced by high-protein diets, which provide excess amino acids to the liver.

19.6   ABNORMAL FUNCTION OF THE PATHWAYS OF NITROGEN METABOLISM

19.6.1   Hyperammonemia

The finding of elevated blood levels of ammonia is evidence that the conversion of ammonia to urea is impaired in some way. Hyperammonemia in adults is usually the consequence of impaired liver function, secondary to liver disease (e.g., cirrhosis), organ transplantation, or chemotherapy. Transient hyperammonemia is often seen in premature neonates with immature liver function and/or inadequate hepatic blood flow. Impaired urea synthesis may also be the result of a genetic defect in one of the enzymes of the urea cycle. Regardless of its etiology, hyperammonemia is usually accompanied by increased plasma levels of glutamine, the amino acid that the brain uses as a vehicle to export excess ammonium ions.

Ammonia is toxic to the central nervous system, where it can cause both acute encephalopathy and long-term irreversible brain damage; however, the pathophysiologic mechanisms are not fully understood. One possible cause is the increased synthesis of the neurotransmitters glutamine and γ-aminobutyric acid (GABA) and subsequent derangements of neurotransmission. Another possible mechanism for ammonia toxicity in the brain involves the depletion of TCA-cycle intermediates by diversion of α-ketoglutarate to glutamate and glutamine synthesis, which would compromise the ability of the neural cells to generate ATP.

Treatment for hyperammonemia involves hemodialysis or peritoneal dialysis to remove the excess ammonia. In acute cases, oral sodium benzoate and sodium phenylbutyrate are sometimes administered to provide alternate pathways for nitrogen excretion as hippurate and phenylacetylglutamine, respectively (Fig. 19-10). Protein intake should also be severely restricted in patients with hyperammonemia. At the same time, it is important to provide adequate intake of carbohydrates to minimize further catabolism of endogenous protein.

19.6.2 Ornithine Transcarbamoylase Deficiency

Many inborn errors of urea-cycle metabolism have been described in the literature. All share a common set of biochemical symptoms, including hyperammonemia, respiratory alkalosis, and low blood urea nitrogen (BUN). Neonatal patients present with lethargy, irritability, and hypotonia. Prompt treatment is critical because prolonged hyperammonemia can result in irreversible brain damage, coma, and even death.

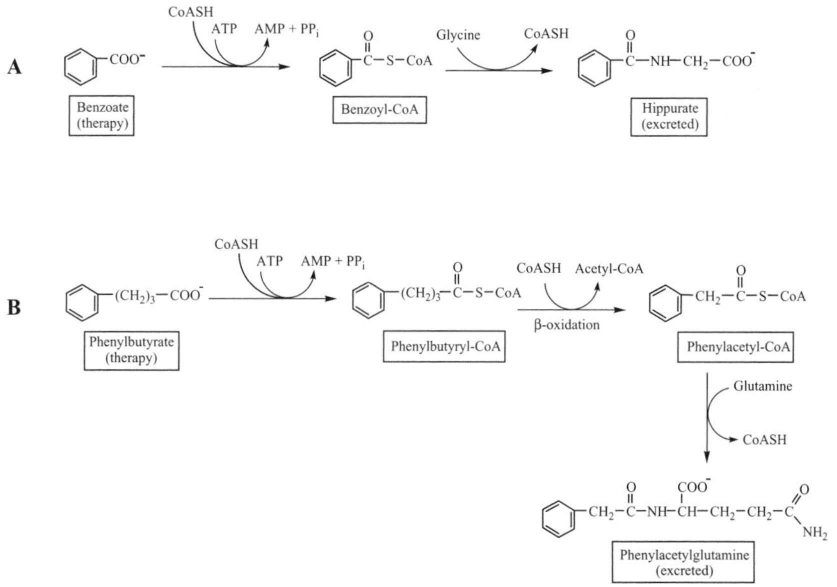

FIGURE 19-10 Alternate mechanisms of nitrogen excretion initiated by pharmacological therapy.

Older children may show a variety of neurological symptoms, including psycho-motor retardation and recurrent cerebellar ataxia, with or without hyperammonemia. The differential diagnosis of inborn errors of the urea cycle is usually made from quantitative analysis of plasma amino acids, which identifies increased levels of urea-cycle intermediates upstream of the specific enzyme blockage.

The most common inborn error of the urea cycle is a deficiency of ornithine trans-carbamoylase, an X-linked disorder with wide genotypic and phenotypic variability in hemizygous males. Heterozygous females are also affected; however, the severity is dependent on the pattern of X-chromosome inactivation. People with ornithine transcarbamoylase deficiency characteristically exhibit decreased plasma levels of citrulline and arginine. They also have elevated levels of orotic acid in their urine, which is a useful diagnostic indicator to rule out deficiency of carbamoyl phosphate synthetase or N-acetylglutamate synthetase. The increased synthesis of orotic acid is the result of the accumulation of intracellular carbamoyl phosphate, which bypasses the regulated step catalyzed by carbamoyl phosphate synthetase II and provides excess substrate for pyrimidine synthesis.

19.6.3 Metabolic Acidosis

The ketone bodies, β-hydroxybutyrate and acetoacetate, that accumulate during both starvation and poorly regulated type I diabetes mellitus, are organic acids. In addition, many inborn errors of metabolism (i.e., medium-chain acyl-CoA dehydrogenase deficiency) and pharmacological therapies (e.g., aspirin) increase renal acid excretion. Excretion of organic acids by the kidney is accompanied by increased excretion of ammonium ions which buffer the urine, resulting in an increase in the ratio of nitrogen excreted as ammonia relative to that excreted as urea.

19.6.4 Elevated BUN

Urea is synthesized in the liver and excreted primarily by the kidneys. Elevated BUN is an indication of a posthepatic failure in nitrogen excretion, the most common causes of which are impaired renal function and poor renal perfusion secondary to congestive heart failure or hypovolemic shock. Severe dehydration, with accompanied decreases in urinary output, may also result in an elevated BUN value.

19.6.5 Hypercatabolic States

Trauma, burns, and sepsis are characterized by increased fuel utilization and a negative nitrogen balance in which excretion of nitrogen—primarily as urea—exceeds dietary intake. These metabolic changes are mediated primarily by hydrocortisone, with contributions by inflammatory cytokines. In these hypermetabolic conditions there is an increased rate of protein degradation, primarily in skeletal muscle, which provides amino acids to support the biosynthetic needs associated with the immune response and wound healing. At the same time, there is increased synthesis and excretion of urea.

CHAPTER 20

AMINO ACIDS

20.1 FUNCTIONS OF AMINO ACID METABOLISM

Since 20 common amino acids, some with cyclic and branched structures, are utilized for protein synthesis, the synthesis and catabolism of amino acid carbon skeletons can be a complex and daunting subject with a myriad of details. Nonetheless, there are a number of common themes that are of major importance in understanding the overall metabolism of the body.

20.1.1 Synthesis of Amino Acids

Some of the 20 common amino acids can be synthesized in the body. The aminotransferase reactions that remove the amino group from most of these amino acids are readily reversible and can therefore be utilized to synthesize amino acids. For example, aspartate aminotransferase can be used to synthesize aspartate from the TCA-cycle intermediate oxaloacetate:

$$\text{oxaloacetate} + \text{glutamate} \rightleftharpoons \text{aspartate} + \alpha\text{-ketoglutarate}$$

Since the glutamate dehydrogenase reaction, too, is reversible, it can be used to incorporate NH_4^+ into α-ketoglutarate, generating glutamate. Glutamate, in turn, can donate its amino group for the synthesis of other amino acids:

$$\alpha\text{-ketoglutarate} + NH_4^+ + NADH^+ + H^+ \rightleftharpoons \text{glutamate} + NAD^+$$

Medical Biochemistry: Human Metabolism in Health and Disease   By Miriam D. Rosenthal and Robert H. Glew
Copyright © 2009 John Wiley & Sons, Inc.

TABLE 20-1   Essential and Nonessential Amino Acids in Humans

Essential	Nonessential
Histidine	Alanine
Isoleucine	Arginine[a]
Leucine	Asparagine
Lysine	Aspartate
Methionine	Cysteine[b]
Phenylalanine	Glutamate
Threonine	Glutamine
Tryptophan	Glycine
Valine	Proline
	Serine
	Tyrosine[b]

[a]Essential in infants but not in adults.
[b]Conditionally essential. Synthesis of cysteine and tyrosine is dependent on adequate dietary intake of methionine and phenylalanine, respectively.

Thus, humans can synthesize a particular amino acid if they can synthesize its corresponding α-ketoacid carbon skeleton.

20.1.1.1 Essential Amino Acids. Certain amino acids, however, cannot be synthesized in the body; these are the essential amino acids and they must be obtained from the diet. The amino acids that are essential in adults are listed in Table 20-1. Two other amino acids, tyrosine and cysteine, can be synthesized from the essential amino acids phenylalanine and methionine, respectively. In addition, although arginine is not an essential amino acid in adults, its rate of synthesis in neonates is not adequate to meet their requirements for optimal growth. Exogenous arginine also becomes essential in cases of sepsis, when there is both a decrease in endogenous synthesis of arginine and an increased requirement of arginine for the synthesis of protein and nitric oxide.

20.1.1.2 Protein Quality. Protein synthesis requires that all 20 of the common amino acids be present, including the essential ones. Dietary proteins of both animal and vegetable origin usually provide all of the essential amino acids, but not necessarily in the proportions required to meet the body's needs for the synthesis of proteins and specialized nitrogen-containing molecules such as neurotransmitters (e.g., dopamine) and polyamines (e.g., spermine).

Nutritionists evaluate the quality of a particular dietary protein by comparing its amino acid composition to that of a reference protein which has the optimum proportions of all the essential amino acids. In general, animal proteins such as beef, fish, and egg white approximate those standards. Although vegetable proteins are sometimes referred to as "incomplete" proteins, they too usually contain all the essential amino acids, but not in the ideal proportions. For example, proteins in cereals such as wheat and rice are relatively deficient in lysine. The amino acid score of rice protein is approximately 50, indicating that only half of the constituent amino acids can be utilized for protein synthesis before the supply of lysine is exhausted; the other 50% of the dietary amino acids must then be catabolized. By contrast, whereas the proteins of legumes such as peas and beans actually contain a higher level of lysine than the reference protein, they are relatively deficient in the sulfur-containing amino acids (methionine plus cysteine). However, when these two types of proteins are mixed as they would be in a meal of rice and beans, they complement each other to provide high-quality dietary protein. Improvement of the quality of vegetable proteins can also be achieved by including a small quantity of animal protein in the meal.

20.1.2 Glucogenic and Ketogenic Amino Acids

One function of amino acid catabolism is to provide their carbon skeletons as substrates for gluconeogenesis. Alanine and glutamine are major gluconeogenic substrates. In addition, the carbon skeletons of 11 of the other common amino acids are readily utilized for the synthesis of glucose. For example, catabolism of cysteine and serine yields pyruvate. The carbon skeletons of other glucogenic amino acids enter the TCA cycle; these include aspartate and asparagine, which are metabolized to oxaloacetate, and arginine and histidine, which are catabolized to α-ketoglutarate.

Two amino acids, lysine and leucine, cannot be utilized for glucose synthesis. Instead, the pathways that catabolize their carbon skeletons generate acetyl-CoA and acetoacetate. Since hepatocytes use acetyl-CoA to synthesize ketone bodies under fasting conditions, lysine and leucine are classified as ketogenic amino acids.

Five of the 20 common amino acids are both glucogenic and ketogenic, in that some of their carbon atoms can be utilized to produce glucose whereas the other carbon atoms generate ketogenic acetyl-CoA. These mixed glucogenic/ketogenic amino acids include tryptophan, isoleucine, threonine, phenylalanine, and tyrosine.

20.1.3 Amino Acids as Precursors of Neurotransmitters and Other Specialized Nitrogen-Containing Compounds

Amino acids provide nitrogen for the synthesis of purines, pyrimidines, nitrogen-containing phospholipids (e.g., phosphatidylcholine), and amino sugars (e.g., glucosamine). Tyrosine is the precursor of the catecholamine neurotransmitters (dopamine, norepinephrine, and epinephrine), the thyroid hormones (thyroxine and triiodothyronine), and the melanin pigments. Tryptophan is the precursor of serotonin, melatonin, and niacin. Other important amino acid–derived products include γ-aminobutyric acid, histamine, nitric oxide, and polyamines.

20.2 LOCALIZATION OF AMINO ACID METABOLISM

20.2.1 Synthesis of Nonessential Amino Acids

Although the liver is the major site of synthesis of nonessential amino acids, other tissues can synthesize and degrade amino acids. For example, the synthesis of arginine for protein synthesis occurs by a two-step interorgan process: Glutamate is converted to ornithine and then to citrulline in the intestinal mucosa. The kidney then utilizes citrulline to synthesize arginine.

20.2.2 Synthesis of Conditionally Essential Amino Acids

The synthesis of tyrosine from the essential amino acid phenylalanine occurs exclusively in the liver. The transsulfuration pathway by which the sulfur atom of methionine is utilized to synthesize cysteine occurs primarily in liver, kidney, and intestinal mucosa.

20.2.3 Synthesis of Specialized Amino Acid–Related Products

Much of the synthesis of specialized metabolites of amino acids is localized to a few cell types. This is particularly true of neurotransmitters such as dopamine, norepinephrine, and serotonin, each of which is produced by specific groups of neurons. The adrenal medulla converts tyrosine to epinephrine, and the thyroid glands synthesize the thyroid hormones thyroxine and 3,5,3'-triiodothyronine from tyrosine.

20.2.4 Catabolism of Amino Acids

The liver is the major site for the catabolism of aromatic amino acid carbon skeletons (phenylalanine, tyrosine, and tryptophan) as well as for the synthesis of urea from amino acid nitrogen. By contrast, the initial transamination of the branched-chain amino acids leucine, isoleucine, and valine, and the subsequent oxidation of their carbon skeletons, occur primarily in skeletal muscle. As pointed out above, the carbon skeletons of alanine and glutamine are major substrates for gluconeogenesis in liver and kidney, respectively. Glutamine is a significant fuel for rapidly dividing cells such as lymphocytes, macrophage, and enterocytes, as well as for the kidney.

20.3 PHYSIOLOGICAL CONDITIONS DURING WHICH THE AMINO ACID CATABOLIC PATHWAYS ARE ESPECIALLY ACTIVE

There are two very different physiological conditions under which there is a high rate of catabolism of amino acids: following intake of a protein-rich diet and during fasting. In the fed state, there is active protein synthesis by muscle, liver, and other tissues; excess amino acids are oxidized for energy. In the fasting state, there is increased catabolism of muscle proteins. Most of the carbon skeletons of the branched-chain amino acids are utilized as fuel. The amino groups of the branched-chain amino acids are exported from muscle as alanine and glutamine, which are transported to the liver and kidney to provide substrates for gluconeogenesis.

20.4 PATHWAYS OF AMINO ACID METABOLISM

Human cells use a multiplicity of metabolic pathways to catabolize the 20 common amino acids and to synthesize the carbon skeletons of the nonessential amino acids. Rather than catalog all of these pathways, specific examples are cited to illustrate the types of reactions involved in amino acid metabolism.

20.4.1 Aminotransferase Reactions

Aminotransferases transfer the α-amino group from a donor amino acid to an acceptor α-ketoacid (see Fig. 19-3). This process initiates amino acid catabolism by transferring the α-amino group of various amino acids to α-ketoglutarate, thereby generating glutamate. Glutamate, in turn, serves as a nitrogen donor for urea synthesis. Transfer of the α-amino group of glutamate to various α-ketoacids can also be utilized to synthesize other nonessential amino acids (e.g., conversion of pyruvate to alanine or oxaloacetate to aspartate).

20.4.2 Synthesis of Glutamine

Glutamine is synthesized by the ATP-driven glutamine synthetase–catalyzed incorporation of amino acid–derived NH_4^+ into glutamate (Fig. 20-1):

$$\text{glutamate} + NH_4^+ + ATP \rightarrow \text{glutamine} + ADP + P_i$$

This reaction is irreversible. The opposite reaction, the hydrolytic removal of the amide group of glutamine, is catalyzed by glutaminase:

$$\text{glutamine} + H_2O \rightarrow \text{glutamate} + NH_4^+$$

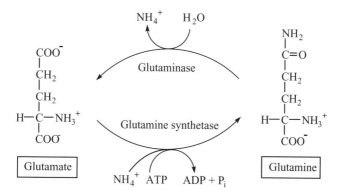

FIGURE 20-1 Synthesis and catabolism of glutamine.

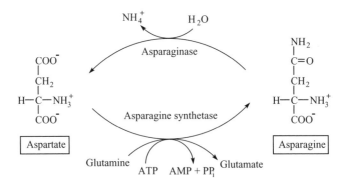

FIGURE 20-2 Synthesis and catabolism of asparagine.

20.4.3 Synthesis of Aspartate And Asparagine

An example of the process of endogenous amino acid synthesis is the synthesis of aspartate and its subsequent conversion to asparagine (Fig. 20-2). The precursors for this pathway are glucose, which is used to generate pyruvate, and amino acids, which donate the α-amino groups needed for glutamate synthesis. Pyruvate is first converted to oxaloacetate by pyruvate carboxylase:

$$\text{pyruvate} + CO_2 + \text{ATP} \rightarrow \text{oxaloacetate} + \text{ADP}$$

Transamination of oxaloacetate by aspartate aminotransferase (AST) in turn generates aspartate:

$$\text{oxaloacetate} + \text{glutamate} \rightleftharpoons \text{aspartate} + \alpha\text{-ketoglutarate}$$

Finally, asparagine synthetase uses the amide nitrogen of glutamine to provide the amide nitrogen of asparagine:

$$\text{aspartate} + \text{glutamine} + \text{ATP} \rightarrow \text{asparagine} + \text{glutamate} + \text{AMP} + PP_i$$

20.4.4 Synthesis of Arginine and Proline

Arginine is synthesized from ornithine, which is generated from glutamate semialdehyde (see Fig. 9.8). Synthesis of glutamate semialdehyde by reduction of the γ-carboxyl group of glutamate to an aldehyde also serves as the initial step in the pathway of proline synthesis (Fig. 20-3):

$$\text{glutamate} + \text{ATP} + \text{NADPH} + H^+ \rightarrow \text{glutamate semialdehyde} + \text{NADP}^+$$
$$+ \text{ADP} + P_i$$

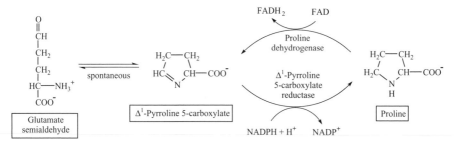

FIGURE 20-3 Synthesis and catabolism of proline.

The next step in the synthesis of proline is the spontaneous reversible cyclization of glutamate semialdehyde:

$$\text{glutamate semialdehyde} \rightleftharpoons \Delta^1\text{-pyrroline 5-carboxylate}$$

Δ^1-pyrroline 5-carboxylate is then reduced further to generate proline:

$$\Delta^1\text{-pyrroline 5-carboxylate} + \text{NADPH} + \text{H}^+ \rightarrow \text{proline} + \text{NADP}^+$$

Catabolism of proline involves essentially this same pathway, operating in reverse except that different enzymes catalyze the two oxidation steps. The enzyme that oxidizes proline to Δ^1-pyrroline 5-carboxylate uses FAD as a cofactor; the one that oxidizes glutamate semialdehyde to glutamate is NAD^+-dependent:

$$\text{proline} + \text{FAD} \rightarrow \Delta^1\text{-pyrroline 5-carboxylate} + \text{FADH}_2$$

$$\Delta^1\text{-pyrroline 5-carboxylate} \rightleftharpoons \text{glutamate semialdehyde}$$

$$\text{glutamate semialdehyde} + \text{NAD}^+ \rightarrow \text{glutamate} + \text{NADH} + \text{H}^+$$

20.4.5 Oxidation of Branched-Chain Amino Acids

The pathways by which the carbon skeletons of the three branched-chain amino acids—leucine, isoleucine, and valine—are oxidized nicely illustrate the general principles of amino acid catabolism and the ultimate metabolic fate of ketogenic and glucogenic amino acids (Fig. 20-4).

The first step in the catabolism of the branched-chain amino acids is the transfer of the α-amino group of each to α-ketoglutarate, generating the corresponding branched-chain α-ketoacid. Thus, for leucine,

$$\text{leucine} + \alpha\text{-ketoglutarate} \rightleftharpoons \alpha\text{-ketoisocaproate} + \text{glutamate}$$

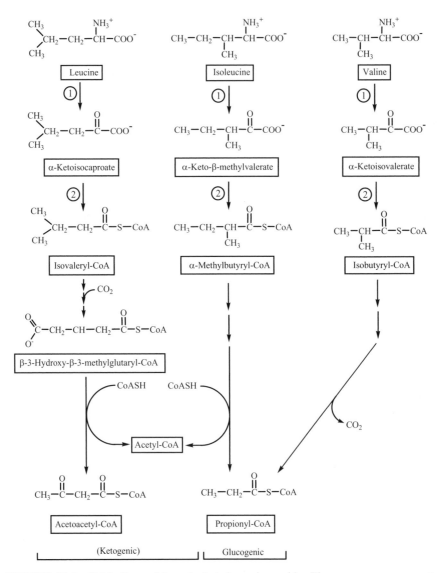

FIGURE 20-4 Catabolism of branched-chain amino acids. The enzymes common to catabolism of all three amino acids are ① branched-chain amino acid transaminase and ② α-ketoacid dehydrogenase.

The next step in the catabolism of the three branched-chain amino acids is an oxidative decarboxylation reaction catalyzed by the branched-chain α-ketoacid dehydrogenase complex. For example, α-ketoisocaproate is oxidized and decarboxylated to form isovaleryl-CoA:

$$\alpha\text{-ketoisocaproate} + NAD^+ + CoASH \rightarrow isovaleryl\text{-}CoA + NADH + H^+ + CO_2$$

This enzyme complex is similar to the pyruvate dehydrogenase and α-ketoglutarate dehydrogenase complexes in that it contains three catalytic components: branched-chain α-ketoacid decarboxylase, dihydrolipoamide branched-chain transacylase, and dihydrolipoamide dehydrogenase. Like the other α-ketoacid dehydrogenases, this enzyme complex catalyzes an irreversible reaction; it thus represents the committed step in the catabolism of branched-chain amino acids. Branched-chain α-ketoacid decarboxylase is similar to the other α-ketoacid dehydrogenases in that the E1 enzyme component requires thiamine pyrophosphate as a cofactor.

The subsequent steps in the oxidation of the three products of the branched-chain α-ketoacid decarboxylase reaction—isovaleryl-CoA (from leucine), α-methylbutyryl-CoA (from isoleucine), and isobutyryl-CoA (from valine)—are outlined in Figure 20-4. Although the specific details about the respective intermediates are beyond the scope of this book, the nature of the end products of the three branched-chain amino acids is illustrative of the ultimate metabolic fates of their amino acid carbon skeletons.

20.4.5.1 *Leucine Is Ketogenic.* Further catabolism of isovaleryl-CoA in mitochondria generates β-hydroxy-β-methylglutaryl-CoA (HMG-CoA). Mitochondrial HMG-CoA is an intermediate in the synthesis of ketone bodies; specifically, it is hydrolyzed by HMG-CoA lyase to yield acetoacetate:

$$\beta\text{-hydroxy-}\beta\text{-methylglutaryl-CoA} \rightarrow \text{acetoacetate} + \text{acetyl-CoA}$$

In peripheral tissues such as muscle, acetoacetate can be activated to acetoacetyl-CoA and then cleaved with the addition of a second molecule of CoASH to produce two molecules of acetyl-CoA. Since the carbon atoms of acetyl-CoA cannot be utilized for glucose synthesis, leucine is classified as a ketogenic rather than a glucogenic or mixed-type (glucogenic and ketogenic) amino acid.

20.4.5.2 *Valine Is Glucogenic.* The successive transamination and decarboxylation of valine generates isobutyryl-CoA. Subsequent oxidation of isobutyryl-CoA results in release of a second molecule of CO_2 and the generation of propionyl-CoA. As discussed in Chapter 10, the metabolism of propionyl-CoA involves carboxylation to methylmalonyl-CoA, which, in turn, is converted to succinyl-CoA. Since succinyl-CoA is a TCA-cycle intermediate that can be further metabolized to oxaloacetate, succinyl-CoA can be used to synthesize glucose. Valine is therefore a glucogenic amino acid.

20.4.5.3 *Isoleucine Is Both Ketogenic and Glucogenic.* Amino transfer and decarboxylation of isoleucine generate α-methylbutyryl-CoA, which is then oxidized to α-methylacetoacetyl-CoA and finally cleaved to produce propionyl-CoA plus acetyl-CoA:

$$\alpha\text{-methylacetoacetyl-CoA} + \text{CoA} \rightarrow \text{propionyl-CoA} + \text{acetyl-CoA}$$

As in the case of valine catabolism, the propionyl-CoA moiety can be used to synthesize glucose. By contrast, the acetyl-CoA moiety cannot be used for gluconeogenesis. The metabolic fate of isoleucine is thus mixed, being part glucogenic and part ketogenic.

20.4.6 Products of Phenylalanine Metabolism

20.4.6.1 *Synthesis of Tyrosine from Phenylalanine.* Phenylalanine, an essential amino acid, is the immediate precursor of tyrosine (Fig. 20-5). The reaction is catalyzed by phenylalanine hydroxylase with tetrahydrobiopterin (BH_4) as cofactor:

$$\text{phenylalanine} + O_2 + BH_4 \rightarrow \text{tyrosine} + BH_2 + H_2O$$

This pathway provides tyrosine for both protein synthesis and the synthesis of catecholamines and thyroid hormones. It is also the major pathway for the catabolism of excess phenylalanine.

The liver also contains a phenylalanine-specific aminotransferase:

$$\text{phenylalanine} + \alpha\text{-ketoglutarate} \rightleftharpoons \text{phenylpyruvate} + \text{glutamate}$$

Phenylalanine aminotransferase requires a higher concentration of phenylalanine for activity than does phenylalanine hydroxylase and is therefore unable to prevent accumulation of deleterious levels of phenylalanine in people who are phenylalanine hydroxylase–deficient. Furthermore, the aromatic metabolites of phenylalanine represent a metabolic dead-end.

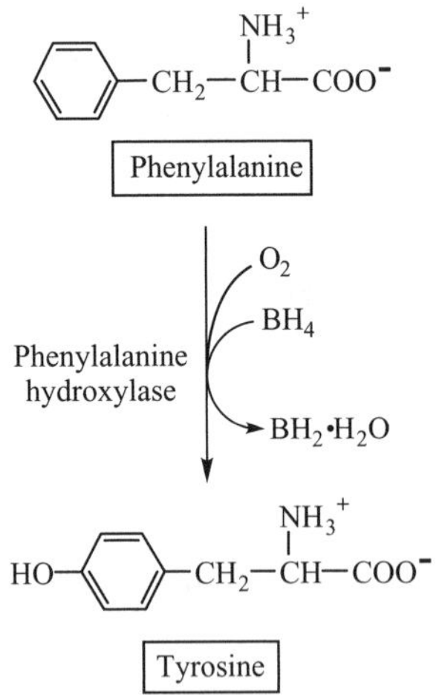

FIGURE 20-5 Synthesis of tyrosine from phenylalanine.

Tetrahydrobiopterin (BH$_4$) is a Cofactor for the Hydroxylation of Aromatic Amino Acids. Unlike so many of the other cofactors in intermediary metabolism, tetrahydrobiopterin, the cofactor for phenylalanine hydroxylase, is not a vitamin. Instead, BH$_4$ is synthesized from GTP (Fig. 20-6).

Recycling of Tetrahydrobiopterin. Phenylalanine hydroxylase is a mixed-function oxidase, which simultaneously oxidizes phenylalanine and removes two hydrogen atoms from tetrahydrobiopterin. The resulting pterin 4α-carbinolamine is then recycled to tetrahydrobiopterin, with NADPH serving as the reductant (Fig. 20-6):

$$\text{pterin } 4\alpha\text{-carbinolamine} \rightarrow \text{dihydrobiopterin} + H_2O$$

$$\text{dihydrobiopterin} + NADPH + H^+ \rightarrow \text{tetrahydrobiopterin} + NADP^+$$

20.4.6.2 *Synthesis of Catecholamines.* Dopamine and norepinephrine are neurotransmitters synthesized in the brain. Dopamine, norepinephrine, and epinephrine, collectively called *catecholamines*, have two hydroxyl groups on the aromatic or phenolic ring (catechol is *o*-dihydroxybenzene). The catecholamines are synthesized through a common pathway that starts with tyrosine (Fig. 20-7). The complete pathway, which generates epinephrine, occurs primarily in the adrenal gland.

Synthesis of DOPA (3.4-Dihydroxyphenylalanine). The first step in the synthesis of the catecholamines is the hydroxylation of tyrosine by tyrosine hydroxylase, which like phenylalanine hydroxylase is a mixed-function oxidase that also reduces BH$_4$:

$$\text{tyrosine} + O_2 + BH_4 \rightarrow \text{dihydroxyphenylalanine} + BH_2 + H_2O$$

Synthesis of Dopamine. Dopamine, the prominent neurotransmitter in the substantia nigra and several other parts of the brain, is synthesized by decarboxylation of DOPA. DOPA decarboxylase, like many other amino acid decarboxylases, utilizes pyridoxal phosphate (PLP) as a cofactor:

$$\text{dihydroxyphenylalanine} \rightarrow \text{dopamine} + CO_2$$

Norepinephrine. Norepinephrine is synthesized by oxidizing dopamine. Unlike the reactions catalyzed by phenylalanine hydroxylase and tyrosine hydroxylase, dopamine β-hydroxylase oxidizes the side chain rather than the phenyl ring and utilizes ascorbic acid rather than tetrahydrobiopterin as the cofactor and reducing agent:

$$\text{dopamine} + O_2 + \text{ascorbate} \rightarrow \text{norepinephrine} + \text{dehydroascorbate} + H_2O$$

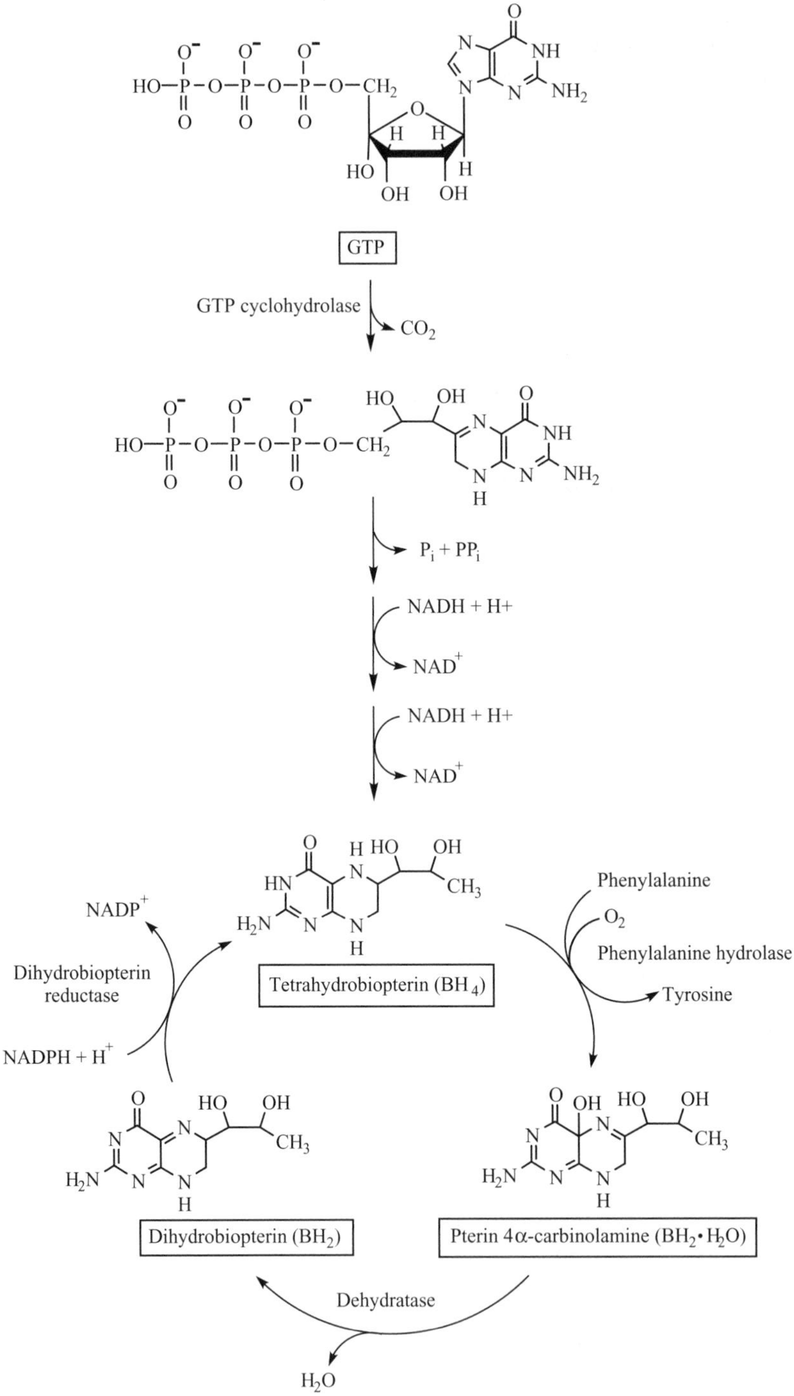

FIGURE 20-6 Synthesis and recycling of the cofactor tetrahydrobiopterin.

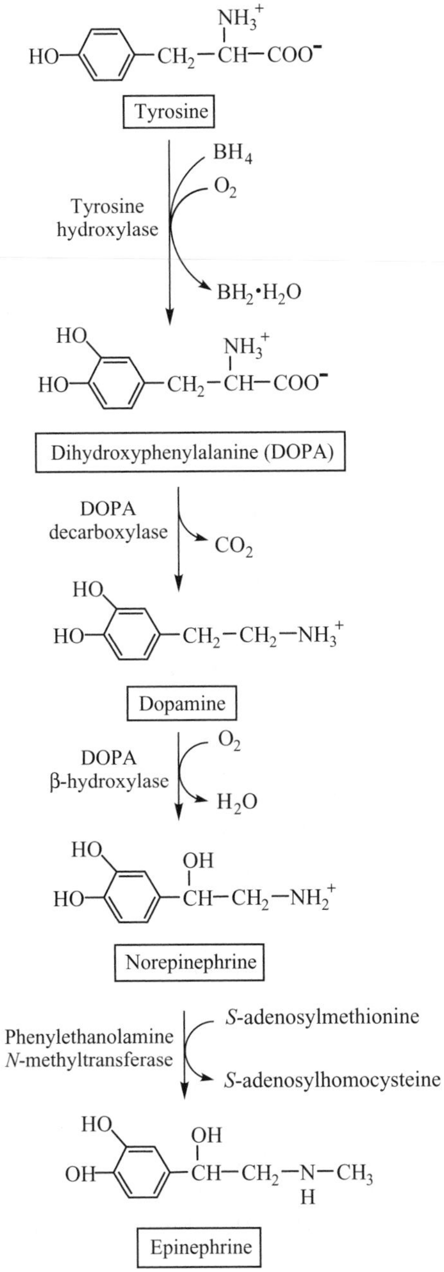

FIGURE 20-7 Synthesis of catecholamines.

Ascorbate can be regenerated from dehydroascorbate by dehydroascorbate reductase, which uses reduced glutathione as the reductant.

Epinephrine. Epinephrine is synthesized by methylating norepinephrine:

norepinephrine + S-adenosylmethionine $\rightarrow$ epinephrine + S-adenosylhomocysteine

The methyl donor for this reaction, S-adenosylmethionine (SAM), represents the activated form of methionine.

20.4.6.3 Inactivation of Catecholamines.
The two enzymes that inactivate the catecholamines are catechol O-methyltransferase (COMT) and monoamine oxidase (MAO). COMT uses S-adenosylmethionine to methylate the hydroxyl group in the $2'$-position on the phenyl ring. MAO catalyzes the removal of the terminal amino group of a catecholamine such as dopamine, generating an aldehyde which is then oxidized further to a carboxyl group. The reactions catalyzed by COMT and MAO can occur in either order; the resulting degradation products (i.e., vanillylmandelic acid from norepinephrine, homovanillic acid from dopamine) are excreted in the urine (Fig. 20-8).

20.4.6.4 Synthesis of Melanin.
Melanin is synthesized by specialized cells called *melanocytes*, located in the skin, hair roots, and iris and retina of the eye. Melanocytes contain tyrosinase, a copper-dependent tyrosine hydroxylase that converts tyrosine first to DOPA quinone and then to a family of bicyclic molecules called *indoles*. Subsequent oxidation and polymerization of the indoles results in the formation of melanins, whose multiple aromatic rings account for the pigmentation for the skin and hair. Synthesis of tyrosinase in melanocytes is induced by exposure to UV light.

20.4.6.5 Synthesis of Thyroid Hormones.
Tyrosine is also the precursor of the thyroid hormones T_4 ($3,5,3',5'$-tetraiodothyronine or thyroxine) and T_3 ($3,5,3'$-triiodothyronine). The molecular iodine (I_2) required in this reaction is synthesized by thyroid peroxidase, which catalyzes the reaction of iodide ions with hydrogen peroxide and subsequent incorporation of iodine atoms into tyrosine residues of thyroglobulin to form mono- and diiodotyrosine (Fig. 20-9). Thyroid peroxidase also catalyzes the coupling of two diiodinated or one iodinated plus one diiodinated tyrosine residue of thyroglobulin. Subsequent lysosomal hydrolysis of the thyroglobulin releases the active hormones thyroxine and triiodothyronine.

20.4.7   Metabolism of Tryptophan

Tryptophan is the precursor of the neurotransmitter serotonin (Fig. 20-10). Serotonin, in turn, is utilized by the pineal gland to synthesize melatonin, which regulates seasonal and circadian rhythms.

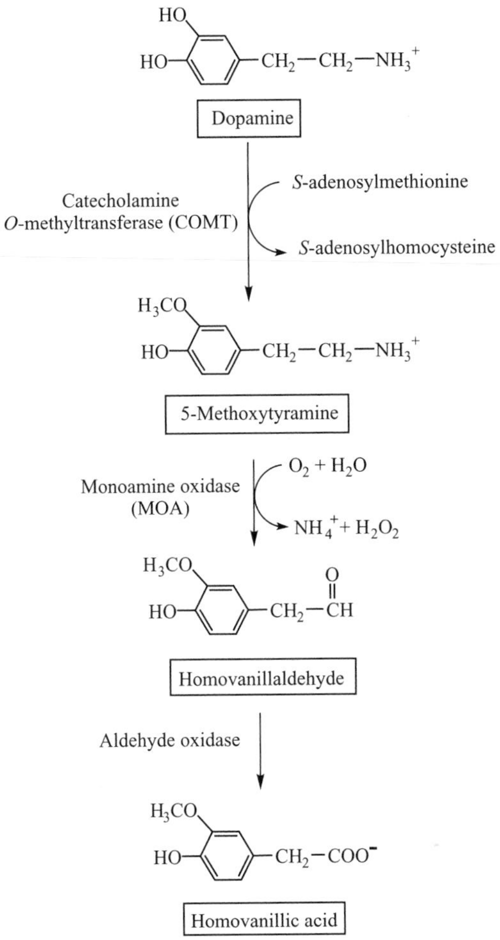

FIGURE 20-8   Inactivation of dopamine.

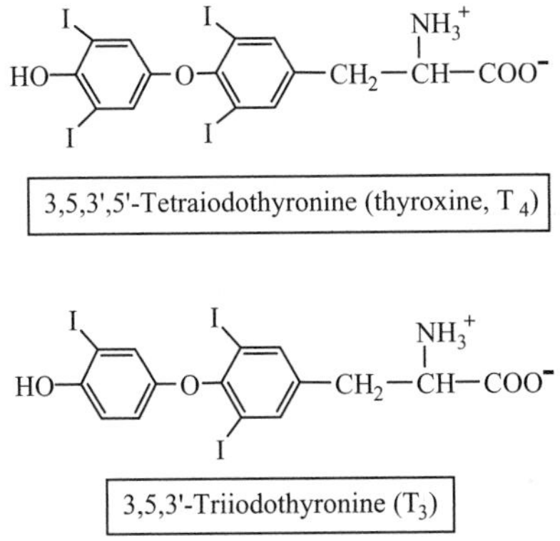

FIGURE 20-9   Structures of the thyroid hormones derived from tyrosine.

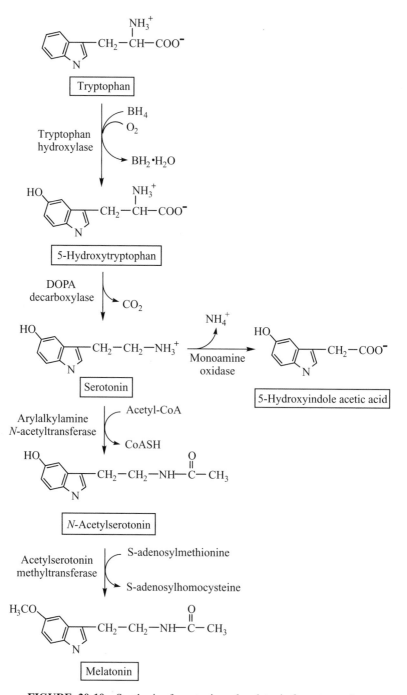

FIGURE 20-10 Synthesis of serotonin and melatonin from tryptophan.

20.4.7.1 Synthesis of Serotonin. The pathway for the synthesis of serotonin is similar to that which generates dopamine. The first step is catalyzed by tryptophan hydroxylase, which, like tyrosine hydroxylase, is a tetrahydrobiopterin (BH_4)-dependent enzyme:

$$\text{tryptophan} + O_2 + BH_4 \rightarrow \text{5-hydroxytryptophan} + BH_2 + H_2O$$

DOPA decarboxylase, the enzyme that catalyzes the decarboxylation of DOPA to produce dopamine, then decarboxylates 5-hydroxytryptophan:

$$\text{5-hydroxytryptophan} \rightarrow \text{serotonin} + CO_2$$

20.4.7.2 Inactivation of Serotonin. Monoamine oxidase, which inactivates catecholamines, also catalyzes the oxidative deamination of serotonin to produce 5-hydroxyindole acetic acid (Fig. 20-10). However, unlike the pathway that inactivates catecholamines, the hydroxyl group on the ring of serotonin is not methylated.

20.4.7.3 Synthesis of Melatonin. Melatonin is synthesized from serotonin by the following two-step reaction sequence (Fig. 20-10):

$$\text{serotonin} + \text{acetyl-CoA} \rightarrow N\text{-acetylserotonin} + \text{CoASH}$$

$$N\text{-acetylserotonin} + S\text{-adenosylmethionine} \rightarrow \text{melatonin} + S\text{-adenosylhomocysteine}$$

20.4.7.4 Synthesis of Niacin. One of the alternative pathways for tryptophan catabolism produces niacin, which is the precursor of the nicotinamide component of NAD^+ and $NADP^+$. Niacin synthesis, however, represents a minor pathway for the catabolism of tryptophan; only about 3% of the tryptophan that is metabolized actually follows the pathway to nicotinamide adenine nucleotide synthesis. Niacin is therefore an essential dietary requirement and is designated as a vitamin (B_3).

20.4.8 Decarboxylation of Other Amino Acids

We have seen that decarboxylation of tyrosine results in the synthesis of DOPA and other catecholamines (Fig. 20-7), while decarboxylation of 5-hydroxytryptophan generates serotonin (Fig. 20-10). Several other important bioactive amines are also synthesized by the pyridoxal phosphate (PLP)–dependent decarboxylation of amino acids. Among these are the neurotransmitter γ-aminobutyric acid (GABA) and histamine, which is synthesized by mast cells as part of the allergic response:

$$\text{glutamate} \rightarrow \gamma\text{-aminobutyric acid} + CO_2$$
$$\text{histidine} \rightarrow \text{histamine} + CO_2$$

20.5 ABNORMAL FUNCTIONING OF AMINO ACID METABOLISM

20.5.1 Maple Syrup Urine Disease

Maple syrup urine disease (MSUD, a.k.a. branched-chain α-ketoaciduria) results from a genetic deficiency of branched-chain α-ketoacid dehydrogenase. Thus, it is an inborn error in the metabolism that impairs the catabolism of the branched-chain amino acids leucine, isoleucine, and valine. Affected persons exhibit hyper-aminoacidemia and excrete excessive amounts of branched-chain α-ketoacids and related side products in their urine. As a result, the urine has a characteristic odor, which Western physicians find reminiscent of maple syrup: Mediterranean physicians report that the odor is similar to that of fenugreek seeds.

Maple syrup urine disease can result from mutations in any of the genes that encode the enzymes of the α-ketoacid dehydrogenase enzyme complex, with the severity of the disease being inversely related to the level of residual enzyme activity. The major clinical features of MSUD are mental and physical retardation. The classic, most severe form of MSUD results in lethargy, weight loss, and encephalopathy in infancy. Untreated, MSUD can result in seizures and coma followed by death. Treatment involves restricting dietary intake of the branched-chain amino acids to the minimum amount necessary to sustain growth.

20.5.1.1 Thiamine-Responsive MSUD. Some MSUD patients have mutant forms of α-ketoacid dehydrogenase that have a lower affinity for thiamine pyrophos-phate or are more subject to proteolysis when the cofactor is not bound to the enzyme. For these patients, the hyperaminoacidemia can usually be corrected with moderate levels of dietary protein restriction plus very large doses of thiamine hydrochloride.

20.5.1.2 MSUD with Lactic Acidosis. MSUD type III is the result of a de-ficiency in dihydrolipoamide dehydrogenase, which is the common E3 compo-nent of three mitochondrial thiamine-dependent enzyme complexes: branched-chain α-ketoacid dehydrogenase, pyruvate dehydrogenase, and α-ketoglutarate dehydro-genase. Because multiple dehydrogenase activities are affected, the clinical conse-quences are more severe than in classic MSUD.

20.5.2 Phenylketonuria

Phenylketonuria is another relatively common inborn error of amino acid metabolism (1/20,000 in the United States). People with PKU lack phenylalanine hydroxylase, which catalyzes the reaction

$$phenylalanine + O_2 + BH_4 \rightarrow tyrosine + BH_2 + H_2O$$

Lacking hepatic phenylalanine hydroxylase activity, plasma phenylalanine increases to the point where phenylalanine is metabolized by phenylalanine transaminase, with a resulting plasma accumulation and urinary excretion of phenylpyruvate, phenyl-acetate, and other aromatic metabolites of phenylalanine.

Phenylketonuria was one of the first genetic diseases for which neonatal screening was established. The initial Guthrie test employed β-2-thienylalanine, which inhibits the growth of the bacterium *Bacillus subtilis* on minimal culture medium; addition of phenylalanine restores growth. As a result, growth of a bacterial colony in response to application of a blood or urine sample is a positive indicator for PKU in a patient. The Guthrie test has now generally been replaced with tandem mass spectroscopy, which permits simultaneous screening for multiple inborn errors of metabolism.

Untreated, severe forms of PKU result in progressive and severe mental retarda-tion, with other neurological manifestations, including seizures, gait abnormalities, and unstable temperature regulation. Fortunately, PKU is one of a number of genetic diseases in which neonatal screening and rapid medical intervention, ideally within the first week of life, result in successful outcomes. Treatment involves severe restric-tion of dietary phenylalanine, which requires elimination of protein-rich foods and reliance on semisynthetic food substitutes. Patients must also avoid products con-taining the artificial sweetener aspartame (the methyl ester of aspartylphenylalanine) since hydrolysis of aspartame in the liver generates phenylalanine.

Phenylalanine is an essential amino acid and as such should not be completely eliminated from the diet. In the absence of adequate phenylalanine hydroxylase activity, tyrosine also becomes an essential amino acid. Indeed, the hypopigmentation observed in PKU patients is due to decreased availability of tyrosine for melanin synthesis as well as competitive inhibition of tyrosinase by the elevated levels of phenylalanine.

20.5.2.1 *Maternal PKU.*

Fetuses with genetic PKU do not develop neurological degeneration in utero because maternal metabolism of phenylalanine prevents exces-sive accumulation of the amino acid in fetal tissue. However, when a woman with PKU becomes pregnant, her fetus is at risk because, if maternal phenylalanine levels are not tightly controlled during pregnancy, the mother with PKU will expose her fetus to elevated levels of phenylalanine and its toxic metabolites during in utero development. Such infants are born with preexisting neurological damage, micro-cephaly, and congenital heart malformations. For this reason, females with PKU are strongly encouraged to maintain their phenylalanine-restricted diets into adulthood.

20.5.2.2 *Variant PKU.*

Tetrahydrobiopterin is the cofactor for phenylalanine hy-droxylase. Some people with PKU have a deficit in the ability to either synthesize or recycle the cofactor, and therefore also have impaired synthesis of catecholamines and serotonin. Administration of BH_4 is effective in treating the hyperphenylalane-mia of these patients; however, since BH_4 does not cross the blood–brain barrier, this therapy does not restore neurotransmitter synthesis in the CNS. These patients therefore require treatment with L-DOPA and 5-hydroytryptophan, which do cross the blood–brain barrier and are precursors for dopamine, epinephrine, and serotonin syn-thesis. Therapy of variant PKU also includes administration of carboxy-DOPA, an in-hibitor of dopamine decarboxylase which prevents excessive synthesis of epinephrine by the adrenals. Carboxy-DOPA does not cross the blood–brain barrier and thus does not inhibit neural synthesis of catecholamines.

20.5.2.3 BH₄-Responsive PKU. Some people with deficiencies in phenylalanine hydroxylase deficiency also respond to BH_4 therapy. It appears that the mutant enzyme in these individuals has either a reduced affinity for BH_4 or is stabilized by the presence of additional cofactor. For such persons, supplementation with BH_4 permits some relaxation of the stringent dietary exclusion of phenylalanine. A stereoisomer of BH_4 called Kuvan has recently become available for the treatment of BH_4-responsive PKU.

20.5.3 Pyridoxine-Dependent Epilepsy

People with pyridoxine-dependent epilepsy (EPD) exhibit seizures in the first hours of life and are unresponsive to standard anticonvulsant therapy. The genetic defect had long been thought to reside in the gene for glutamate decarboxylase, the enzyme that converts glutamate to the neurotransmitter γ-aminobutyric acid (GABA). However, recent studies indicate that patients with EPD are deficient in α-aminoadipic semialdehyde dehydrogenase, an intermediate in a minor (pipecolic acid) pathway of lysine catabolism. As a result, there is accumulation of a metabolic intermediate (piperidine-6-carboxylate) that reacts with and sequesters pyridoxal phosphate (PLP). Since PLP is an essential cofactor in neurotransmitter metabolism, those affected are dependent on relatively high doses of pyridoxine hydrochloride to prevent recurrence of seizures.

20.6 PARKINSON'S DISEASE

The characteristic tremors of the shaking palsy of Parkinson's disease are the result of gradual loss of dopamine-producing neurons in the substantia nigra. Dramatic symptomatic relief is obtained by administering DOPA, which is decarboxylated to dopamine by α-amino acid decarboxylase in the brain. Side effects such as gastrointestinal disturbances are prevented by addition of a peripheral inhibitor such as carboxy-DOPA.

20.7 ALBINISM

Albinism is a genetic disorder in which production of the photoprotective pigment melanin is impaired. One cause of the disease is a defect in the gene for tyrosinase, the enzyme that catalyzes several steps in the pathway of melanin synthesis, including the conversion of tyrosine to dopaquinone. Defects in several other proteins also cause albinism, but the role of each is not well understood. Affected persons have visual impairment and are at increased risk for skin cancer.

CHAPTER 21

SULFUR AMINO ACID METABOLISM

21.1 FUNCTIONS OF METHIONINE

Methionine is unique among the essential amino acids in that it is the only one that contains a sulfur atom (Fig. 21-1). Furthermore, in addition to its role as an amino acid constituent of proteins, different portions of the methionine molecule serve as precursors for a variety of other key molecules.

21.1.1 Methionine Provides Sulfur Atoms for Other Organic Molecules in the Body

Methionine provides the sulfur atom for the synthesis of cysteine, which is not an essential amino acid. Like methionine, cysteine is a constituent of proteins; in most proteins the sulfhydryl groups of particular cysteines form intra- and interpolypeptide disulfide bridges. Cysteine is also a component of glutathione (Fig. 21-1), which in its reduced form is a major component of the intracellular defense system that protects cells against oxidative damage from hydrogen peroxide and other reactive oxygen species.

Cysteine is the precursor of taurine, an unusual amino acid whose acidic nature derives from a sulfate rather than a carboxyl group. Although not a component of proteins, taurine plays many essential roles: It is one of two amino acids (the other being glycine) used to form conjugated bile salts in adults, and the one most often used to form conjugated bile salts in neonates (see Fig. 3-5). Taurine is also an abundant

Medical Biochemistry: Human Metabolism in Health and Disease   By Miriam D. Rosenthal and Robert H. Glew
Copyright © 2009 John Wiley & Sons, Inc.

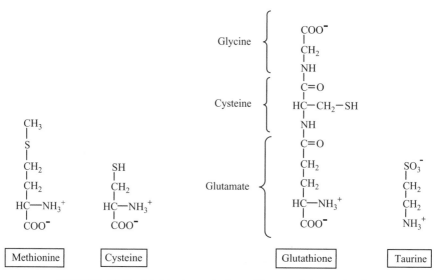

FIGURE 21-1   Methionine and other sulfur-containing molecules.

intracellular free amino acid which is involved in osmoregulation, particularly in the brain and neural retina.

Catabolism of cysteine produces free sulfate ions, which provide the sulfur moieties for the sulfation reactions required to synthesize proteoglycans and sulfated steroids. Many hormones, including catecholamines, steroids, and thyroid hormones, can be sulfated in vivo. Sulfation of both thyroxine (T_4) and triiodothyronine (T_3) inactivates these hormones. Sulfotransferases also play an important role in drug metabolism and detoxification of xenobiotics by rendering organic molecules more soluble and thus more readily excretable in the urine.

21.1.2   Methionine Donates Methyl Groups

S-Adenosylmethionine (SAM) (Fig. 21-2) is the major donor of methyl groups in a variety of biosynthetic pathways. These include biosynthesis of a number of small molecules, including epinephrine, creatine, melatonin, and phosphatidylcholine as well as methylation of proteins (see Chapter 20). In addition, SAM-dependent methylation of cytosine residues in DNA provides a mechanism for regulating gene expression, particularly during fetal development.

21.1.3   Methionine Donates Aminopropyl Groups

S-Adenosylmethionine can also donate three carbons plus an amino group to putrescine, generating spermidine and then spermine, which are two polyamines that stabilize DNA during cell growth and proliferation (Fig. 21-3).

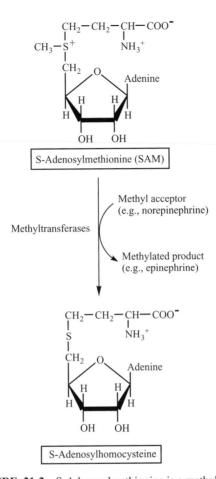

FIGURE 21-2   S-Adenosylmethionine is a methyl donor.

21.2 LOCALIZATION

Nearly all cells of the body use S-adenosylmethionine as a methyl donor. By contrast, the transsulfuration pathway, which utilizes the sulfur atom of methionine in the synthesis of cysteine, occurs primarily in the liver, kidney, and gastrointestinal tract.

21.3 METHIONINE METABOLISM

21.3.1 Activation of Methionine

S-Adenosylmethionine is synthesized from methionine plus ATP (Fig. 21-4). This ATP-dependent reaction is unusual in that all three of the phosphate atoms of

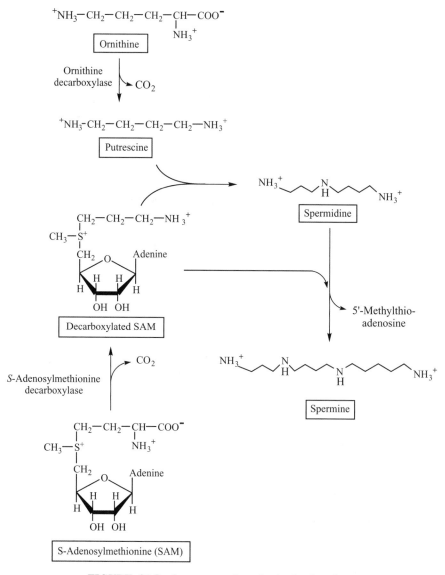

FIGURE 21-3   Structures and synthesis of polyamines.

ATP are cleaved from the ATP molecule:

$$\text{methionine} + \text{ATP} \rightarrow S\text{--adenosylmethionine} + PP_i + P_i$$

The reaction occurs in two steps: Methionine adenosyltransferase first forms SAM and a triphosphate and then the triphosphate is cleaved into $PP_i + P_i$.

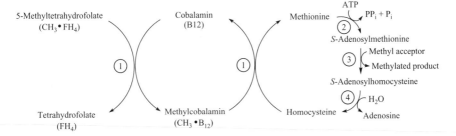

FIGURE 21-4   Regeneration of methionine from homocysteine: ① methionine synthase (methyltetrahydrofolate homocysteine methyltransferase); ② methionine adenosyltransferase; ③ SAM-dependent methyltransferase; ④ S-adenosylhomocysteine hydrolase.

21.3.2   Donation of Methyl Groups

SAM is the major methyl donor in a number of biosynthetic reactions, each of which is catalyzed by a specific SAM-dependent methyltransferase:

$$norepinephrine + S\text{-adenosylmethionine} \rightarrow epinephrine + S\text{-adenosylhomocysteine}$$

$$acetylserotonin\ S\text{-adenosylmethionine} \rightarrow melatonin + S\text{-adenosylhomocysteine}$$

$$guanidinoacetate + S\text{-adenosylmethionine} \rightarrow creatine + S\text{-adenosylhomocysteine}$$

Three successive methyl transfer steps are involved in the synthesis of phosphatidyl-choline from phosphatidylethanolamine (see Fig 14-10):

$$phosphatidylethanolamine + 3S\text{-adenosylmethionine} \rightarrow$$

$$phosphatidylcholine + 3S\text{-adenosylhomocysteine}$$

SAM-dependent methyltransferases also methylate both DNA and RNA, and catalyze the methylation of specific lysine, arginine, and glutamine residues in proteins. Subsequent proteolysis of trimethyllysine-containing proteins releases the trimethyllysine, which is a precursor for the synthesis of carnitine, the molecule that serves to transport long-chain fatty acids into the mitochondrion.

In all cases, the donation of a methyl group from SAM produces S-adenosylhomo-cysteine, which is subsequently hydrolyzed by S-adenosylhomocysteine hydrolase to release free homocysteine (Fig. 21-3). Homocysteine is a homolog of cysteine in that it has an additional $-CH_2-$ group in its carbon backbone; however, homocysteine is not utilized in protein synthesis. There are two pathways that metabolize homocysteine: remethylation and transsulfuration.

21.3.3   Remethylation of Homocysteine

Homocysteine is remethylated by methionine synthase (Fig. 21-4):

$$homocysteine + methylcobalamin \rightarrow methionine + cobalamin$$

The major source of the methyl group of methylcobalamin is 5-methyltetrahydrofolate ($FH_4 \cdot CH_3$), which is discussed in greater detail in Chapter 22. A minor pathway for remethylating homocysteine, which is present only in liver and kidney, utilizes betaine instead of 5-methyltetrahydrofolate as the methyl donor. Betaine (*N*-trimethylglycine) is not an essential nutrient since it can be synthesized in the liver from choline. Remethylation of homocysteine regenerates methionine, which can be activated at the expense of ATP to provide *S*-adenosylmethionine. The remethylation pathway thus completes a metabolic cycle in which *S*-adenosylmethionine plays a catalytic role in methylation reactions.

21.3.4   The Transsulfuration Pathway and Cysteine Synthesis

The other route for metabolizing homocysteine is the transsulfuration pathway, which transfers the sulfur atom of homocysteine to serine, generating cysteine in the process (Fig. 21-5). This pathway consists of two reactions, which form and then cleave cystathionine. The first reaction is catalyzed by cystathionine synthase and utilizes pyridoxal phosphate (PLP) as its cofactor:

$$\text{homocysteine} + \text{serine} \rightarrow \text{cystathionine} + H_2O$$

Cystathionase, another PLP-dependent enzyme, then catalyzes the hydrolytic deamination of cystathionine.

$$\text{cystathionine} + H_2O \rightarrow \text{cysteine} + \alpha-\text{ketobutyrate} + NH_4^+$$

Note in Figure 21-5 that the newly formed cysteine molecule contains a sulfur atom that is derived from homocysteine but none of the carbon atoms of homocysteine. α-Ketobutyrate, which is the product of the transsulfuration reaction, is oxidatively decarboxylated to form propionyl-CoA.

21.3.5   Synthesis of Taurine

Taurine is synthesized from cysteine (Fig. 21-6). The first step in the pathway oxidizes the sulfhydryl group of cysteine to generate cysteinosulfinate. The second step is a decarboxylation reaction that produces hypotaurine. The sulfite group of hypotaurine is then oxidized to taurine.

21.3.6   Release of Inorganic Sulfate from Sulfur-Containing Amino Acids

Catabolism of cysteine results in the release of the sulfur atom as SO_4^{2-}. The initial step is the generation of cysteinosulfinate, as occurs in taurine synthesis. Cysteine

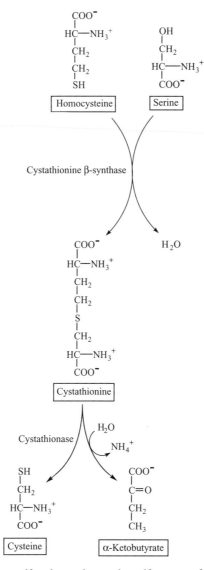

FIGURE 21-5 The transsulfuration pathway; the sulfur atom of homocysteine is used to synthesize cysteine.

aminotransferase then releases inorganic sulfite (Fig. 21-6):

$$\text{cysteinosulfinate} + \alpha\text{-ketoglutarate} \rightarrow HSO_3^- + \text{glutamate} + \text{pyruvate}$$

Sulfite oxidase then generates sulfate:

$$HSO_3^- + O_2 + H_2O \rightarrow SO_4^{2-} + H_2O_2 + H^+$$

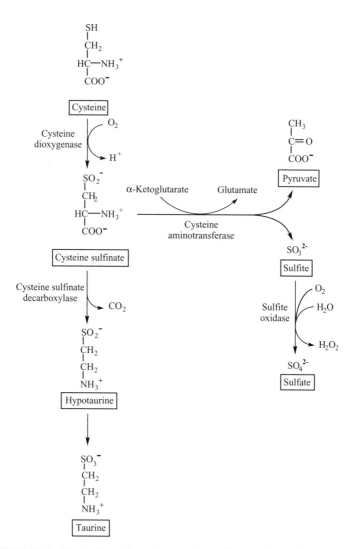

FIGURE 21-6   Metabolism of cysteine produces either taurine or inorganic sulfate.

21.3.7 Activation of Inorganic Sulfate

Inorganic sulfate can be activated at the expense of ATP to form $3'$-phosphoadenosine $5'$-phosphosulfate (PAPS), which serves as the sulfur donor for sulfation reactions. PAPS synthesis involves two successive ATP-dependent reactions that release PP_i and P_i, respectively:

$$SO_4^{2-} + ATP \rightarrow \text{adenosine phosphosulfate} + PP_i$$

$$\text{adenosine phosphosulfate} + ATP \rightarrow \text{phosphoadenosine phosphosulfate (PAPS)}$$
$$+ ADP$$

21.3.8 Synthesis of Polyamines

Polyamines have multiple positive charges that stabilize DNA during cell division and are therefore essential for cell survival (Fig. 21-3). Putrescine, the simplest of the polyamines, is produced by decarboxylation of ornithine. The larger, more positively charged polyamines, spermidine and spermine, are synthesized by means of the transfer of aminopropyl groups to putrescine. In this pathway, SAM is first decarboxylated to decarboxylated-SAM (S-adenosylmethylthiopropylamine). Transfer of an aminopropyl group from decarboxylated SAM to putrescine generates spermidine; transfer of a second aminopropyl group to spermidine generates spermine.

21.4 REGULATION OF SULFUR AMINO ACID METABOLISM

As described above, homocysteine is generated when S-adenosylmethionine serves as a methyl donor. There are two competing pathways that metabolize homocysteine: remethylation to regenerate methionine, and transsulfuration, which synthesizes cysteine and leads to the catabolism of the carbon skeleton of the homocysteine. Flux through these two completing pathways is determined by the concentration of S-adenosylmethionine. Since SAM is an activator of cystathionine β-synthase (Fig. 21-5), high concentrations of SAM tend to favor the transsulfuration pathway. In addition, SAM is an allosteric inhibitor of methylenetetrahydrofolate reductase, which synthesizes 5-methyltetrahydrofolate (N^5-methyl-F_4). N^5-methyl-F_4 donates its methyl group to cobalamin to form methylcobalamin, which in turn serves as the methyl donor for the remethylation of homocysteine (Fig. 21-4). Thus, high concentrations of SAM will inhibit the regeneration of methionine from homocysteine. The transsulfuration pathway is also regulated by end-product inhibition, whereby cysteine allosterically inhibits cystathionine β-synthase.

21.5 ABNORMAL FUNCTION

21.5.1 Deficiency of Cystathionine β-Synthase

Homocystinuria is an autosomal recessive genetic disease that results from a deficiency of cystathionine β-synthase (CBS). Patients with homocystinuria have marked elevations of both methionine and total homocysteine (tHcy = homocysteine + homocystine) in their blood. The clinical features of CBS deficiency include skeletal deformities, abnormalities of the ocular lens, and mental retardation. When untreated, patients with CBS deficiency also have a 50% chance of developing a myocardial infarction, stroke, or serious blood clot before the age of 30 years.

CBS is a pyridoxal phosphate–dependent enzyme, and many patients with CBS deficiency respond to pharmacological doses of pyridoxine (vitamin B_6). Other therapies have included restriction of dietary methionine, and supplementation with folate and vitamin B_{12} to enhance activity of the remethylation pathway. Patients with homocystinuria may also benefit from treatment with betaine, which is the alternate methyl donor for methyl cobalamin synthesis.

21.5.2 Hyperhomocysteinemia

Mild or moderate hyperhomocysteinemia (plasma tHcy concentration 15 to 100 μmol/L; normal < 12 μmol/L) is a multifactorial condition associated with a variety of different nutritional and genetic factors, impaired kidney function, or excessive alcohol intake. Vitamins required for homocysteine metabolism include pyridoxine, folate, and vitamin B_{12}. Genetic causes of hyperhomocysteinemia include polymorphisms in any of the enzymes involved in either the transsulfuration pathway or the remethylation pathway and related enzymes of folate metabolism [e.g., methylenetetrahydrofolate reductase (MTHFR)]. Regardless of etiology, even mild hyperhomocysteinemia confers an increased risk of adverse cardiovascular events. Although the mechanism by which homocysteine causes endothelial cell dysfunction is not fully understood, it is thought that homocysteine, which is a potent oxidizing agent, inactivates the vasoprotective agent nitric oxide. Alternatively, the impairment of sulfur amino acid metabolism and consequent decreased intracellular SAM/SAH ratio may result in hypomethylation of DNA and increased expression of particular genes.

21.5.3 Cirrhosis of the Liver

One of the features of cirrhosis of the liver is impaired synthesis of S-adenosylmethionine due to abnormally low activity of methionine adenosyltransferase. The enzyme is inactivated by a variety of reactive oxygen species which overwhelm the normal enzyme reactivation by reduced glutathione (GSH). As a result, intrahepatic concentrations of SAM are reduced under conditions of chronic oxidative stress such as those induced by alcohol or hepatitis C. The lack of SAM, in turn, exacerbates the liver injury. Pharmacological treatment with stable salts of SAM appears to be beneficial, especially for patients with less advanced liver disease.

CHAPTER 22

FOLATE AND VITAMIN B$_{12}$ IN ONE-CARBON METABOLISM

22.1 FUNCTIONS OF ONE-CARBON METABOLISM

Many reactions in human metabolism involve the transfer of an activated one-carbon group from a donor molecule to an acceptor molecule. Some of these reactions function in catabolic pathways, for example in the breakdown of serine and histidine, whereas others occur in anabolic processes, such as in the pathway of purine synthesis or the conversion of deoxyuridine monophosphate (dUMP) to deoxythymidine monophosphate (dTMP).

One-carbon units can exist in various oxidation states. As discussed previously, biotin is the cofactor that carboxylases use to transfer CO_2, the most oxidized of the one-carbon units, to substrates such as pyruvate (to produce oxaloacetate). At the most reduced end of the oxidation–reduction spectrum, S-adenosylmethionine is the main donor of activated methyl groups for biosynthetic reactions. The body uses tetrahydrofolate (FH$_4$), the most reduced form of the vitamin folic acid, as the carrier of one-carbon groups in all the intermediate oxidation states of carbon and as the substrate for the oxidation or reduction of these one-carbon groups (Table 22-1). In addition, 5-methyl-FH$_4$ provides a source of methyl groups for the synthesis of methionine from homocysteine, thus providing a mechanism for replenishing the pool of methyl groups for methylation reactions that require S-adenosylmethionine. Transfer of a methyl group from methyl-FH$_4$ to homocysteine represents one of the two cofactor roles that vitamin B$_{12}$ plays in humans; the other is as the cofactor in the reaction that converts methylmalonyl-CoA to succinyl-CoA.

Medical Biochemistry: Human Metabolism in Health and Disease   By Miriam D. Rosenthal and Robert H. Glew
Copyright © 2009 John Wiley & Sons, Inc.

TABLE 22-1   One-Carbon Group Carried by Tetrahydrofolate

Oxidation Level	Group		Cofactor
Formic acid	Formyl	$\overset{\displaystyle O}{\overset{\displaystyle \|}{-CH-}}$	N^5-Formyl-FH$_4$, N^{10}-formyl-FH$_4$
	Methenyl	$-CH=$	N^5, N^{10}-Methenyl-FH$_4$
	Formimino	$-CH=NH$	N^5-Formimino-FH$_4$
Formaldehyde	Methylene	$-CH_2-$	N^5, N^{10}-Methylene-FH$_4$
Methanol	Methyl	$-CH_3$	N^5-Methyl-FH$_4$

FH$_4$, tetrahydrofolate.

22.2   LOCALIZATION OF REACTIONS INVOLVING ONE-CARBON TRANSFER

Although reactions involving one-carbon transfer occur in essentially all cells, they are especially prominent in the liver which is the major site of purine synthesis. Relatively high levels of enzymes that use tetrahydrofolate are also found in the brain, where one-carbon groups are used to maintain the pool of S-adenosylmethionine for the methylation reactions involved in both catecholamine synthesis and inactivation as well as to synthesize tetrahydrobiopterin, the cofactor for hydroxylation reactions of catecholamine and serotonin synthesis.

22.3   PHYSIOLOGICAL CONDITIONS IN WHICH ONE-CARBON METABOLISM IS ESPECIALLY ACTIVE

Because of the central role one-carbon metabolism plays in the synthesis of purines that are components of RNA and DNA and in the generation of thymidylate for DNA synthesis, it is most active during periods of rapid cellular growth, including embryogenesis and early postnatal development, and in rapidly dividing cells such as the intestinal epithelium and stem cells of both the erythropoietic and immune cell lineages.

22.4   REACTIONS OF ONE-CARBON METABOLISM

22.4.1   Absorption of Folate and Its Conversion to the Active Cofactor Form

Folic acid consists of three components: a pteridine ring, *para*-aminobenzoic acid (PABA), and glutamate (Fig. 22-1). Although humans can synthesize all of these components of the vitamin, they lack the enzyme required to join PABA to the pteridine ring. Natural forms of folic acid have a polyglutamate tail that contains an additional one to five glutamates joined in γ-peptide linkages. All but one of these glutamate residues is hydrolyzed sequentially in the small intestine by mucosal folate conjugase. Inside the enterocyte, folic acid is reduced to FH$_4$. The pathway

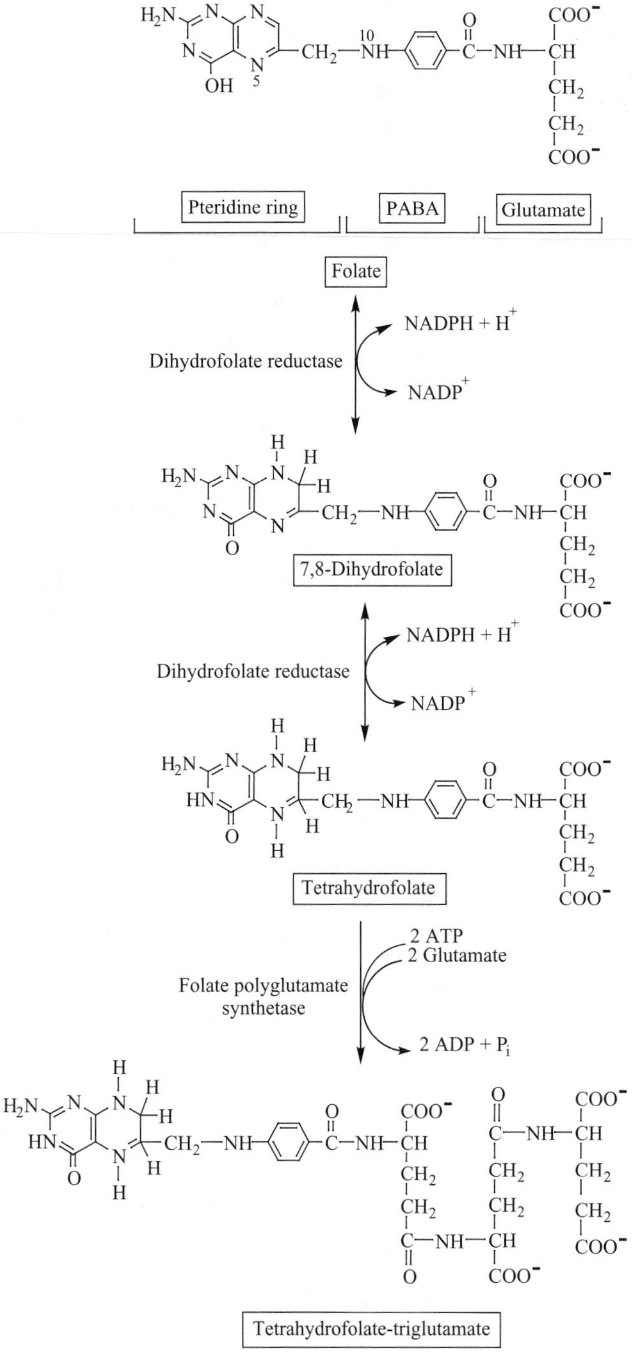

FIGURE 22-1 Structure of folate and its conversion to FH$_4$, the active form, and FH$_4$ polyglutamate, the form in which it is stored. PABA, *para*-aminobenzoic acid.

involves two successive reductions, both of which are catalyzed by dihydrofolate reductase, and both of which use NADPH as the reductant or hydrogen donor. After absorption from the intestine and transport of folate to the liver and other cells, the polyglutamate chain is restored by polyglutamate synthetase, trapping the active form of the cofactor within the cell. Subsequent release of the vitamin from hepatic stores into the blood requires hydrolysis of these additional glutamate residues by folate conjugase.

22.4.2   Oxidation States of the Folate One-Carbon Pool

FH_4 can carry one-carbon groups in several different oxidation states, including methyl ($-CH_3$), methylene ($-CH_2-$), and formyl ($-CHO$). Thus, there are several activated one-carbon carrying forms of FH_4, where the one-carbon group is covalently attached either to the N^5 or the N^{10} atom of FH_4, or forms a bridge between the N^5 and N^{10} atoms. Note that the reduction of N^5,N^{10}-methylene-FH_4 is irreversible:

$$N^5,N^{10}\text{-methylene-}FH_4 + NADPH + H^+ \rightarrow N^5\text{-methyl-}FH_4 + NADP^+$$

N^5-Methyl-FH_4 is the donor of methyl groups to vitamin B_{12}. Whenever transfer of a methyl group to vitamin B_{12} is blocked, FH_4 accumulates in the N^5-methyl-FH_4

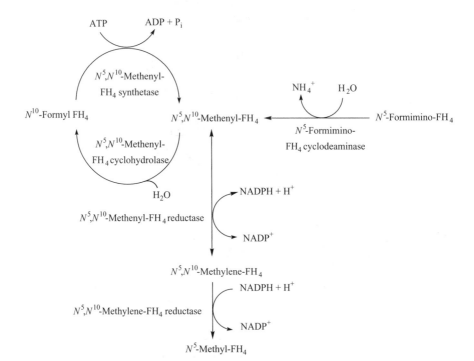

FIGURE 22-2   Interconversions of the one-carbon group attached to FH_4. The structures of the one-carbon units are shown in Table 22-1.

form and cannot be utilized for reactions requiring donation of one-carbon groups at other oxidation states. This phenomenon, referred to as the *methyltetrahydrofolate trap*, results in the depletion of the pool of FH_4 available for use in other pathways, such as purine synthesis.

22.4.3   Donors of One Carbon Groups to FH_4

22.4.3.1   *Serine.* The β-carbon atom of serine, which bears the hydroxyl group, is the major carbon source for the tetrahydrofolate one-carbon pool. Humans can synthesize serine from 3-phosphoglycerate, an intermediate in glycolysis, thereby providing a pathway for the utilization of carbohydrate-derived carbon atoms for biosynthetic reactions requiring one-carbon fragments (Fig. 22-3). Serine synthesis

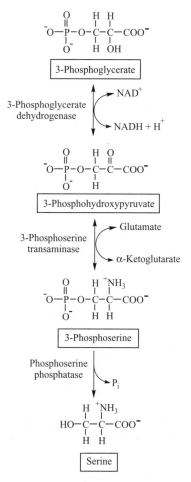

FIGURE 22-3   Synthesis of serine from the glycolytic intermediate 3-phosphoglycerate.

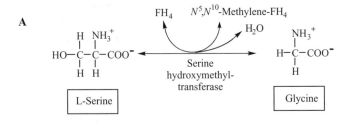

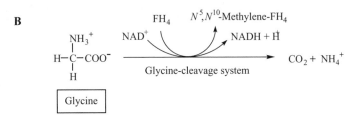

FIGURE 22-4   Reactions by which serine and glycine donate one-carbon units to tetrahydrofolate: (A) serine hydroxymethyltransferase; (B) the glycine-cleavage system.

involves the NAD^+-dependent reduction of 3-phosphoglycerate, transamination of 3-phosphohydroxypyruvate with glutamate serving as the amino group donor, and finally, hydrolysis of the phosphate from 3-phosphoserine.

The transfer of the carbon 3 of serine to tetrahydrofolate is catalyzed by serine hydroxymethyltransferase, which utilizes pyridoxal phosphate as a cofactor (Fig. 22-4A):

$$serine + FH_4 \rightleftharpoons N^5, N^{10}\text{-methylene-FH}_4 + glycine$$

22.4.3.2 *Glycine.*
Glycine can donate a one-carbon unit to FH$_4$ in a reaction catalyzed by the glycine-cleavage system (Fig. 22-4B):

$$glycine + FH_4 + NAD^+ \rightleftharpoons N^5, N^{10}\text{-methylene-FH}_4 + CO_2 + NH_4^+ + NADH + H^+$$

The net effect of the two reactions shown in Figure 22-4 is to use two carbon atoms of serine to synthesize N^5, N^{10}-methylene-FH$_4$. Glycine can also be transaminated to form glyoxylic acid, which is then oxidized to oxalic acid (see Fig. 13-2). Excessive production of oxalic acid leads to the deposition of oxalate stones in the kidney.

22.4.3.3 *Formate.*
Formic acid (HCOOH) is generated during the catabolism of tryptophan. It is also produced from ingested methanol by alcohol dehydrogenase

and aldehyde dehydrogenase. Formate can be incorporated into the one-carbon pool by formyltetrahydrofolate ligase:

$$\text{formate} + FH_4 + ATP \rightarrow N^{10}\text{-formyl-}FH_4 + ADP + P_i$$

22.4.3.4 Histidine. The pathway for the catabolism of histidine to N-formiminoglutamate (FIGLU) is shown in Figure 22-5. The formimino group of FIGLU is transferred to FH_4 by glutamate formiminotransferase:

$$N\text{-formiminoglutamate} + FH_4 \rightarrow N^5\text{-formimino-}FH_4 + \text{glutamate}$$

N^5, N^{10}-methenyl-FH_4 is then generated by deamination of N^5-formimino-FH_4 (Fig. 22-2):

$$N^5\text{-formimino-}FH_4 \rightarrow N^5, N^{10}\text{-methenyl-}FH_4 + NH_4^+$$

The complete pathway of histidine catabolism to N-formiminoglutamate is confined to the liver. Histidase, which catalyzes the first step of the pathway, is also present in skin, and urocanate (Fig. 22-5), the product of this reaction, is present in sweat.

22.4.4 Synthetic Reactions Utilizing One-Carbon Groups from Tetrahydrofolate Derivatives

22.4.4.1 Purines. A major role of FH_4 is to provide one-carbon units for synthesis of the ring structures of adenosine and guanine, the purine bases that are constituents of DNA and RNA. Synthesis of inosine monophosphate, the precursor of both AMP and GMP, involves donation of two separate formyl groups from N^{10}-formyl-FH_4 (see Chapter 23).

22.4.4.2 Deoxythymidylate (dTMP). The principal pyrimidine bases of RNA are uracil and cytosine. In DNA, however, the pyrimidine base thymine replaces uracil. Synthesis of the thymine base of deoxythymidylate is catalyzed by thymidylate synthase in a reaction that transfers a methyl group from N^5, N^{10}-methylene-FH_4 to deoxyuridylate (dUMP) (Fig. 22-6):

$$\text{dUMP} + N^5, N^{10}\text{-methylene-}FH_4 \rightarrow \text{dTMP} + FH_2$$

Note that in this reaction the methylene group is reduced to a methyl group and the FH_4 moiety is oxidized to FH_2. Dihydrofolate reductase regenerates FH_4:

$$FH_2 + NADPH + H^+ \rightarrow FH_4 + NADP^+$$

22.4.4.3 Serine. As indicated above, serine can donate a carbon atom to the one-carbon tetrahydrofolate pool, generating glycine (Fig. 22-4A). Since this reaction

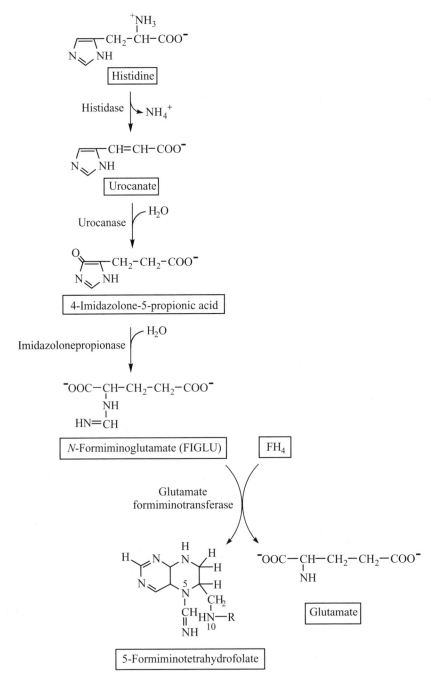

FIGURE 22-5 Metabolism of histidine results in the transfer of a formimino group to tetrahydrofolate. R, PABA-glutamate.

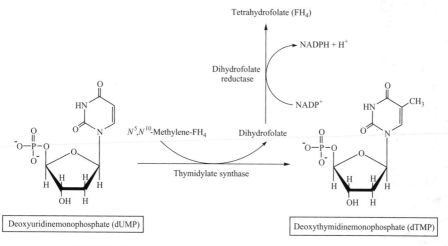

FIGURE 22-6 Synthesis of deoxythymidine monophosphate.

is reversible, when the level of N^5, N^{10}-methylene-FH_4 becomes high, it can also generate serine from glycine.

22.4.4.4 Methionine. Regeneration of methionine from homocysteine by methionine synthase (Fig. 22-7) requires methyl-B_{12} as a cofactor:

$$\text{homocysteine} + \text{methyl-}B_{12} \rightarrow \text{methionine} + \text{vitamin } B_{12}$$

Vitamin B_{12}, in turn, receives its methyl group from N^5-methyl-FH_4:

$$N^5\text{-methyltetrahydrofolate} + \text{vitamin } B_{12} \rightarrow \text{tetrahydrofolate} + \text{methyl-}B_{12}$$

This is the only reaction in the body that uses N^5-methyl-FH_4. Since the reduction of N^5, N^{10}-methylene-FH_4 to N^5-methyl-FH_4 is irreversible, lack of vitamin B_{12} to regenerate free tetrahydrofolate will "trap" the tetrahydrofolate as N^5-methyl-FH_4, thereby limiting the availability of tetrahydrofolate for other biosynthetic reactions.

22.4.5 Absorption of Vitamin B$_{12}$

Vitamin B_{12} (Fig. 22-8) is composed of a planar corin ring that is similar to the porphyrin ring structure of heme but contains cobalt instead of iron. In its active forms, the cobalt atom of vitamin B_{12} binds either a methyl or a deoxyadenosyl group. Most commercial vitamin B_{12} supplements have cyanide bound to the cobalt; they are rapidly converted in the body first to hydroxy-B_{12} and then ultimately to the active deoxyadenosyl-B_{12} or methyl-B_{12} coenzyme forms.

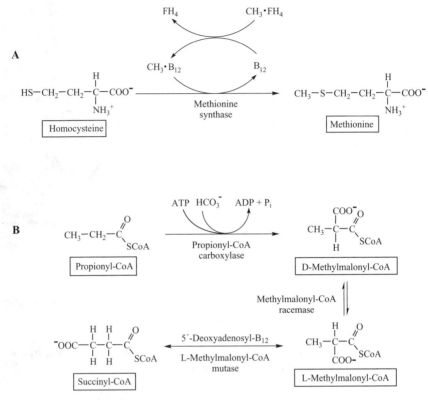

FIGURE 22-7 Two reactions catalyzed by vitamin B$_{12}$.

Utilization of dietary vitamin B$_{12}$ is dependent on both gastric HCl and two specialized proteins, R proteins and intrinsic factor. Dietary vitamin B$_{12}$ is covalently bound to polypeptides; release of vitamin B$_{12}$ normally occurs in the stomach through the combined hydrolytic actions of HCl and pepsin. *R proteins* (also designated *haptocorrins* or *cobalophilins*) are present in both saliva and gastric juice. They bind the vitamin B$_{12}$ prior to its release from the polypeptides, and remain associated with vitamin B$_{12}$ until the R proteins are hydrolyzed in the small intestine.

Intrinsic factor (IF), a glycoprotein produced by the parietal cells of the stomach, is essential for the absorption of vitamin B$_{12}$. Intrinsic factor is so named because early studies demonstrated that both a dietary (extrinsic) factor and a protein produced by the normal stomach (intrinsic) were necessary for the prevention of pernicious anemia. As soon as vitamin B$_{12}$ is released from the R proteins, it binds to intrinsic factor. The IF-B$_{12}$ complex is then recognized by specific receptors, called *cubulins*, located primarily in the distal ileum. The major protein that transports vitamin B$_{12}$ from the intestine to other tissues is transcobalamin II, which is a member of the R-protein family.

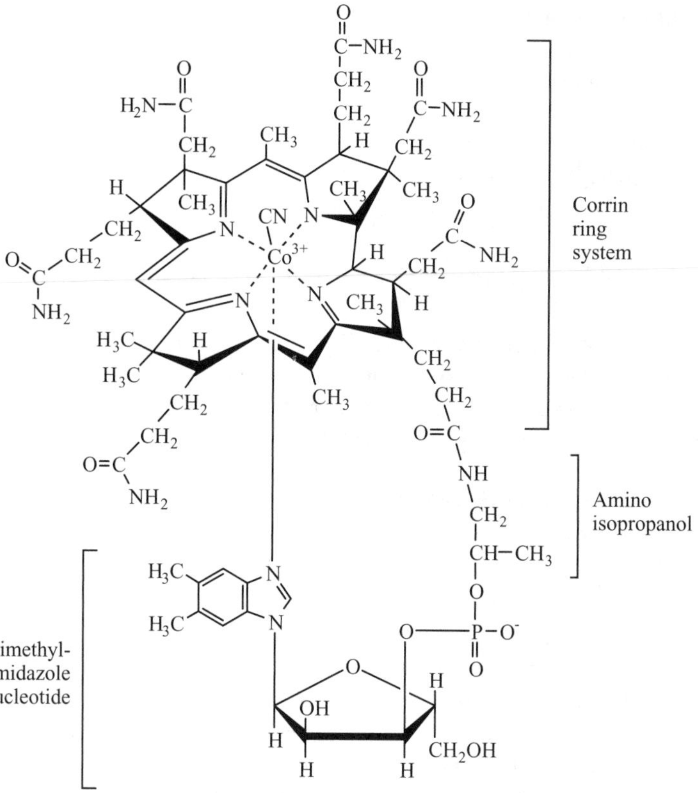

FIGURE 22-8   Structure of dietary vitamin B_{12}.

22.4.6   Reactions Utilizing Vitamin B_{12} as a Cofactor

Vitamin B_{12} is the precursor of two different cofactor forms that are involved in two very different metabolic reactions (Fig. 22-7). One, catalyzed by methionine synthase (a.k.a. homocysteine methyltransferase), uses methyl-B_{12} as a one-carbon donor. The other reaction is catalyzed by methylmalonyl-CoA mutase, which utilizes 5′-deoxyadenosyl-B_{12} as a cofactor and involves the transfer of a one-carbon unit within the molecule rather than between reactants.

22.4.6.1   *Resynthesis of Methionine from Homocysteine.*   As described above, vitamin B_{12} is the cofactor through which the methyl group of N^5-methyl-FH_4 is transferred to homocysteine to regenerate methionine. Vitamin B_{12} is active in its reduced form, Cob(I). Over time a fraction of the Cob(I) oxidizes spontaneously to Cob(II). The catalytic activity of methionine synthase is regenerated by methionine synthase reductase, which uses S-adenosylmethionine to catalyze the reductive

methylation of vitamin B_{12}:

$$Cob(II) + S\text{-adenosylmethionine} \rightarrow methyl\text{-}Cob(I) + S\text{-adenosylhomocysteine}$$

In most cells, N^5-methyl-FH_4 is the only methyl donor for the synthesis of methyl-B_{12}. Liver cells, however, have a second form of homocysteine methyltransferase which can use betaine (trimethylglycine) as the methyl donor:

$$betaine + homocysteine \rightarrow methionine + dimethylglycine$$

Betaine is derived from choline when the latter is oxidized by choline oxidase. The betaine pathway is normally of minor importance in humans because we have a low level of choline oxidase. Betaine supplements, however, are effective in promoting methionine regeneration in patients with hyperhomocysteinemia.

22.4.6.2 *Methylmalonyl-CoA Mutase.* Deoxyadenosyl-B_{12} is a cofactor for methylmalonyl-CoA mutase, which catalyzes the reaction (Fig. 22-7B)

$$methylmalonyl\text{-}CoA \rightleftharpoons succinyl\text{-}CoA$$

This reaction is a key component of the pathway by which the carbon skeleton of propionyl-CoA is metabolized. Propionyl-CoA is generated when odd-chain fatty acids are oxidized and during the catabolism of the carbon skeletons of valine, isoleucine, and cysteine. The pathway by which propionyl-CoA is converted to succinyl-CoA is shown in Figure 22-7B. Subsequent metabolism of succinyl-CoA by means of the TCA cycle generates oxaloacetate, which can be used to synthesize glucose.

22.5   REGULATION OF ONE-CARBON METABOLISM

22.5.1   Regulation of N^5,N^{10}-Methylenetetrahydrofolate Reductase

As described in Chapter 21, S-adenosylmethionine (SAM) is an allosteric inhibitor of the reduction of N^5,N^{10}-methylene-FH_4 to N^5-methyl-FH_4. High concentrations of SAM inhibit the irreversible conversion of the folate one-carbon pool to a form that can only be used to regenerate methionine.

22.5.2   Serine Hydroxymethyltransferase

Because of the reversibility of the serine hydroxymethyltransferase (SHMT) reaction, it can either generate or consume N^5,N^{10}-methylene-FH_4. There are two SHMT isozymes, one cytosolic and the other mitochondrial. The cytosolic isozyme preferentially supplies one-carbon units for thymidylate synthesis. The cytosolic isozyme also binds 5-methyl-FH_4, thereby limiting the activity of methionine synthase. Thus, the level of expression of the cytosolic SHMT gene appears to modulate competition between nucleotide synthesis and S-adenosylmethionine synthesis for the one-carbon units carried on FH_4.

22.6 ABNORMAL FUNCTIONING OF THE PATHWAYS OF ONE-CARBON METABOLISM

22.6.1 Folate Deficiency

A dietary deficiency of folate impairs one-carbon metabolism and preferentially impacts rapidly dividing cells, including the stem cells that generate erythrocytes, enterocytes, and cells of the immune system. The typical clinical presentation of folate deficiency is megaloblastic anemia. Lack of adequate nucleic acid synthesis results in decreased red cell number and release into the circulation of normochromic red blood cells, which are larger than normal (megaloblasts) due to impaired cell division. Persons with folate deficiency also often have decreased white cell counts. One measure of folate insufficiency is increased urinary excretion of N-formiminoglutamate (FIGLU) because of the absence of sufficient FH_4 for removal of the formimino group of FIGLU and formation of glutamine.

Inadequate folate levels during the first weeks of pregnancy increase the risk for congenital neural tube defects such as spina bifida. The fortification of grains with folate in both the United States and Canada is credited with a 15 to 30% decrease in the incidence of neural tube defects in recent years. Women who are able to become pregnant are advised to obtain 400 µg of folic acid daily from fortified foods, supplements, or both; the synthetic form of folate is preferred because it is more bioavailable than folate in foods.

Epidemiological studies indicate that lack of adequate dietary folate is also associated with increased incidence of other pathological conditions, including cardiovascular disease and colon cancer. As discussed in Chapter 21, hyperhomocysteinemia is a risk factor for cardiovascular disease. Lack of sufficient folate is one of the factors that can contribute to hyperhomocysteinemia; a deficiency of vitamin B_{12} (cobalamin) or vitamin B_6 (pyridoxine) can also result in hyperhomocysteinemia due to impaired remethylation of homocysteine or transsulfuration of homocysteine to generate cysteine, respectively. The increased cancer risk associated with folate deficiency appears to be the result of decreased methylation of DNA and to increased DNA damage secondary to incorporation of deoxyuridylate into DNA in the absence of an adequate supply of thymidylate for DNA synthesis.

22.6.2 Genetic Polymorphisms and Increased Risk of Neural Tube Defects

Mutations affecting many of the genes related to one-carbon metabolism have been described. For example, there are rare cases of methylenetetrahydrofolate reductase (MTHFR) deficiency which are characterized by severe hyperhomocysteinemia and homocystinuria. There are also more common genetic polymorphisms of MTHFR that are associated with only partial loss of enzymatic activity. The most prevalent of these is C677T, where there is an alanine-to-valine substitution in MTHFR, resulting in a thermolabile enzyme. The 10 to 20% of the population in the United States who are homozygous for the 677TT genotype have approximately 33% of normal enzyme activity and exhibit only mild to moderate hyperhomocysteinemia. This particular polymorphism is associated with an increased risk of neural tube defects, which can be prevented by supplementation with larger amounts of folate than is required by people with the CC or CT genotype.

22.6.3 Use of Folate Analogs for Chemotherapy

Rapidly dividing cells require high levels of DNA and RNA synthesis. For this reason, drugs that reduce the availability of FH$_4$ for one-carbon metabolism have proven useful chemotherapeutic agents for treating cancers. Methotrexate and aminopterin are two structurally similar analogs of FH$_2$ that are potent competitive inhibitors of dihydrofolate reductase. Dihydrofolate reductase inhibitors also impair the functioning of normal, rapidly dividing cells such as the stem cells in bone marrow, hair follicles, and enterocytes, thus producing a variety of adverse side effects.

22.6.4 Pernicious Anemia

Pernicious anemia is a severe megaloblastic anemia that results from inadequate tissue levels of vitamin B$_{12}$. The underlying problem is a lack of intrinsic factor production by the stomach. Pernicious anemia is due principally to an autoimmune gastritis in which the blood contains antibodies against intrinsic factor and other proteins of the parietal cells. These antibodies damage the patient's mucosa and abolish the secretion of both intrinsic factor and HCl.

The hematological presentation of pernicious anemia is indistinguishable from that of folic acid deficiency. Vitamin B$_{12}$ deficiency results in an inability to transfer the methyl group from 5-methyl-FH$_4$ to homocysteine to form methionine. Since the methyl group of 5-methyl-FH$_4$ cannot be oxidized to 5′,10′-methylene-FH$_4$ or other one-carbon folate derivatives, the FH$_4$ pool becomes "trapped" as 5-methyl-FH$_4$ in vitamin B$_{12}$–deficient persons. This, in turn, diminishes the availability of FH$_4$ for nucleotide synthesis, resulting in megaloblastic anemia.

With time, pernicious anemia can result in progressive neurological degeneration, which in its later stages may present with tingling or numbness of the extremities and diminished reflexes. Neurological damage may be due to insufficient 5′-deoxyadenosyl-B$_{12}$ to support the methylmalonyl-CoA mutase reaction, resulting in a buildup of methylmalonic acid. When present at elevated concentrations, methylmalonyl-CoA can substitute for malonyl-CoA in fatty acid synthesis, leading to the synthesis of branched-chain fatty acids, which are incorporated into phospholipids of the myelin sheath. Alternatively, the neurological symptoms may be due to inadequate amounts of *S*-adenosylmethionine in neural tissues.

Urinary excretion of methylmalonic acid and propionic acid provides a laboratory means for distinguishing vitamin B_{12} deficiency from folic acid deficiency. Vitamin B_{12} deficiency can also be detected by abnormally low serum vitamin B_{12} levels or by elevated blood levels of methylmalonic acid. For many years the Schilling test was used to distinguish dietary vitamin B_{12} deficiency from intrinsic factor deficiency. In this test, the patient is given an oral dose of radioactive cobalt-labeled vitamin B_{12}, followed by an intramuscular injection of nonradiolabeled vitamin B_{12}. The injected vitamin B_{12} saturates the vitamin B_{12}–binding sites in tissues and promotes urinary excretion of absorbed radioactive vitamin B_{12}; excretion of 10% or more of the labeled vitamin B_{12} in 24 hours indicates normal absorption, whereas less than 5% excretion indicates malabsorption.

Since the underlying deficit in pernicious anemia is malabsorption of vitamin B_{12}, treatment involves intramuscular vitamin B_{12} injections rather than oral vitamin therapy. The hematological symptoms of pernicious anemia can also be ameliorated with folate supplements, which replenish the FH_4 pool. Therapeutic doses of folate are, however, ineffective in preventing the ongoing and eventually irreversible neurological degeneration. Thus, folic acid supplements may mask the underlying vitamin B_{12} deficiency and, in some cases, even exacerbate it by promoting utilization of vitamin B_{12} by cells in the bone marrow rather than in neural tissues.

22.6.5   Other Causes of Vitamin B_{12} Deficiency

The term *pernicious anemia* is usually reserved for the pathology that results from a lack of gastric intrinsic factor. There are, however, other conditions that can result in poor vitamin B_{12} status, as described below.

22.6.5.1   *Dietary Deficiency of B_{12}.*   Nutritional deficiency of vitamin B_{12} is relatively rare but can occur in people who consume vegan diets. Plants do not contain vitamin B_{12} unless they are contaminated by bacteria. For people who consume a normal diet with ample intake of animal products, the large stores of vitamin B_{12} in the liver provide protection for up to 5 to 7 years of reduced vitamin B_{12} intake or absorption.

22.6.5.2   *Insufficient Gastric Acid.*   The proteolytic activity of pepsin is required to release food-bound vitamin B_{12}. Many forms of gastritis, including those resulting from *Helicobacter pylori* infection and AIDS, impair both pepsinogen secretion and the availability of acid for the conversion of pepsinogen to pepsin. Hypochlorhydria (decreased stomach acid) can also result from prolonged use of proton-pump inhibitors.

22.6.5.3   *Absent or Diseased Ileal Mucosa.*   As described earlier, the vitamin B_{12}–intrinsic factor complex is absorbed by a receptor-mediated process in the terminal ileum. Ilial resection or mucosal disease such as Crohn's disease will impair absorption of the vitamin.

22.6.5.4 Impaired Translocation of Vitamin B$_{12}$. Transcobalamin II (TCII), the protein that transports newly absorbed vitamin B$_{12}$, is synthesized in both the ileum and the liver. Genetic defects resulting in lack of functional TCII produce biochemical and hematological signs of vitamin B$_{12}$ deficiency in early infancy.

CHAPTER 23

PURINES AND PYRIMIDINES

23.1 FUNCTIONS OF PURINES AND PYRIMIDINES

23.1.1 Purines and Pyrimidines Play a Key Structural and Informational Role in DNA and RNA

Both DNA (deoxyribonucleic acid) and RNA (ribonucleic acid) contain purine and pyrimidine bases attached to sugar phosphates. The backbones of DNA and RNA are composed of sugar phosphate polymers (deoxyribose phosphate and ribose phosphate, respectively), while the purine and pyrimidine bases provide the specific information of the genetic code.

23.1.1.1 Structures of Purines. Purines (adenine and guanine) contain two fused heterocyclic, nitrogen-containing rings (Fig. 23-1). They are components of both DNA and RNA.

23.1.1.2 Structures of Pyrimidines. Pyrimidines are six-membered, heterocyclic, nitrogen-containing carbon structures. The pyrimidine bases cytosine and thymine are found in DNA. RNA also contains cytosine but has uracil in lieu of thymine (Fig. 23-2).

23.1.1.3 Nucleosides and Nucleotides. When purine or pyrimidine bases are attached to pentoses (ribose or deoxyribose), they are called *nucleosides* (Fig. 23-3), and when they are attached to ribose or deoxyribose that has one or more

Medical Biochemistry: Human Metabolism in Health and Disease   By Miriam D. Rosenthal and Robert H. Glew
Copyright © 2009 John Wiley & Sons, Inc.

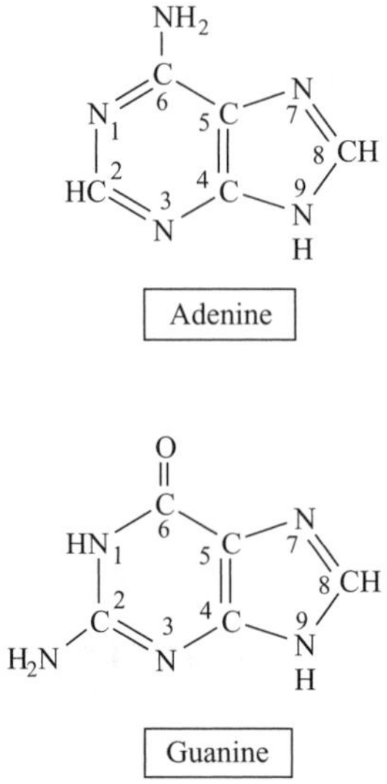

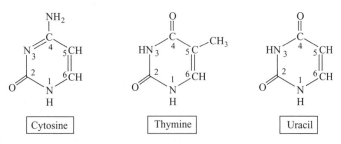

FIGURE 23-1 Purine bases.

phosphates attached to the 5′-hydroxyl group of the pentoses, they are called *nucleotides*. Thus, the nucleoside adenosine is formed by attachment of adenine to ribose, and deoxyadenosine is formed by attachment of adenine to deoxyribose. Adenine linked to ribose 5-phosphate is adenosine 5′-phosphate (AMP), and adenine linked to deoxyribose 5′-phosphate is deoxyadenosine 5′-phosphate (dAMP). The adenine nucleotides are sometimes called *adenylates*.

23.1.2 Energy Metabolism

Purine nucleoside triphosphates are at the heart of energy metabolism. ATP is the high-energy phosphoanhydride product of both glycolysis and oxidative phosphorylation. ATP is also the main source of energy for metabolic work, including biosynthesis, active ion transport, muscle contraction, and detoxification of xenobiotics.

FIGURE 23-2 Pyrimidine bases.

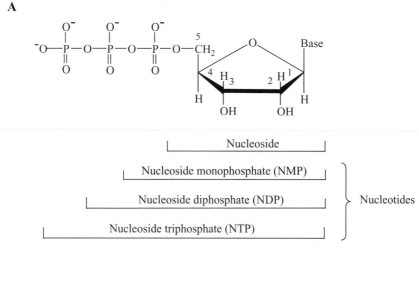

FIGURE 23-3 Structures of nucleoside and nucleotides: (A) structures of a generic ribonu-cleoside and related phosphorylated ribonucleotides; (B) structure of the deoxynucleotide, deoxyadenosine 5′-phosphate.

Other ribonucleoside triphosphates also participate directly in metabolic pathways. GTP, which is produced directly in the tricarboxylic acid cycle, provides energy for protein synthesis and for certain biosynthetic reactions, such as that catalyzed by phosphoenolpyruvate carboxykinase during gluconeogenesis. Uridine nucleotide-activated monosaccharides such as UDP-glucose and UDP-glucuronic acid are in-volved in the synthesis of polysaccharides, lactose, and the oligosaccharide chains of glycoproteins, and in glucuronidation reactions involved in detoxification processes (e.g., bilirubin conjugation; Chapter 24). Several CTP-activated substrates, including CDP-choline, CDP-ethanolamine, and CDP-diacylglycerol, are intermediates in glycerophospholipid synthesis.

The GTP, CTP, and UTP required for energy metabolism are generated from ATP by reactions catalyzed by nucleoside diphosphate kinase: for example,

$$GDP + ATP \rightleftharpoons GTP + ADP$$

23.1.3 Components of Cofactors

Adenosine monophosphate (AMP) is a core element in the structure of the major enzymatic cofactors of oxidation–reduction reactions, including $NAD^+/NADH$, $NADP^+/NADPH$, and $FAD/FADH_2$. $FAD/FADH_2$ and $FMN/FMNH_2$ are also key components of several of the complexes of the mitochondrial electron transport chain. Adenine nucleotides are components of coenzyme A, which is involved in the synthesis of bile acids and the catabolism of branched-chain amino acids as well as providing the activated form of acetate and fatty acids for both biosynthetic and catabolic reactions. In addition, S-adenosylmethionine (SAM) is the activated form of methionine, which functions as a methyl donor in the synthesis of epinephrine, phosphatidylcholine, and other methyl group–containing compounds.

23.1.4 Activation of G Proteins

Guanine nucleotides bind to and regulate the activity of G proteins, which are important components of intracellular signal-transduction cascades. The heterotrimeric G proteins are activated by G protein–coupled receptors, such as the receptors for glucagon, epinephrine, and various prostaglandins. The smaller monomeric G proteins, such as Ras, are components of the MAP kinase (mitogen-activated protein kinase) cascade, which is linked to receptor tyrosine kinases such as the epidermal growth factor receptor. Both classes of G proteins become active when the GDP bound to them is exchanged for a GTP molecule. Hydrolysis of the bound GTP to GDP (and P_i) returns the G protein to its inactive state.

23.1.5 Activation of Protein Kinases

The cyclic purine nucleotides cAMP and cGMP are second messengers that regulate numerous metabolic processes. cAMP-dependent protein kinase A (PKA) is the major mediator by which glucagon and epinephrine regulate multiple aspects of energy metabolism, including gluconeogenesis, glycogen mobilization, and lipolysis in adipocytes. cGMP activates protein kinase G and thus provides the mediator by which nitric oxide (NO) triggers muscle relaxation.

AMP-activated protein kinase (AMPK) plays an important role in energy homeostasis. Whereas cAMP and cGMP are generated in response to hormonal and autocoid signals, increases in cellular AMP occur when the cell becomes depleted of ATP. AMPK-mediated protein phosphorylation inhibits pathways that require ATP, such as fatty acid and cholesterol synthesis, and activate pathways such as fatty acid oxidation that generate ATP. The activation of AMPK also exerts long-term effects on gene expression and protein synthesis.

23.1.6 Extracellular Signaling Molecules

Adenosine, ATP, and diadenosine polyphosphates (two adenosines coupled via ester bonds to three to six phosphates, Ap_nA) are neurotransmitters that activate certain classes of G protein–coupled receptors. ATP also activates cation-permeable ligand-gated ion channels. Although these various receptors are all called purinergic receptors, some are also activated by UTP, a pyrimidine triphosphate.

23.2 LOCALIZATION OF PURINE AND PYRIMIDINE METABOLISM

Pyrimidine synthesis occurs in a variety of tissues, including spleen, thymus, testes, and intestinal enterocytes. By contrast, de novo synthesis of purines is active primarily in liver. Nonhepatic tissues generally rely on preformed purines salvaged from intracellular turnover and purines synthesized by the liver. The *purine salvage pathway* is particularly important in the brain, where it serves to regenerate adenine nucleotides from the neurotransmitter adenosine.

The end product of purine catabolism in humans is uric acid, a relatively insoluble substance that is excreted in the urine. The final steps of uric acid synthesis occur only in the liver and intestine.

23.3 PATHWAYS OF NUCLEOTIDE METABOLISM

Purine and pyrimidine bases are mostly synthesized in the body, with dietary sources contributing less than 1% to the body's needs. The reactions involved in the de novo synthesis of purines and pyrimidines take place in the cytosol and mitochondria, and the enzymes of each pathway tend to function as components of multienzyme complexes. The two pathways differ primarily in that the purine base is constructed sequentially on phosphoribosyl-1-pyrophosphate (PRPP) (Fig. 23-4), whereas the pyrimidines are synthesized as free bases which are then attached to PRPP.

23.3.1 Synthesis of the Sugar Moieties

The two pentoses found in nucleotides are ribose and deoxyribose. Ribose, generated as ribose 5-phosphate via the pentose phosphate pathway, is activated to 5-phosphoribosyl 1-pyrophosphate (PRPP), which is used to synthesize ribonucleotides. Deoxyribose is synthesized by oxidizing the ribose moiety of ribonucleotides.

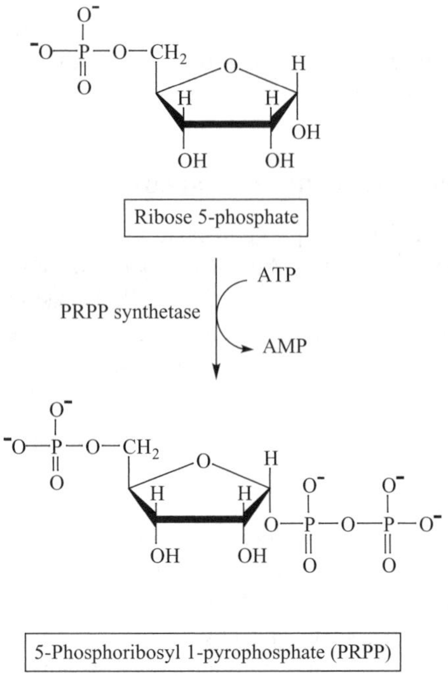

Ribose 5-phosphate

5-Phosphoribosyl 1-pyrophosphate (PRPP)

FIGURE 23-4 Synthesis of 5-phosphoribosyl 1-pyrophosphate.

23.3.1.1 Synthesis of 5-Phosphoribosyl 1-Pyrophosphate.
PRPP synthetase, also called ribose phosphate pyrophosphokinase, catalyzes the reaction that attaches a pyrophosphate moiety to carbon-1 of the ribose molecule (Fig. 23-4):

$$\text{ribose 5-phosphate} + \text{ATP} \rightarrow \text{5-phosphoribosyl 1-pyrophosphate} + PP_i$$

23.3.1.2 Synthesis of Deoxyribose.
The typical cell contains 5 to 10 times as much total RNA (mRNAs, rRNAs, and tRNAs) as DNA. Therefore, the bulk of nucleotide biosynthesis has as its purpose the production of ribonucleoside triphosphates (NTPs). However, because proliferating cells must replicate their genomes, the production of deoxyNTPs (dNTPs) is also necessary for the synthesis of DNA. Nucleoside diphosphates are the substrate for the reduction of carbon 2 of ribose to deoxyribose; the reaction is catalyzed by ribonucleotide reductase (Fig. 23-5):

$$\text{NDP} + \text{NADPH} + H^+ \rightarrow \text{deoxy-NDP} + \text{NADP}^+$$

The subsequent phosphorylation of dNDPs to dNTPs is catalyzed by nucleoside diphosphate kinases that use ATP as the phosphate donor:

$$\text{deoxy-NDP} + \text{ATP} \rightleftharpoons \text{deoxy-NTP} + \text{ADP}$$

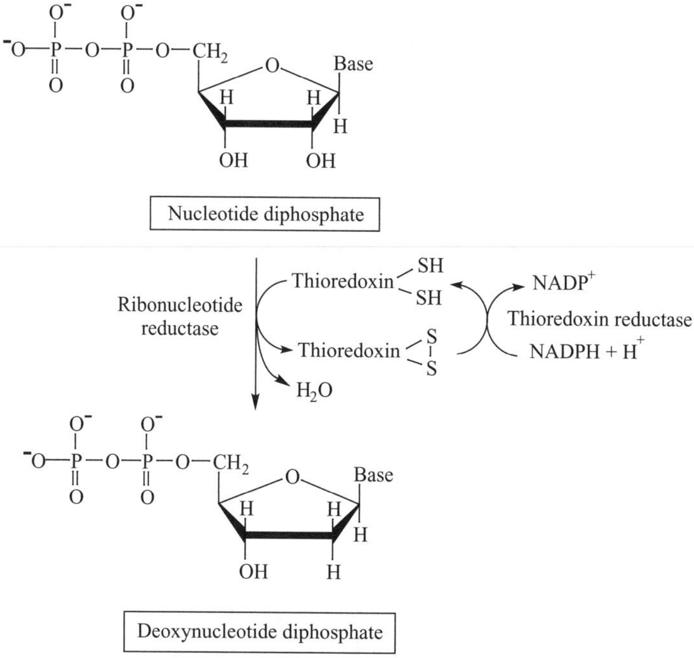

FIGURE 23-5 Role of ribonucleotide reductase in the generation of deoxynucleotides. The reaction results in the oxidation of thioredoxin (shown) or glutaredoxin (not shown).

Ribonucleotide reductase is a multifunctional enzyme complex that contains redox-active thiol groups. In the process of reduction of NDPs, the thiol groups become oxidized. They are then reduced by either one of two low-molecular weight proteins, thioredoxin or glutaredoxin. Oxidized thioredoxin, in turn, is reduced by the flavoprotein thioredoxin reductase (Fig. 23-5), while oxidized glutaredoxin is reduced by the combined action of glutathione and glutathione reductase. In both cases, NADPH is the ultimate reductant.

23.3.2   De Novo Synthesis of Purines

23.3.2.1   *Synthesis of Inosine Monophosphate.* Starting with PRPP, 10 steps are required for the de novo synthesis of inosine monophosphate (IMP), which is the common precursor for both adenosine monophosphate (AMP) and guanosine monophosphate (GMP) (Fig. 23-6). The multienzyme complex that carries out de novo purine synthesis is called the *purinosome*. The names of the enzymes and intermediates in the de novo purine pathway are cumbersome and complex; therefore, except for the steps that are regulated or the ones that are implicated in a particular human disease, there is little need for a medical student to focus on all the reactions.

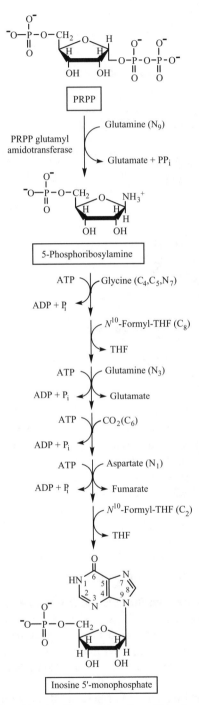

FIGURE 23-6 Synthesis of inosine 5′-monophosphate. The subscripts after C and N indicate the location of the atom in the purine ring.

However, it is worth noting several important features of the de novo synthesis of IMP from PRPP:

- Four of the 10 enzymes utilize ATP, thereby underscoring the fact that de novo purine synthesis is a costly process. Additional ATP is required for the synthesis of PRPP from ribose 5-phosphate.
- Two of the four nitrogen atoms in the ring structure of inosine are derived from glutamine and one each are derived from glycine and aspartate (Fig. 23-6).
- Two enzymes utilize N^{10}-formyl-FH$_4$ to donate carbon atoms to the purine rings. These reactions underscore the dependency of nucleic acid synthesis on adequate dietary folate. Inadequate folate impairs cell division, as exemplified by the megaloblastic anemia associated with folate deficiency. Inhibitors of dihydrofolate reductase such as methotrexate and aminopterin are useful cancer chemotherapeutic agents.

23.3.2.2 Synthesis of AMP and GMP from IMP.

Synthesis of both the adenine and guanine ring structures from IMP involves addition of an amino group to the ring structure (Fig. 23-7). The amino group of adenine is derived from aspartate, whereas glutamine is the source of the amino group of guanine. Both of these reactions require energy: GTP is required for the conversion of IMP to AMP, and ATP is consumed in the conversion of IMP to GMP.

23.3.3 Catabolism of Purines

The catabolism of purines is an especially important process in humans, in part at least, because the pathway ends in uric acid, the accumulation of which causes gout. Unlike most other mammals, humans lack the enzyme urate oxidase, which converts uric acid to allantoin, a more soluble end product than urate.

GMP catabolism is initiated by a 5'-nucleotidase, which converts GMP to guanosine:

$$GMP + H_2O \rightarrow guanosine + P_i$$

The release of ribose from the purine base is catalyzed by purine nucleoside phosphorylase:

$$guanosine + P_i \rightarrow guanine + ribose\ 1\text{-phosphate}$$

Guanine is then deaminated to xanthine (Fig. 23-8):

$$guanine \rightarrow xanthine + NH_4^+$$

There are two alternative pathways for catabolism of AMP (Fig. 23-8), both of which involve deamination of the adenosine moiety to inosine and hydrolysis of

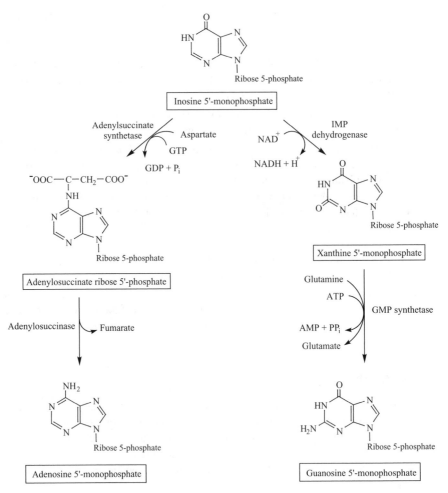

FIGURE 23-7 Synthesis of adenosine 5′-monophosphate and guanosine 5′-monophosphate from inosine 5′-monophosphate.

the nucleoside monophosphate to a nucleoside. The major route of AMP catabolism involves generation of inosine monophosphate (IMP) by AMP deaminase (a.k.a. adenosine monophosphate deaminase), followed by removal of the phosphate group by 5′-nucleotidase. The AMP deaminase reaction is also part of the purine nucleotide cycle discussed below and shown in Figure 23-9. In the alternative pathway of AMP catabolism, 5′-nucleotidase converts AMP to adenosine, which is then deaminated by adenosine deaminase. Purine nucleoside phosphorylase then catalyzes release of the ribose moiety:

$$\text{inosine} + P_i \rightarrow \text{hypoxanthine} + \text{ribose 5-phosphate}$$

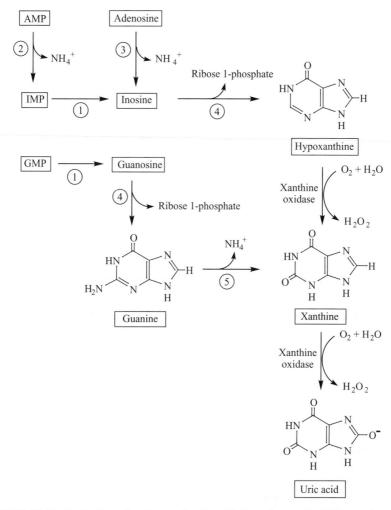

FIGURE 23-8 Catabolism of purine nucleotides: ①, 5′-nuclease; ②, AMP deaminase; ③, adenosine deaminase; ④, purine nucleoside phosphorylase; ⑤, guanine deaminase.

The catabolism of the deoxynucleotides dAMP and dGMP follow the same pathways as those that catabolize AMP and GMP, respectively, with the release of deoxyribose rather than ribose.

The final enzyme in the pathway of purine catabolism is xanthine oxidase, a molybdenum-containing flavoenzyme that uses molecular oxygen and produces hydrogen peroxide. Xanthine oxidase oxidizes both hypoxanthine and xanthine (Fig. 23-8):

$$\text{hypoxanthine} + O_2 \rightarrow \text{xanthine} + H_2O_2$$

$$\text{xanthine} + O_2 \rightarrow \text{uric acid} + H_2O_2$$

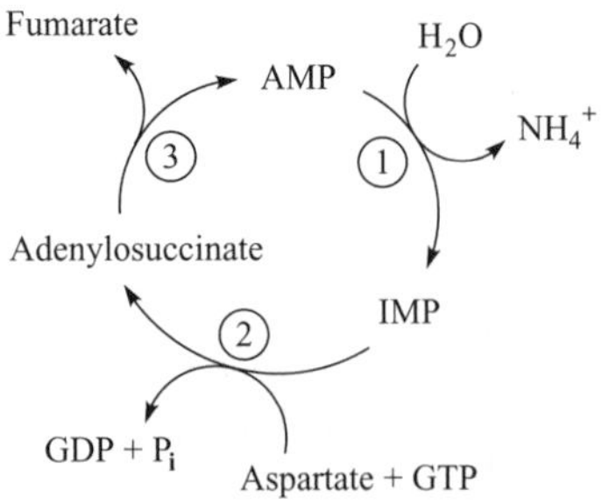

FIGURE 23-9 Purine nucleotide cycle: ①, AMP deaminase; ②, adenylosuccinate synthase; ③, adenylosuccinase. AMP, adenosine 5'-monophosphate; IMP, inosine 5-monophosphate.

23.3.4 Salvage of Purines

As described above, de novo purine synthesis involves the stepwise assembly of the purine base on PRPP. When free purines are released during nucleotide catabolism, they can be salvaged by reattachment of the bases to PRPP. The key enzyme in the salvage of purines is hypoxanthine-guanine phosphoribosyltransferase (HGPRTase), which utilizes both guanine and hypoxanthine, the product of adenine deamination:

$$\text{guanine} + \text{PRPP} \rightarrow \text{guanosine monophosphate} + \text{PP}_i$$

$$\text{hypoxanthine} + \text{PRPP} \rightarrow \text{inosine monophosphate} + \text{PP}_i$$

The IMP generated by HGPRTase is then converted to AMP by enzymes of the de novo pathway.

A second phosphoribosyl transferase, adenine phosphoribosyltransferase, uses adenine as a substrate:

$$\text{adenine} + \text{PRPP} \rightarrow \text{adenosine monophosphate} + \text{PP}_i$$

However, since the major pathway for the catabolism of adenosine generates inosine and then hypoxanthine, salvage of adenine is of only minor physiological importance. Unlike guanosine, adenosine can also be phosphorylated directly back to the level of a nucleotide by adenosine kinase:

$$\text{adenosine} + \text{ATP} \rightleftharpoons \text{AMP} + \text{ADP}$$

23.3.5 Purine Nucleotide Cycle

During exercise, there is increased catabolism of AMP to IMP in skeletal muscle. Multiple bouts of sprint exercise result in the loss of adenine nucleotides from muscle and subsequent urinary excretion of increased amounts of purine metabolites, including inosine, hypoxanthine, and uric acid. However, when oxidative metabolism is not stressed, most of the IMP is converted back to AMP and then to ADP and ATP (Fig. 23-9). This process, called the *purine nucleotide cycle*, serves several interrelated functions, discussed below.

23.3.5.1 Generation of ATP. During high-intensity exercise, cytosolic ATP is rapidly converted to ADP. Additional ATP can be generated directly from ADP by myokinase (adenylate kinase):

$$2ADP \rightleftharpoons ATP + AMP$$

The reversible myokinase reaction is driven to the right by the action of AMP deaminase, which removes AMP:

$$AMP + H_2O \rightarrow IMP + NH_3$$

AMP deaminase is expressed at high levels in skeletal muscle and is associated with myofibrils.

23.3.5.2 Generation of Ammonium Ions for Glutamine Synthesis. Aspartate is the nitrogen donor in the pathway that generates AMP from IMP:

$$IMP + aspartate + GTP \rightarrow adenylosuccinate + GDP + P_i$$

$$adenylosuccinate \rightarrow AMP + fumarate$$

Fumarate can then be hydrated to malate by fumarase and transported into the mitochondrion, where it is converted to oxaloacetate and transaminated to regenerate aspartate. Some of the fumarate generated in the purine nucleotide cycle is metabolized and thus represents a source of energy for muscle.

Aspartate provides the amino group for the regeneration of AMP, and thus for subsequent release of ammonium ions in the AMP deaminase reaction. The ammonium ions are utilized for the synthesis of glutamine from glutamate. Although the immediate nitrogen donor for the transamination of oxaloacetate to aspartate is glutamate, most of the amino groups are ultimately derived from the catabolism of branched-chain amino acids.

23.3.6 De Novo Synthesis of Pyrimidines

The pathway for pyrimidine synthesis differs from the de novo purine synthesis pathway in that the pyrimidine ring is assembled first and then combined with PRPP to form the initial pyrimidine nucleotide, UMP (Fig. 23-10).

23.3.6.1 Synthesis of Uracil-Containing Nucleotides. The uracil synthesis pathway (Fig. 23-10) starts with the synthesis of carbamoyl phosphate from glutamic

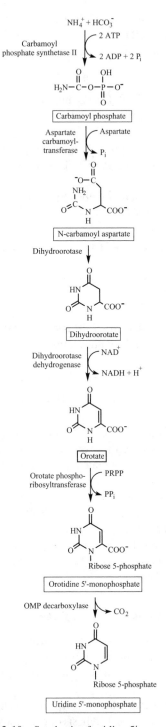

FIGURE 23-10 Synthesis of uridine 5'-monophosphate.

acid and CO_2 (HCO_3^-), catalyzed by carbamoyl phosphate synthetase II (CPS II):

$$glutamine + CO_2 + ATP \rightarrow carbamoyl\ phosphate + ADP + glutamic\ acid$$

This enzyme is distinct from carbamoyl phosphate synthetase I (CPS I), which is localized to mitochondria and catalyzes the first step in urea synthesis. CPS II utilizes glutamine rather than free ammonia as the nitrogen sources, and unlike CPS I is not activated by N-acetylglutamate.

The next reaction in pyrimidine synthesis is both the committed step and the regulated step, and it is catalyzed by aspartate transcarbamoylase:

$$aspartate + carbamoyl\ phosphate \rightarrow N\text{-carbamoyl\ aspartate} + P_i$$

Next, dehydration of N-carbamoyl aspartate by dihydroorotase forms the ring structure dihydroorotic acid (Fig. 23-10). An NAD^+-linked dehydrogenase introduces a double bond into the ring, producing orotate. Orotate is then transferred to PRPP by orotate phosphoribosyltransferase:

$$orotate + PRPP \rightarrow orotidine\ 5\text{-monophosphate} + PP_i$$

Once the initial pyrimidine nucleotide has been synthesized, orotidylate decarboxylase removes the carboxyl group from the ring to form uridylic acid (UMP), the primary product of the de novo pyrimidine pathway. Two kinase reactions then convert UMP to UTP:

$$UMP + ATP \rightarrow UDP + ADP$$

$$UDP + ATP \rightarrow UTP + ADP$$

23.3.6.2 *Synthesis of Cytosine-Containing Nucleotides.* Cytidylate synthetase (CTP synthetase) catalyzes the conversion of UTP to CTP (Fig. 23-11), with glutamine as the nitrogen donor:

$$uridine\ 5'\text{-triphosphate} + glutamine + ATP \rightarrow cytidine\ 5'\text{-triphosphate} + ADP + P_i$$

23.3.6.3 *Synthesis of Deoxythymidylate.* As noted earlier, DNA contains thymine (methyluracil) in place of uracil. The methylation of uracil is catalyzed by thymidylate synthase, which utilizes deoxyuridylate (dUMP) as its substrate (Fig. 23-12):

$$deoxyuridine\ 5'\text{-monophosphate} + N^5, N^{10}-methylene - FH_4 \rightarrow$$

$$deoxythymidine\ 5'\text{-monophosphate} + 7, 8\text{-}FH_2$$

N^5, N^{10}-methylene-FH_4 is regenerated by the sequential action of two enzymes: dihydrofolate reductase and serine hydroxymethyltransferase. Because thymidylate

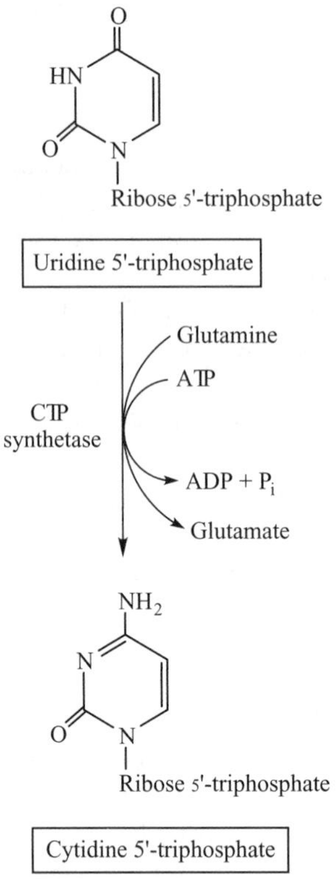

FIGURE 23-11   Synthesis of cytidine 5′-triphosphate from uridine 5′-triphosphate.

synthetase is strongly dependent on an adequate supply of the components of the folate one-carbon pool, cancer chemotherapeutic agents (e.g., methotrexate) that interfere with steps in purine nucleotide synthesis also inhibit the synthesis of dTMP.

23.3.7 Catabolism of Pyrimidines

The ribonucleotides and deoxyribonucleotides are first dephosphorylated to the level of nucleosides by nonspecific phosphatases. Cytidine and deoxycytidine are then deaminated by pyrimidine nucleoside deaminase to uridine and deoxyuridine, respectively:

$$\text{cytidine} + H_2O \rightarrow \text{uridine} + NH_4^+$$

5′-Nucleosidases (nucleoside phosphorylases) then catalyze the phosphorolytic cleavage of uridine, deoxyuridine, and deoxythymidine to generate the free nitrogen bases (uracil and thymine) and ribose 1-phosphate or deoxyribose 1-phosphate:

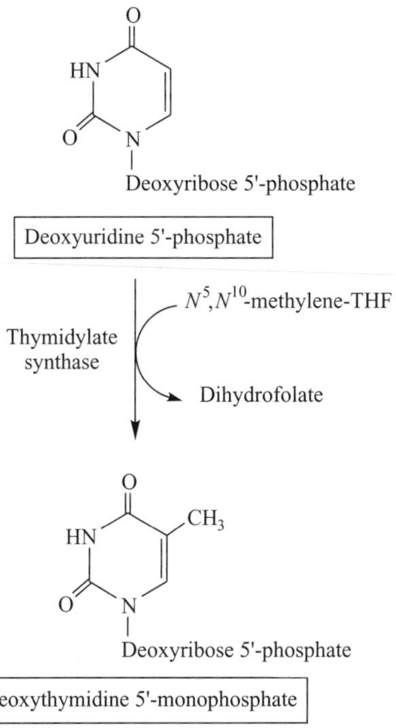

FIGURE 23-12 Synthesis of thymidylate (deoxythymidine 5'-phosphate) from deoxyuridine 5'-phosphate.

for example,

$$uridine + P_i \rightarrow uracil + ribose\ 1\text{-phosphate}$$

In contrast to purines, pyrimidines can undergo ring cleavage. Catabolism of both uracil and thymine involves the same three-enzyme sequence, which releases ammonia and carbon dioxide and generates a β-amino acid from the rest of the ring (Fig. 23-13). β-Aminoisobutyrate, the main end product of thymine metabolism, is mostly excreted in the urine. The end product of uracil metabolism is β-alanine, some of which is incorporated into carnosine (histidine-β-alanine) and anserine (methyl histidine-β-alanine), two dipeptides found in brain and muscle. Excess β-alanine is excreted in the urine.

23.3.8 Salvage of Pyrimidines

Salvage of pyrimidine bases is only a minor pathway in humans. However, the pyrimidine nucleosides can be salvaged by nucleoside kinases:

$$nucleoside + ATP \rightarrow nucleotide + ADP$$

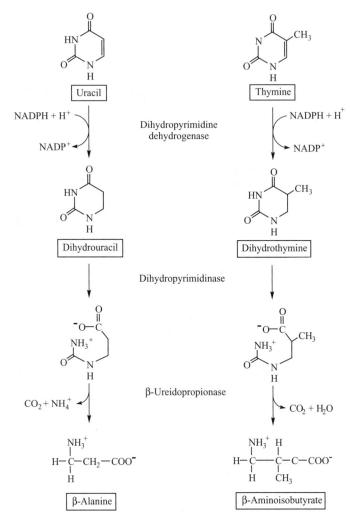

FIGURE 23-13 Catabolism of the pyrimidine bases uracil and thymine.

23.4 REGULATION OF PURINE AND PYRIMIDINE METABOLISM

23.4.1 Purine Synthesis

As is the case with many metabolic pathways, the first committed step in de novo purine synthesis is the regulated step. Glutamine PRPP amidotransferase, which catalyzes the first step in the process of IMP synthesis (Fig. 23-6), is inhibited allosterically by the end products of the pathway, AMP and GMP. A second level of control is exerted at the level of PRPP synthesis, where PRPP synthetase is inhibited by both ADP and GDP.

23.4.2 Balance Between the Generation of Adenine and Guanine Nucleotides

Two different reciprocal regulatory mechanisms act to maintain the balance between synthesis of AMP and GMP (see Fig. 23-7). First, adenylosuccinate synthase and IMP dehydrogenase, which catalyze the initial steps in the conversion of IMP to AMP and GMP, respectively, are inhibited by AMP and GMP, respectively. Second, the synthesis of AMP requires GTP, while that of GMP requires ATP.

23.4.3 Pyrimidine Synthesis

The key control point in pyrimidine synthesis is the reaction catalyzed by carbamoyl phosphate synthetase II. The enzyme is inhibited by UTP, and activated by PRPP. Carbamoyl phosphate synthetase II is also regulated by posttranslational phosphorylation. Phosphorylation of CPS II by MAP kinase results in an enzyme that is more sensitive to activation by PRPP and is not inhibited by UTP, thus increasing carbamoylphosphate synthetase II activity as cells approach the S-phase of the cell cycle.

23.5 ABNORMAL PURINE AND PYRIMIDINE METABOLISM

23.5.1 Gout

Gout occurs when a high concentration of uric acid in the blood (hyperuricemia) results in deposition of monosodium urate crystals in tissues and joints, causing pain and inflammation. Uric acid is poorly soluble in plasma, especially at lower temperatures; thus the hallmark of gout is deposition of tophi or urate crystals under the skin of the ear, fingers, and toes. Gout can be caused by either underexcretion or overproduction of uric acid. Causes of underexcretion of purines include lactic acidosis and drugs such as thiazide diuretics.

Hyperuricemia is one of the hallmarks of tumor lysis syndrome, a cluster of metabolic complications of cancer chemotherapy. The catabolism of purines from large numbers of lysed cancer cells results in increased production of uric acid. One of the genetic causes of increased uric acid synthesis is impaired salvage of purine bases as occurs in people who have a partial deficiency of HRPTase (see below). Gout is also associated with von Gierke disease, the glycogen storage disease that results from a deficiency in glucose 6-phosphatase activity. The increased availability of glucose 6-phosphate increases the rate of flux through the pentose phosphate pathway, yielding an elevation in the level of ribose 5-phosphate, which results in increased synthesis of PRPP and excess purine biosynthesis.

The drug allopurinol, which is used to treat gout, is oxidized by xanthine oxidase to oxypurinol, which is a potent inhibitor of xanthine oxidase. When xanthine oxidase is inhibited, hypoxanthine and xanthine accumulate and the concentration of uric acid is reduced. Hypoxanthine and xanthine are more water-soluble than uric acid, thereby facilitating the urinary excretion of purine degradation products and reducing the likelihood of urate crystal deposition.

23.5.2 Lesch–Nyhan Disease

Lesch–Nyhan disease is a rare, X-linked disorder caused by a genetic defect in hypoxanthine-guanine phosphoribosyltransferase (HGPRTase), the regulated enzyme in the purine salvage pathway that catalyzes the conversion of the purine bases hypoxanthine and guanine to their respective nucleotides, IMP and GMP. Lacking HGPRTase activity, PRPP levels increase and purines are synthesized in excess by the de novo pathway, resulting in high plasma and urine concentrations of urate. Affected children have severe neurologic deficits, retarded motor development, muscle weakness, and self-injurious behavior; the most characteristic feature is loss of tissue from biting themselves. Although the mechanism is not fully understood, it appears that abnormalities in purine metabolism impair dopaminergic function in the basal ganglia. Allopurinol is effective in controlling the hyperuricemia and goutlike symptoms of Lesch–Nyhan disease, but is ineffective in modifying the neurological or behavioral manifestations of the disease. A partial deficiency of HGPRTase (residual activity $> 8\%$) results in gout, renal stones, and uric acid nephropathy without the neurological problems associated with Lesch–Nyhan disease.

23.5.3 Adenosine Deaminase Deficiency

Adenosine deaminase (ADA) deaminates both adenosine to inosine (Fig. 23-8) and deoxyadenosine to deoxyinosine. Deficiency of ADA causes severe combined immunodeficiency disease. People with ADA deficiency have 100-fold increases in the concentration of dATP, which is a potent inhibitor of ribonucleotide reductase. The resulting deficiency in the other deoxynucleotide triphosphates inhibits DNA synthesis and impairs the immune response of both B and T lymphocytes. The disease has been treated successfully by bone marrow transplantation.

Immunodeficiency can also be caused by a deficiency of purine nucleoside phosphorylase, the enzyme that catalyzes the phosphorolytic cleavage of inosine to hypoxanthine. Unlike ADA deficiency, however, purine nucleoside phosphorylase deficiency affects only the T cells; B-cell function remains normal.

23.5.4 Orotic Aciduria

Orotic aciduria is an inborn error of metabolism characterized by large quantities of orotic acid in the urine and by megaloblastic anemia, which is unresponsive to vitamin B_{12} and folic acid. The metabolic defect is in uridine-5-monophosphate synthase, which contains two enzyme activities in a single protein: orotate phosphoribosyltransferase and orotidylate decarboxylase. The disease is treated effectively with high doses of oral uridine, which leads to increased UMP production; the UMP then inhibits carbamoyl phosphate synthetase II, thereby decreasing synthesis of orotic acid.

23.5.5 Cholera and Whooping Cough

Cholera is an acute, life-threatening diarrheal disease caused by *Vibrio cholera* bacteria that are transmitted in water contaminated by human wastes. The bacterium produces a toxin that catalyzes a covalent modification that attaches an ADP-ribose moiety derived from NADH to a specific arginine of the $G_{\alpha s}$ subunit of G proteins.

$$G_\alpha + NAD^+ \rightarrow ADP\text{-ribose-}G_\alpha + niacin$$

ADP ribosylation stabilizes the $G_{\alpha s}$ in the active form by preventing hydrolysis of bound GTP. As a result, intestinal epithelial cells produce excess cAMP, which stimulates the secretion of excess sodium into the intestinal lumen. Water follows the sodium, resulting in diarrhea and dehydration. Treatment requires aggressive rehydration therapy.

The bacterium *Bordetella pertussis*, which causes pertussis or whooping cough, also produces a toxin that catalyzes ADP-ribosylation. In this case, the target is a different Gα subunit called $G_{\alpha i}$, and the result is the stabilization of the inactive GDP-bound form of the G protein.

NAD^+ is also a substrate for transfer of ADP-ribose in a number of physiological reactions. Poly(ADP-ribose) polymerases, a family of cell-signaling enzymes in the nucleus of eukaryotic cells, are involved in poly(ADP-ribosylation) of DNA-binding proteins and use poly(ADP)ribose to repair DNA single-strand breaks. ADP-ribose cyclases generate cyclic AMP-ribose, which is an intracellular messenger involved in Ca^{2+} signaling.

23.5.6 Cancer Chemotherapy

Since tumor cells have a much greater demand for nucleotides for DNA and RNA synthesis, certain enzymes in the nucleotide biosynthetic pathways are major targets of cancer chemotherapy. For example, azaserine and acivicin are analogs of glutamine that inhibit glutamine-PRPP amidotransferases involved in nucleotide synthesis. One of the earliest chemotherapeutic agents, fluorouracil, inhibits thymidylate synthase. Hydroxyurea, which is used as a component of some cancer protocols, inhibits ribonucleotide reductase. Methotrexate and aminopterin inhibit dihydrofolate reductase, thus impairing synthesis of purines as well as thymidylate.

CHAPTER 24

HEME AND IRON

24.1 FUNCTIONS OF HEMOGLOBIN AND OTHER IRON-CONTAINING PROTEINS

24.1.1 Hemoglobin and Myoglobin

Heme (Fe^{2+}-protoporphyrin IX) is the prosthetic group of hemoglobin and myoglobin. Hemoglobin is the molecule in red blood cells that transports oxygen from the lungs to peripheral tissues. Myoglobin is an intracellular protein that extracts oxygen from the red blood cells and stores it until the oxygen is needed in various oxidation–reduction reactions and respiration. Hemoglobin and myoglobin both contain heme, a porphyrin or tetrapyrrole that consists of four pyrrole rings joined by methylene bridges (Fig. 24-1). All of the double bonds in heme are conjugated; that is, double bonds and single bonds alternate throughout the structure.

Porphyrins differ in the side chains attached to the tetrapyrrole ring; in the case of heme, one methyl and one vinyl group ($-CH=CH_2$) are located on each of the A and B rings, and methyl and propionic acid groups are attached to rings C and D. The iron in the center of the porphyrin ring is in the ferrous (Fe^{2+}) state and is the site where molecular oxygen binds. The iron atom also binds noncovalently to a particular histidine in the hemoglobin and myoglobin proteins. Myoglobin contains one polypeptide chain and one protoporphyrin IX ring. Hemoglobin is a tetrameric molecule with four polypeptide chains, two α and two β chains, each with a bound heme molecule.

Medical Biochemistry: Human Metabolism in Health and Disease   By Miriam D. Rosenthal and Robert H. Glew
Copyright © 2009 John Wiley & Sons, Inc.

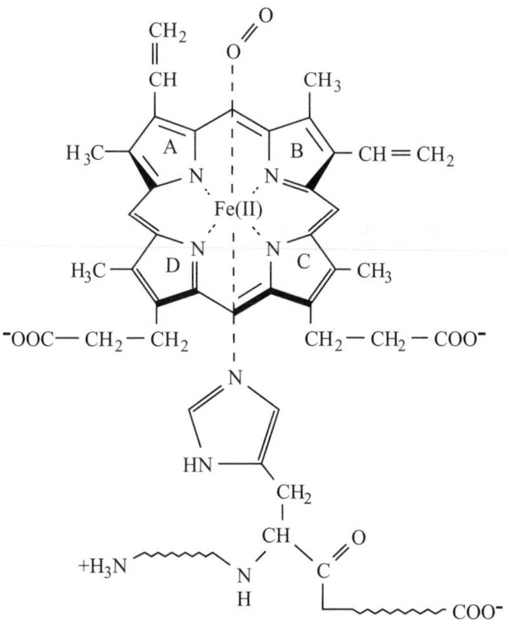

FIGURE 24-1 Structure of oxyheme in oxyhemoglobin. The sawtooth represents the globin chain.

24.1.2 Cytochromes

Heme is also the prosthetic group for a large class of hemoproteins called *cytochromes*. Whereas the iron in hemoglobin and myoglobin remains in the ferrous (Fe^{2+}) state as molecular oxygen is bound successively to and then released from the molecule, the iron in cytochromes accepts and donates electrons by undergoing reversible interconversion between the ferrous (Fe^{2+}, reduced) and ferric (Fe^{+3}, oxidized) forms. Heme-containing enzymes include prostaglandin H synthase (a.k.a. cyclooxygenase), and catalase, which converts hydrogen peroxide to water and O_2. Heme is also a component of members of the superfamily of cytochrome P450–containing enzymes that catalyze the oxidation of many molecules, including steroids and xenobiotics. Cytochromes differ from each other in the nature of the side chains attached to the porphyrin ring. Furthermore, in some cytochromes, such as cytochrome c_3, the porphyrin ring is covalently bound to the protein through the side chains of two cysteine residues of the protein via thioether bonds with vinyl groups of protoporphyrin IV.

24.1.3 Nonheme Iron

There are also some proteins in which nonheme iron participates in enzyme-catalyzed oxidation–reduction reactions. These include the iron–sulfur (Fe–S) clusters in proteins of the electron-transport system and ribonucleotide reductase, which converts ribonucleotides to deoxynucleotides. Iron also plays a central role in the generation of the hydroxyl radical ($OH^\bullet$) in the (nonenzymatic) Fenton reaction:

$$Fe^{2+} + H_2O_2 \rightarrow Fe^{3+} + OH^\bullet + OH^-$$

24.2   LOCALIZATION OF HEME AND IRON METABOLISM

24.2.1   Synthesis of Heme

Heme is synthesized de novo in all cells. Most of heme synthesis takes place in the liver and in erythroid cells within the bone marrow. A total of eight enzymes are required for heme synthesis. The first and last three steps take place in the mitochondrion, whereas the second through fifth steps take place in the cytosol.

24.2.2   Heme Catabolism

When hemoproteins turn over, the heme is degraded rather than salvaged. The iron atom is reutilized, while the porphyrin ring is oxidized and cleaved to produce the breakdown product, bilirubin. The liver converts hydrophobic bilirubin to the more water-soluble bilirubin diglucuronide, which is secreted into the bile and ultimately excreted in the feces.

Most of the bilirubin arising from the degradation of hemoglobin is produced in splenic phagocytes. This means that bilirubin must be transported from nonhepatic phagocytic cells to the liver. It is critical that bilirubin is transported in the blood bound to albumin; when the binding capacity of albumin is exceeded, the unbound bilirubin can be toxic.

24.2.3   Localization of Iron in the Body

The average man and woman contain about 3500 and 2600 mg of iron, respectively (Fig. 24-2). The hemoglobin of red blood cells and the myoglobin of muscle cells account for approximately 2500 and 100 mg of this iron, respectively. Approximately 0.8% of a person's red blood cells are broken down each day by the reticuloendothelial (macrophage) system, which results in the release of 20 mg of iron into the blood. Ninety-five percent of this iron is recycled and reutilized by the bone marrow to synthesize new red blood cells, which replace those that were broken down.

The body is also capable of storing iron, mainly in the liver. Ferritin is a ubiquitous iron-binding protein that is found mostly in the cytosol. Ferritin is composed of two subunits, designated H and L. The apoferritin shell, which is formed by assembly of 24 of these dimers, can accommodate up to 4500 atoms of iron in its core. The ratio of H to L subunits is tissue-dependent: Liver and spleen contain mostly L-subunit ferritin, whereas the H subunit predominates in kidney and heart. In a healthy adult,

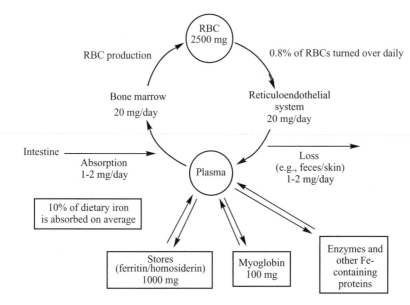

FIGURE 24-2 Disposition of iron in the body. RBC, red blood cells.

approximately 1000 mg of iron is found in ferritin. As ferrous iron enters the pores of the apoferritin shell, it is oxidized to the Fe^{3+} state and deposited as hydrated ferric oxide crystals.

Iron is transported between cells and tissues in the circulation in the ferric form, most of which is complexed to a protein named *transferrin* that is synthesized and secreted by hepatocytes. Dietary iron, absorbed in the proximal small intestine, is also transported on transferrin. Transferrin is normally about one-third saturated with ferric iron. The Fe^{3+}–transferrin complex binds to transferrin receptors on the plasma membrane of iron-utilizing cells and is internalized by receptor-mediated endocytosis (Fig. 24-3). A proton pump in the endosome acidifies the endosome, resulting in the release of iron from transferrin, which is then transported out of the endosome to the cytosol, where it becomes available for heme synthesis or storage as ferritin. The transferrin : transferin receptor complex is recycled to the plasma membrane, whereupon transferrin dissociates from its receptor.

Small amounts of ferritin are also present in plasma, reflecting slow release of this iron protein from storage sites (mostly liver) during normal cellular turnover. Since the plasma ferritin concentration is directly proportional to the intracellular stores of ferritin, determination of the plasma ferritin level in the clinical laboratory provides a convenient way of assessing a person's iron status.

About 1 mg of iron is lost each day through exfoliation of skin and intestinal cells. This iron loss is made up for by the absorption of an equivalent amount of iron from the intestine.

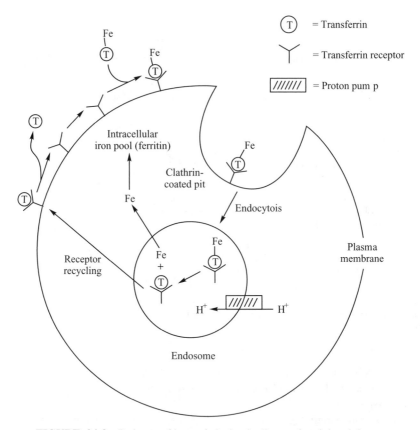

FIGURE 24-3   Pathway of iron uptake by the liver and peripheral tissues.

24.3   CONDITIONS WHERE HEME METABOLISM IS ESPECIALLY ACTIVE

Heme synthesis in erythroid cells is required for erythropoiesis and is therefore most active during periods of growth or pregnancy and when the body is compensating for increased blood loss or destruction of existing red blood cells. Heme synthesis in the liver is most active when there is induction of enzymes of the cytochrome P450 family which participate in the detoxification of drugs and xenobiotics.

24.4   PATHWAYS OF HEME AND IRON METABOLISM

24.4.1   Heme Synthesis

The overall pathway of heme synthesis is summarized in Figure 24-4. The first step is catalyzed by δ-aminolevulinic acid (ALA) synthase and involves the condensation

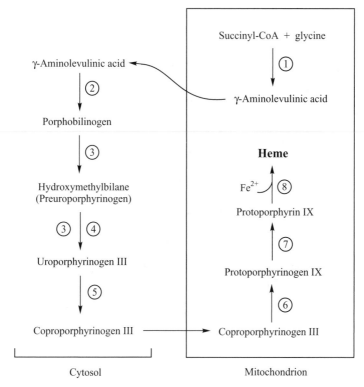

FIGURE 24-4 Synthesis of heme: ①, δ-aminolevulinic acid synthase; ②, porphobilinogen synthase; ③, uroporphrinogen synthase; ④, uroporphyrinogen III cosynthase; ⑤, uropor-phyrinogen decarboxylase; ⑥, coproporphyrinogen oxidase; ⑦, protoporphyrinogen oxidase; ⑧, ferrocheletase.

of glycine with succinyl-CoA (Fig. 24-5):

$$\text{glycine} + \text{succinyl-CoA} \rightarrow \delta\text{-aminolevulinic acid} + CO_2 + \text{CoASH}$$

This reaction is irreversible and represents the committed step in the pathway. ALA synthase is a mitochondrial enzyme and the succinyl-CoA for the reaction is provided by the TCA cycle.

The next step, the condensation of two molecules of δ-aminolevulinic acid to form a pyrrole ring, is catalyzed by ALA dehydratase (also called *porphobilinogen synthase*) (Fig. 24-5):

$$2 \,\delta\text{-aminolevulinic acid} \rightarrow \text{porphobilinogen} + 2H_2O$$

Porphobilinogen deaminase then generates hydoxymethylbilane by the sequential condensation of four porphobilinogen molecules with the concomitant release of ammonia at each step.

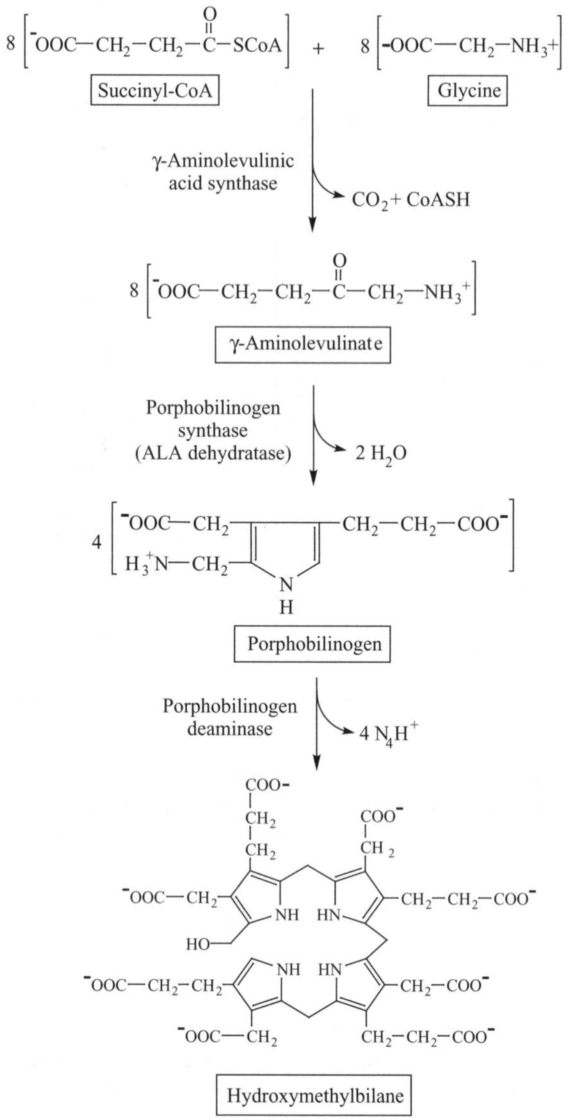

FIGURE 24-5 Synthesis of hydroxymethylbilane.

Next, hydoxymethylbilane is converted to uroporphyrinogen III by the combined action of uroporphyrinogen synthase and uroporphyrinogen III cosynthase (Fig. 24-6). Uroporphyrinogen synthase catalyzes formation of the porphyrinogen ring. The function of uroporphyrinogen III cosynthase is to direct the fourth pyrrole ring into the inverse orientation relative to that of the other three pyrrole rings. Uroporphyrinogen III decarboxylase then converts uroporphyrinogen III to coproporphyrinogen III, which is then transported from the cytosol into the mitochondrion.

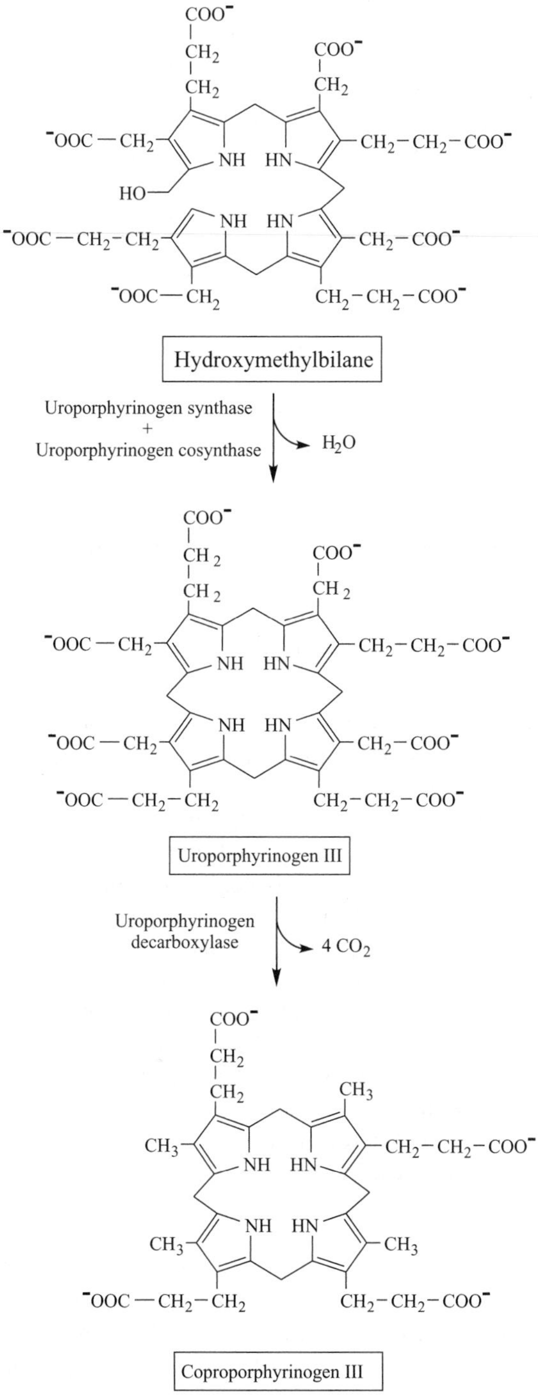

FIGURE 24-6 Synthesis of coproporphyrinogen III from hydroxymethylbilane.

Inside the mitochondrion, the propyl side chains of rings A and B of copro-porphyrinogen III are oxidized to vinyl groups by coproporphyrinogen III oxidase, thereby generating protoporphyrinogen IX (Fig. 24-7). The methylene rings con-necting the pyrrole rings of colorless protoporphyrinogen IX are then oxidized by protoporphyrinogen oxidase to generate protoporphyrin IX. This reaction generates a molecule that has resonance of double bonds around the entire great ring and the characteristic red color of heme. Finally, ferrochelatase (heme synthetase) catalyzes the insertion of ferrous iron (Fe^{2+}) into protoporphyrin IX to form heme.

24.4.1.1 Reduction of Methemoglobin. The iron of heme is normally in the ferrous state. Hemoglobin cannot bind oxygen when the iron of hemoglobin is in the ferric state (a.k.a. *methemoglobin*). Furthermore, methemoglobin is a potent oxidizing agent that can damage red blood cells and shorten their lifespan. Red blood cells contain an enzyme called *methemoglobin reductase* (a.k.a. *diaphorase*), which uses NADH or NADPH to reduce the iron atom of methemoglobin from the ferric to the ferrous state:

$$\text{methemoglobin} + NAD(P)H + H^+ \rightarrow \text{hemoglobin} + NAD(P)^+$$

24.4.2 Heme Catabolism and Bilirubin Excretion

Bilirubin is the breakdown product of heme. About 75% of bilirubin is derived from hemoglobin that has been ingested by phagocytic cells during the process of destroying senescent red cells in the spleen and liver. Bilirubin is also derived from the turnover of other heme-containing proteins (e.g., myoglobin, catalase, cytochromes).

Some free hemoglobin is released into the circulation when senescent red cells are destroyed. This hemoglobin is complexed to haptoglobins, a family of plasma glycoproteins synthesized by the liver. Haptoglobin : hemoglobin complexes are re-moved from the circulation by splenic phagocytes and Kupffer cells in the liver. Any free heme that dissociates from hemoglobin in the circulation is complexed immedi-ately by hemopexin, another liver-synthesized plasma glycoprotein, and transported to the liver. Binding of hemoglobin and heme to haptoglobins and hemopexin, re-spectively, serves both to prevent loss of iron through filtration by the kidney and to protect against oxidative stress. Both the haptoglobins and hemopexin are acute-phase proteins whose synthesis is increased in response to infection and which, by removing hemoglobin and heme from the plasma, prevent iron-utilizing bacteria from benefitting from the iron released by hemolysis.

24.4.3 Catabolism of Heme

The conversion of heme to bilirubin (Fig. 24-8) can be visualized in a bruise that is initially reddish purple (heme) and with time turns yellow-green (biliverdin) and then red-orange (bilirubin). The initial reaction that cleaves the porphyrin ring is catalyzed by heme oxygenase, producing biliverdin IX and carbon monoxide, and concurrently

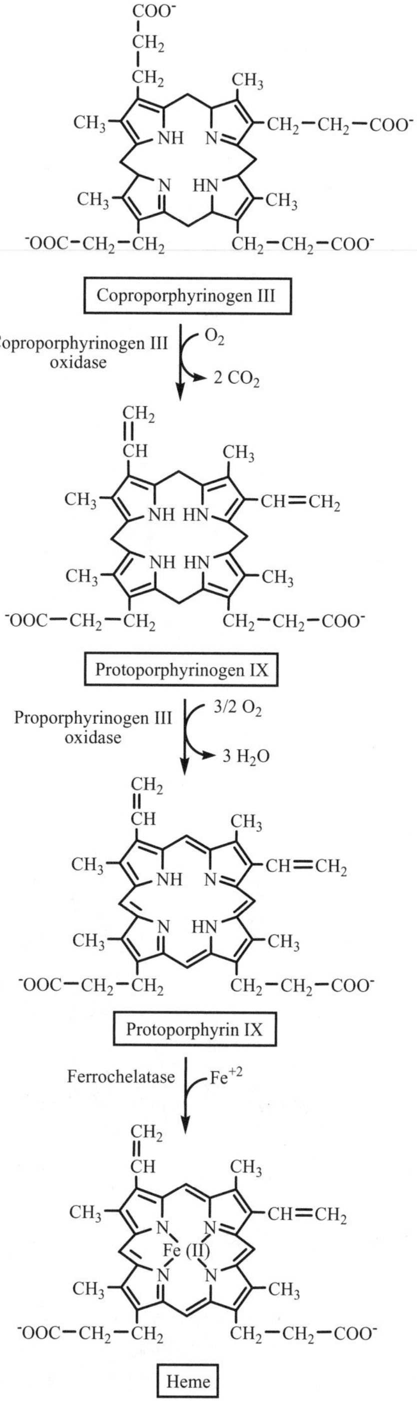

FIGURE 24-7   Synthesis of heme from coproporphyrinogen III.

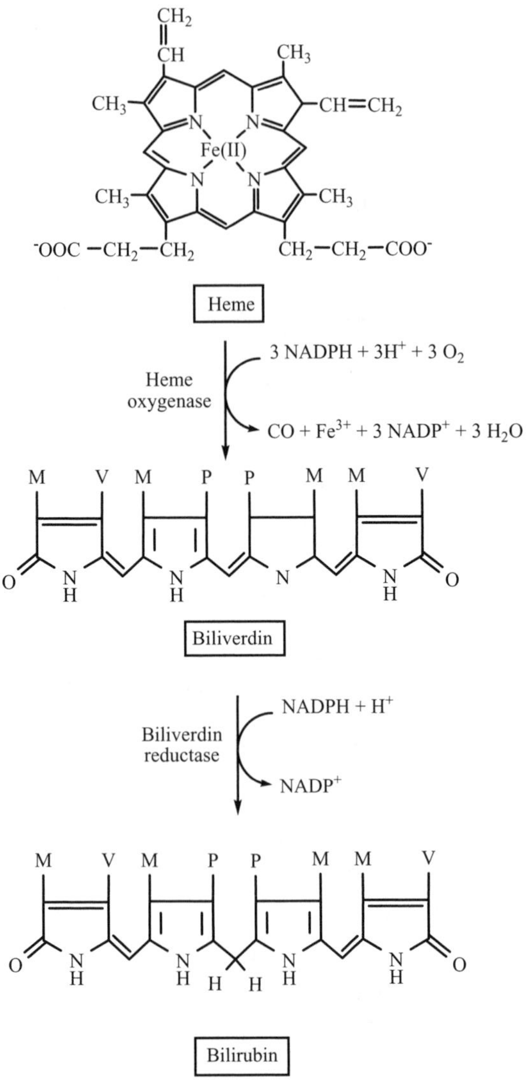

FIGURE 24-8 Metabolism of heme to bilirubin. M, methyl; P, propionyl; V, vinyl.

releasing the oxidized Fe^{3+} ion:

$$\text{heme} + 3O_2 + 3NADPH + 3H^+ \rightarrow \text{biliverdin} + CO + Fe^{3+} + 3NADP^+ + 3H_2O$$

Note that heme oxygenase is a mixed-function oxidase that utilizes both molecular oxygen and NADPH; this reaction is the only known endogenous source of carbon monoxide. The substrate for heme oxygenase is Fe^{2+} heme; any hemin (Fe^{3+} heme) that is produced during the phagocytic process must first be reduced to the Fe^{2+} form.

Biliverdin reductase then reduces the central methene bridge of biliverdin, producing bilirubin (Fig. 24-8):

$$\text{bilirubin} + \text{NADPH} + \text{H}^+ \rightarrow \text{biliverdin} + \text{NADP}^+$$

24.4.3.1 *Conjugation of Bilirubin.* Although all cells contain heme oxygenase and can convert heme generated during turnover of hemoproteins to bilirubin, only the liver is capable of converting bilirubin to the more water-soluble bilirubin diglucuronide. This reaction, catalyzed by UDP-glucuronyl transferase, involves successive transfers of two glucuronic acid residues from UDP-glucuronic acid to form ester linkages with the propionic acid side chains of bilirubin (Fig. 24-9):

$$\text{bilirubin} + \text{UDP-glucuronate} \rightarrow \text{bilirubin glucuronide} + \text{UDP}$$

$$\text{bilirubin glucuronide} + \text{UDP-glucuronate} \rightarrow \text{bilirubin diglucuronide} + \text{UDP}$$

Conjugation with glucuronic acid is also the mechanism the liver uses to increase the solubility of steroids and certain drugs prior to their excretion. UDP-glucuronic acid is synthesized by the oxidation of UDP-glucose by UDP-glucose dehydrogenase (Fig. 24-10):

$$\text{UDP-glucose} + 2\text{NAD}^+ + \text{H}_2\text{O} \rightarrow \text{UDP-glucuronic acid} + 2\text{NADH} + 2\text{H}^+$$

Direct and Indirect Bilirubin. Clinically, conjugated bilirubin or bilirubin diglucuronide is often called *direct-acting bilirubin* and unconjugated bilirubin is called *indirect-acting bilirubin*. This nomenclature is related to the colorimetric reaction, called the van den Bergh reaction, which is commonly used to quantify the two

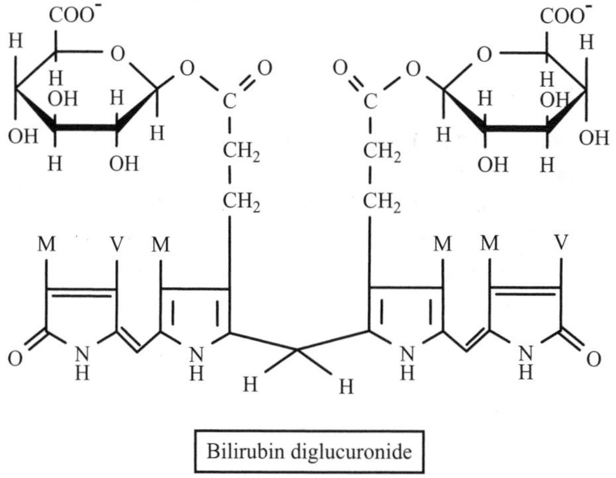

FIGURE 24-9 Structure of bilirubin diglucuronide. M, methyl; P, propionyl; V, vinyl.

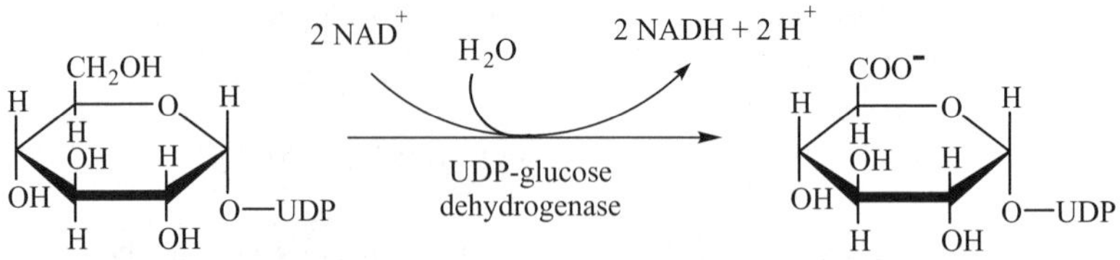

FIGURE 24-10   Synthesis of UDP-glucuronic acid.

forms of bilirubin, which is important for the differential diagnosis of the causes of hyperbilirubinemia. In this assay, conjugated bilirubin reacts readily with an azo dye. Unconjugated bilirubin, on the other hand, is much more lipophilic and tightly bound to serum albumin; it must be released with alcohol before the dye-coupling reaction can occur. The van den Bergh assay first quantifies conjugated bilirubin; then, with the addition of alcohol, the test quantifies total plasma bilirubin. The quantity of unconjugated bilirubin is determined by subtraction, and unconjugated bilirubin is therefore designated *indirect bilirubin*.

24.4.3.2 Metabolism of Excreted Bilirubin Diglucuronide. Conjugated bilirubin is released into the biliary system and delivered into the small intestine when the gallbladder contracts. In the lower intestine and colon, bacterial β-glucuronidases remove glucuronic acid to form unconjugated bilirubin. Further metabolism of bilirubin by bacteria reduces bilirubin to a colorless tetrapyrrolic compound called *urobilinogen*. A small amount of urobilinogen is absorbed and enters into the enterohepatic circulation; a minor fraction of this urobilinogen is ultimately excreted by the kidney, partly as the oxidized, colored compound urobilin, which imparts the characteristic yellow color of urine. Most of the urobilinogen formed in the gut is further metabolized by the enteric bacteria to stercobilinogen and excreted mainly in its oxidized form, stercobilin, which imparts the characteristic color of stool.

24.4.4 Iron Utilization

24.4.4.1 Oxidation and Reduction of Iron. Much of the metabolism of iron is involved with its interconversion between the reduced (Fe^{2+}) and oxidized (Fe^{3+}) forms. Hemoglobin, the major iron-containing molecule of the body, contains ferrous iron, while both plasma transport of iron by transferrin and intracellular storage as ferritin (cytosolic) and frataxin (mitochondrial) require ferric iron. Many enzymes catalyze the oxidation or reduction of molecular iron, including the following three.

Ceruloplasmin. Ceruloplasmin is a blue-colored, copper-containing glycoprotein synthesized by and secreted from the liver. This protein, also known as *plasma ferroxidase*, oxidizes Fe^{2+} to Fe^{3+}. It functions to provide ferric iron for transport by transferrin. The dependency on a copper-containing enzyme for iron transport accounts for why a copper deficiency results in anemia.

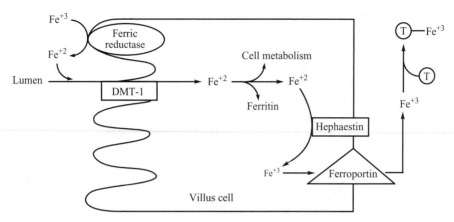

FIGURE 24-11 Pathway of iron uptake by enterocytes. DMT-1, divalent metal transporter-1. ⓉT, transferrin.

Hephaestin. Hephaestin is a membrane-bound protein which, like ceruloplasmin, is a copper-dependent ferroxidase (Fig. 24-11). It is located on the basal–lateral surface of intestinal enterocytes and acts to provide ferric iron to ferroportin, the basolateral transporter that transfers Fe^{3+} from the enterocyte to plasma transferrin.

Ferric Reductase. This enzyme, localized on the luminal surface of duodenal enterocytes, catalyzes the conversion of Fe^{3+} to Fe^{2+}, permitting absorption of dietary nonheme iron.

24.4.4.2 Absorption of Dietary Iron. Normally, about 10% of dietary iron is absorbed into the blood. The efficiency of iron absorption in the small intestine is influenced by a number of factors, including the nature of the dietary source of iron, a person's iron status, and the availability of reducing agents (ascorbic acid, cysteine) and an appropriate pH. In general, heme iron is absorbed more efficiently than nonheme iron.

Absorption of Heme Iron. In North America and Europe, about one-third of dietary iron is heme iron, which enters the enterocyte as the intact metalloporphyrin by a mechanism involving heme carrier protein 1 (HCP1). Inside the enterocyte heme oxygenase cleaves the porphyrin ring and releases the free Fe^{2+} (Fig. 24-11).

Absorption of Nonheme Iron. For dietary nonheme iron to be absorbed, it must first be freed from proteins and other ligands during digestion. The dissociation of iron from iron-binding ligands is facilitated by the acid pH of the stomach, which aids the reduction of ferric iron to ferrous iron; ascorbic acid (vitamin C) enhances this process. Enterocytes absorb nonheme iron in the ferrous form; any iron that has been oxidized to Fe^{3+} in the alkaline intestinal lumen is first reduced to Fe^{2+} by a brush-border ferric reductase (Fig. 24-11). Ferrous iron is then transported into the enterocyte by means of a divalent metal iron transporter called *DMT1*, which also transports copper, zinc, manganese, and lead.

Fate of Iron Within Enterocytes. Iron, whether derived from heme or nonheme sources, has two potential fates within the enterocyte. Some of the iron is reoxidized to Fe^{3+} by hephaestin and exported by ferroportin for transport on transferrin. The remainder is bound to ferritin and trapped within the enterocyte. This ferritin-sequestered iron is ultimately excreted when the enterocytes reach the end of their life cycle and are sloughed into the intestinal lumen.

Delivery of Iron to Cells. The circulating iron : transferrin complex binds to transferrin receptors on the plasma membrane of hepatocytes, cells of the reticuloendothelial system and other cells, and is internalized by receptor-mediated endocytosis.

24.5 REGULATION OF HEME AND IRON METABOLISM

24.5.1 Regulation of Heme Synthesis

ALA synthase catalyzes the regulated step of heme synthesis. Both the synthesis and activity of the enzyme are inhibited by heme and by hemin (which contains ferric iron). There are two isoforms of ALA synthase: ALAS1 in nonerythroid cells and ALAS2 in erythroid cells and their expression is regulated differently. Essentially all of the heme made by erythroid cells is committed to hemoglobin synthesis. Hypoxia and erythropoietin increase heme synthesis in erythroid cells. ALAS2 mRNA contains an iron-responsive element (IRE) in its 5′-untranslated region (UTR) and is responsive to the intracellular availability of iron. Heme synthesis is also coordinated with globin-chain protein synthesis.

By contrast, most of the heme synthesized in hepatocytes is incorporated into cytochromes of the electron-transport chain and P450-type cytochromes involved in the detoxification of xenobiotics. The expression of ALAS1 in hepatocytes is increased in response to many of the drugs and toxins that are metabolized in the liver (e.g., carbon tetrachloride, phenobarbital, acetaminophen, some oral contraceptives).

24.5.2 Iron Homeostasis

Since there are no mechanisms to eliminate excess iron from the body, iron homeostasis is regulated primarily by controlling the uptake of dietary iron by the intestine and release of iron from stores in the liver. Furthermore, all cells regulate their intracellular iron concentrations, a process necessary for preventing toxicity associated with iron-catalyzed free-radical reactions.

Iron metabolism is regulated by two iron-binding proteins, IRP-1 and IRP-2 (Fig. 24-12). IRP-1 is actually a cytosolic form of the iron-containing protein aconitase, the enzyme normally thought of in the context of the tricarboxylic acid cycle and energy metabolism, while IRP-2 lacks aconitase activity. Both IRP-1 and IRP-2

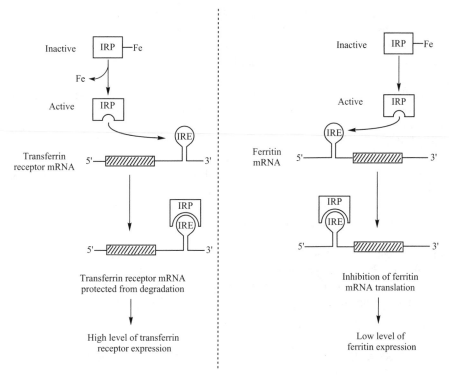

FIGURE 24-12   Regulation of iron stores in hepatocytes when there is iron sufficiency.

act by binding to iron-responsive elements in mRNAs and play similar roles in post-transcriptional gene regulation.

24.5.2.1 How Do Hepatocytes Regulate Their Iron Stores? The synthesis of transferrin and ferritin from their respective mRNA molecules is regulated by the iron-binding proteins IRP-1 and IRP-2. When these proteins bind iron, they are inactive (Fig. 24-12). However, in the iron-depleted state, the apo-IRPs bind to stem–loop structures: in particular, mRNA molecules. Binding of IRP to iron-responsive elements (IRE) in the 5′-untranslated region of ferritin mRNA inhibits initiation of translation, thus decreasing ferritin synthesis. By contrast, the mRNA for the transferrin receptor has IRE elements in the 3′-untranslated region; binding of apo-IRP to these IREs stabilizes the mRNA molecule and protects it from degradation. The result of these two regulatory processes is such that when iron stores are low, hepatocytes synthesize transferrin receptors but not ferritin, thereby enabling iron to be taken up from the blood and made available for heme synthesis (Fig. 24-12). Conversely, when the intracellular iron concentration is high, the two regulatory processes act to inhibit iron uptake (by down-regulating transferrin receptor expression) and increase storage of excess iron in the form of ferritin.

24.5.2.2 How Does the Intestine Regulate the Uptake of Dietary Iron into the Body? As described earlier, uptake of iron into the enterocyte requires the activity of the transporter DMT1. Furthermore, absorbed iron can either be transferred to transferrin in the blood by ferroportin or sequestered in ferritin and eventually excreted. Developing enterocytes in the crypt of the intestinal villus are programmed by the IRP-IRE mechanism to synthesize the appropriate amounts of DMT1, ferritin, and ferroportin. This programming provides a direct response to the levels of dietary iron, since when mucosal cells have accumulated iron, additional uptake into the body is down-regulated.

Furthermore, when iron stores are abundant, hepatocytes synthesize a protein named *hepcidin* and secrete it into the blood. Hepcidin travels to the enterocytes, where it forms a complex with ferroportin. The ferroportin : hepcidin complex is internalized and degraded; the end result of this process is to decrease release of iron from enterocytes into the blood. Hepcidin also binds to ferroportin in the plasma membrane of macrophages, thereby decreasing the release of iron from macrophages that have catabolized hemoglobin contained in senescent red blood cells. An additional, although not yet well-characterized mechanism for regulation of intestinal iron metabolism involves a soluble regulatory molecule produced by erythroid cells in response to anemia.

24.6 DISEASES INVOLVING HEME AND IRON METABOLISM

24.6.1 Iron Deficiency Anemia

Iron deficiency can result from inadequate intake of iron or loss of iron resulting from hemorrhage (e.g., gastrointestinal-tract bleeding) or excessive menstrual blood loss. Globally, inadequate dietary intake of iron remains the major cause of iron deficiency, especially where the diet is largely cereal-based and contains little meat. Iron-deficiency anemia is a major cause of pregnancy-related mortality in developing countries. Infants beyond 1 year of age who are fed solely breast milk or cow's milk without iron supplementation are also at risk of anemia. Although certain foods contain phytates, oxalates, and tannins that chelate nonheme iron, thereby rendering the iron less available for absorption, malabsorption of iron is uncommon in the absence of disease of the small intestine (e.g., celiac disease, enteritis, parasitic infection). Persistent achlorohydria (decreased HCl secretion by the stomach) may result in iron deficiency because low pH increases the solubility of iron and promotes the release of protein- and mucin-bound iron in the gastrointestinal tract.

Hemorrhage is the leading cause of excess loss of body iron. Indeed, in men and postmenopausal women, the differential diagnosis of iron deficiency starts with gastrointestinal cancer. Younger women tend to have lower body stores of iron than men and lower plasma ferritin levels; this difference is due to menstrual blood loss. Excessive iron loss during menstruation, or the iron requirements of frequent pregnancies, can also result in iron deficiency and anemia.

In iron-deficiency anemia, red blood cells are small (microcytic) and pale (hypochromic). Prior to development of frank anemia, the most sensitive clinical indicator of emerging iron deficiency is a plasma ferritin concentration that falls below the reference range. When iron stores become so low as to compromise erythropoiesis, there is an increase in the serum concentration of transferrin (increased iron binding capacity) and a decrease in transferrin saturation (less than the normal 30% of the iron-binding sites of transferrin occupied by iron atoms). Iron deficiency is commonly treated by supplementing the diet with ferrous salts (e.g., ferrous sulfate) and juices containing ascorbic acid and citric acid which enhance iron absorption.

24.6.2 Hereditary Hemochromatosis

Hereditary hemochromatosis (HH), a relatively common autosomal recessive disease in Caucasians, is characterized by progressive accumulation of excess iron in cells of the liver, heart, pancreas, and other organs. Menstrual blood loss explains why women are usually less severely affected than males. The clinical hallmarks of HH include cirrhosis, diabetes mellitus, cardiac failure, and bronzing of the skin. In people with HH, the total serum iron concentration and the ferritin level are increased, and the transferrin saturation is usually >50% and often >90%.

Classic hemochromatosis is the result of mutations in the gene for a protein called *HFE*. Normally, HFE binds to the transferrin receptor and reduces the affinity of the receptor for transferrin. The HH mutations cause loss of function of HFE and increased absorption of iron by the liver. This, in turn, down-regulates hepcidin synthesis and ultimately increases uptake of iron from the intestine into the blood. Interestingly, a genetic deficiency of hepcidin results in a juvenile form of hemochromatosis that is associated with a severe form of iron overload.

The only therapy recommended for HH is routine phlebotomy, with the removal of one unit (0.5 L) of whole blood (approximately 250 mg of iron) weekly or biweekly until iron stores return to normal levels. This therapy is quite effective if initiated prior to end-organ damage.

24.6.3 Hemosiderosis

Hemosiderin is an intracellular iron-storage complex that contains denatured ferritin (Fig. 24-2). Hemosiderin accumulates in macrophage following the increased phagocytosis of red blood cells associated with hemorrhage. Accumulation of hemosiderin in liver is associated with conditions of iron overload, as can occur in people who receive frequent blood-transfusion therapy for sickle cell anemia or thalassemia, as well as in patients with hemochromatosis.

24.6.4 Lead Poisoning

Lead inhibits both ALA dehydratase and ferrochelatase, thereby reducing heme synthesis and resulting in microcytic, hypochromic anemia. Plasma ALA and erythrocyte protoporphyrin concentrations are increased in people with lead poisoning.

Furthermore, since heme is the prosthetic group of many enzymes and proteins, including the cytochromes of the mitochondrial electron-transport chain, lead poisoning can also have detrimental effects on energy metabolism. Lead is especially toxic to the nervous system, probably due to accumulation of δ-aminolevulinic acid as well as to impaired energy metabolism.

24.6.5 Porphyrias

Porphyrias are a family of genetic diseases caused by deficiencies of the various enzymes in the pathway of heme synthesis, with resulting accumulation of intermediate metabolites. Depending on the particular gene affected, porphyrias can affect heme synthesis in all cells or be primarily either hepatic or erythropoietic. Most porphyrias are associated with nervous system pathology. Accumulation of porphyrinogens in the skin can lead to photosensitivity.

24.6.5.1 Acute Intermittent Porphyria. Acute intermittent porphyria is caused by a defect in *porphobilinogen deaminase* (also called *hydroxymethylbilane synthase*). It is associated with elevated δ-aminolevulinic acid and PBG levels and results in severe neurological symptoms. People with acute intermittent porphyria have periodic crises that are usually precipitated by drugs such as barbiturates and sulfonamides that induce the synthesis of heme-containing cytochrome P450 enzymes.

24.6.6 Jaundice

Jaundice (also known as *icterus*) is a condition of impaired heme catabolism. It is characterized by a yellow color of the skin and sclerae of the eyes that is the result of an elevated plasma concentration of bilirubin. Bilirubin toxicity or kernicterus occurs when the plasma level of bilirubin is high enough to result in transfer of excess bilirubin to membrane lipids, particularly in the brain. Jaundice can be a symptom of many different clinical problems.

24.6.6.1 Prehepatic Jaundice. In hemolytic anemias, the excess breakdown of red blood cells results in the production of abnormally large quantities of bilirubin, which may overload the liver's capacity to conjugate bilirubin. As a result, the plasma concentration of unconjugated bilirubin rises. Unconjugated bilirubin may also spill over into bile and increase the risk of developing pigmented gallstones (calcium bilirubinate).

Genetic causes of hemolytic anemia include sickle cell anemia (especially during sickling crises), thalassemia, and hereditary spherocytosis, which is a defect in a red cell membrane protein that results in premature breakdown of the red cells. Inborn errors of metabolism that cause hemolytic anemia include inadequate activities of pyruvate kinase (required for the generation of ATP) and glucose 6-phosphate dehydrogenase (required to generate NADPH and thus maintain adequate levels of reduced glutathione). Hemolytic anemia can also result from infections (particularly malaria), certain drugs (e.g., primaquine), autoimmune reactions, and poisons (e.g., paraquat).

Low plasma levels of haptoglobin are diagnostic for hemolytic anemia. Excess hemoglobin released into the blood binds to haptoglobin and the resulting hemoglobin–haptoglobin complex is cleared by the reticuloendothelial system, thus removing haptoglobin from the circulation. An older diagnostic approach relies on finding increased urinary and fecal levels of urobilinogen, reflecting increased secretion of conjugated bilirubin by the liver and subsequent bacterial metabolism of that bilirubin in the intestine.

24.6.6.2 Hepatic Jaundice. Impaired liver function is one of the major causes of jaundice. Hepatitis and cirrhosis impair the ability of hepatic UDP-glucuronyl transferase to conjugate bilirubin. The secretion of conjugated bilirubin into the bile is also compromised. As a result, both unconjugated and conjugated bilirubin accumulate in the blood, while fecal and urinary urobilinogen levels are decreased.

Hepatic jaundice can also result from deficiency of one or more of the enzymes involved in the metabolism and excretion of bilirubin. Three of these diseases are related to defects in the activity of UDP-glucuronyl transferase (also called UGT1A1) and result in elevated levels of unconjugated bilirubin. The most common of these diseases is Gilbert syndrome, which is generally benign, with affected persons exhibiting mild elevations in unconjugated bilirubin. By contrast, Crigler–Najjar (CN) syndrome type I patients have essentially no UGT1A1 activity and are unable to conjugate any bilirubin. CN type II patients produce a mutated protein that retains residual glucuronyl transferase activity.

24.6.6.3 Posthepatic Jaundice. Obstruction of the common bile duct due to a stone or (less commonly) a tumor results in posthepatic jaundice. The backup of conjugated bilirubin in the liver results in abnormal spillage of conjugated bilirubin into the blood and its excretion in the urine, thereby imparting a dark color. By contrast, lack of biliary excretion results in pale stools that lack normal pigmentation. Prolonged obstruction of the biliary system can lead to liver damage and result in increased plasma levels of both unconjugated and conjugated bilirubin.

Dubin–Johnson syndrome and Rotor syndrome are two rare but benign genetic diseases associated with elevated levels of conjugated bilirubin. They are caused by reduced function of the ATP-dependent pump that transports conjugated bilirubin across the bile canaliculus from hepatocytes into bile.

24.6.6.4 Neonatal Jaundice. Jaundice is common in infants, but it is usually benign and self-limiting. Hyperbilirubinemia often occurs because of a combination of two factors: breakdown of fetal hemoglobin as it is replaced by adult hemoglobin, and limited hepatic conjugation of bilirubin.

Premature infants often develop more severe jaundice, due primarily to hepatic immaturity. It is important that the gene for UDP-glucuronyl transferase not be activated early in fetal development, when it would be undesirable to produce conjugated bilirubin and secrete it into the biliary system; instead, bilirubin produced by the fetus is transferred to the maternal circulation. The activity of UDP-glucuronyl transferase normally increases just before birth and continues to increase after birth. In premature infants, the activity of UDP-glucuronyl transferase is often low, leading to impaired conjugation and secretion of bilirubin and elevated plasma levels of unconjugated bilirubin.

Neonatal jaundice can also result from hemolytic diseases of the newborn, usually associated with blood-group (i.e., ABO, Rh) incompatibilities between mother and child. If during pregnancy a small amount of fetal blood enters the maternal circulation, the mother may produce antibodies against fetal blood cells. If maternal antibodies then pass through the placenta to the fetus, they will produce hemolysis in the fetus or neonate.

Phototherapy continues to be the standard treatment of neonatal hyperbilirubinemia. The effectiveness of this form of therapy is based on the ability of photons to convert bilirubin IXα into photoisomers that are more water-soluble. Because of their increased solubility, these bilirubin photoproducts can also be excreted by way of the liver without requiring glucuronic acid conjugation. Blue light (approximately 450 nm), longer-wavelength green light, and more commonly, fluorescent white light have been used for phototherapy.

CHAPTER 25

INTEGRATION OF METABOLISM

25.1 BALANCE BETWEEN ENERGY PRODUCTION AND UTILIZATION IN HUMAN METABOLISM

The major role of metabolism is to capture chemical energy from foodstuffs as ATP and utilize that ATP for a variety of essential functions, including synthesis of cellular components, active transport of ions and solute, and muscle work. Humans can generate ATP by oxidizing carbohydrates, fatty acids, and amino acids. At the simplest level, energy homeostasis involves a balance between dietary fuel intake and energy expenditure so that the body is neither fuel-depleted (starvation) nor storing excess triacylglycerol (obesity). Since humans do not eat continuously, dietary fuels in excess of immediate needs are therefore processed and stored for subsequent use. Consequently, specific metabolic pathways must be regulated and the activities of different organs coordinated to satisfy the needs of the body.

25.1.1 How Much Energy Can One Get From Different Metabolic Fuels?

The chemical energy in foods is measured in kilocalories (kcal), sometimes denoted Calories (1 Calorie = 1 kcal). Since 1 kcal is defined as the energy required to raise the temperature of 1 kilogram of water by 1 degree Celsius, one can determine the energy content of different chemicals by measuring their heat of combustion in vitro in a calorimeter or by determining physiologically how much heat a person

Medical Biochemistry: Human Metabolism in Health and Disease   By Miriam D. Rosenthal and Robert H. Glew
Copyright © 2009 John Wiley & Sons, Inc.

generates when a particular fuel is being oxidized. The values used by nutritionists and nutritional biochemists are:

Carbohydrates	4 kcal/g
Triacylglycerol	9 kcal/g
Protein	4 kcal/g
Ethanol	7 kcal/g

For carbohydrates, triacylglycerol (fat), and ethanol, the physiological fuel values are approximately equal to their respective heats of combustion. The actual heat of combustion of protein is close to 5.4 kcal/g; the lower physiological value reflects the energy cost of excreting nitrogen as urea. (The synthesis of urea from ammonia, CO_2, and the amino group of aspartate requires four high-energy phosphate bonds.)

25.1.2   What Are the Fuel Stores of a Normal Person?

25.1.2.1   Carbohydrates. In the fed state, the reference standard 70-kg male has about 300 g of glycogen stored in his muscles and 100 g in his liver, with only minor quantities in adipose tissue and the brain.

25.1.2.2   Triacylglycerols. Since triacylglycerols (TAG) have a higher energy content than carbohydrates (9 kcal/g vs. 4 kcal/g) and are stored without hydration, they provide a much more compact form of energy storage than glycogen. Normal body stores of TAG total approximately 15 kg, or 135,000 kcal, compared to only 1600 kcal for glycogen. Although nearly all of this fat is stored in adipocytes, skeletal muscle and liver each contain about 50 g of triacylglycerol, and trained endurance athletes have even greater amounts of intramuscular triacylglycerol. Unlike the storage of carbohydrate as glycogen, the body has a virtually unlimited capacity to store TAG. An imbalance between energy intake and energy expenditure underlies the current epidemic in obesity.

25.1.2.3   Protein. Although there are no stores of proteins as such in the body, some of the normal cellular proteins are mobilized when amino acids are required for other needs, such as synthesis of new protein and providing carbon skeletons for gluconeogenesis. Most of the mobilizable proteins are found in skeletal muscle (6 kg) and in liver (0.1 kg). In cases of starvation and severe negative nitrogen balance, heart muscle proteins may also be degraded.

25.1.3   The Respiratory Quotient Can Be Used to Assess Which Fuels Are Being Utilized at a Particular Time

The nature and quantities of fuels being utilized at a particular time by an organism can be estimated using "indirect calorimetry," which measures oxygen consumption and carbon dioxide production rather than heat generation during a defined interval. The overall equations for the complete oxidation of glucose and a typical triacylglycerol molecule, in this case triolein, are used to determine the *respiratory quotient* ($RQ = CO_2/O_2$) for each reaction:

$$RQ = CO_2/O_2$$

$$C_6H_{12}O_6 + 6O_2 \rightarrow 6CO_2 + 6H_2O \qquad 1.0$$

$$C_{57}H_{104}O_6 + 80O_2 \rightarrow 57CO_2 + 52H_2O \qquad 0.71$$

25.1.3.1 What Can We Learn from RQ Data?

A fasted person at rest has an RQ of approximately 0.75. Based on the equations above, the RQ value indicates that this person is primarily oxidizing fat. By contrast, when the same person begins running rapidly, say on a treadmill, the RQ value will rise to nearly 1.0, indicating that he or she is utilizing mostly carbohydrates (i.e., glycogen, blood glucose). An RQ of 0.85 indicates that a person is utilizing a mixture of carbohydrates and fats.

25.1.3.2 Don't These Calculations Ignore Amino Acid Oxidation?

Many studies use first approximations to estimate the proportions of the various fuels being oxidized and ignore the contribution of protein as a fuel, primarily because considering protein oxidization complicates the picture. Consider the oxidation of alanine:

$$RQ$$

$$C_3H_7O_2N + 6O_2 \rightarrow (NH_2)_2CO + 5CO_2 + 5H_2O \qquad 0.83$$

The only way to accurately estimate the extent to which amino acids are oxidized is to measure urinary excretion of urea (usually over a 24-hour period). By subtraction, one then corrects for the amounts of O_2 consumed and the CO_2 produced during generation of the measured quantity of urea. The remaining O_2 and CO_2 are then use to calculate a "nonprotein" RQ value that can be used to determine the contributions of carbohydrate and fat to total metabolism.

For normal persons consuming a typical American diet, protein usually provides 12 to 20% of metabolic fuel. Thus, an RQ of 0.83 indicates that a person is obtaining approximately half of his or her calories from carbohydrate and half from fat rather than 100% from protein.

25.2 WHAT ARE THE MAJOR PHYSIOLOGICAL CONDITIONS THAT AFFECT FUEL UTILIZATION?

25.2.1 Fasting or Basal State

The *basal metabolic rate* (BMR) is the minimum energy expenditure required for involuntary work of the body (e.g., pumping of the heart, maintenance of ion gradients, protein turnover). BMR is measured in the morning, while the subject is in a prone position and has fasted for at least 12 hours. The measurement is made at an ambient temperature, where shivering thermogenesis and sweating is minimized. For convenience, the *resting energy expenditure* (REE) is usually measured instead of BMR; measurement of REE requires a less stringent fast (2 to 4 hours) and gives slightly higher values. One can approximate the BMR for a given person as 1 kcal/kg per hour for men and 0.9 kcal/kg per hour for women, or 1680 and 1200 kcal per day, respectively, for a 70-kg (154-lb) man and a 56-kg (124-lb) woman. The gender difference in BMR is due to the relatively greater adipose stores and lower muscle mass of women than men. Indeed, the BMR correlates primarily with lean body mass and can be increased by exercise, which promotes accrual of muscle.

25.2.2 Fed State

The resting metabolic rate is higher when measured in a person who has recently eaten a meal. The difference, sometimes referred to as the *thermic effect* of food, reflects the extra energy required for the digestion, transport, and storage of dietary fuels, including the active transport of solutes into cells and the activation of molecules (i.e., glucose to glucose 6-phosphate, fatty acids to acyl-CoAs). The thermic effect of food increases energy expenditure over BMR by 10 to 15%, depending on the person and the diet, with protein-rich foods requiring the greatest amount of energy to process and dietary TAG the least.

25.2.3 Physical Activity

Voluntary movement, including normal daily activities, fidgeting, and purposeful exercise, increases energy expenditure. Physical activity is the most variable component of a person's daily energy expenditure, and represents 20 to 40% of the total for the average person. Physical activity is also the only component of total energy expenditure that is easily altered. Energy expenditure during exercise is affected by the nature of the activity itself (running vs. walking); the intensity, duration, and efficiency of the activity; and the person's body mass.

25.3 ROLES OF DIFFERENT ORGANS IN THE INTEGRATION OF METABOLISM

The metabolism of certain cells changes little between physiological states. Red blood cells, for example, depend exclusively on glycolysis in both fasted and fed states. Similarly, the brain is dependent primarily on glucose as a fuel and does not begin to use significant amounts of ketones until day 3 or 4 of a fast. By contrast, most other cells and organs alter their pattern of fuel utilization, fuel storage, and fuel export to meet the current needs of those cells and of the body as a whole in various physiological states (Table 25-1).

TABLE 25-1   Major Organs Involved in Integration of Fuel Metabolism

Organ	Major Fuel Store	Fuel Exported	When Exported
Adipocytes	TAG	Free fatty acids and glycerol	Fasting, moderate-intensity exercise
Liver	Glycogen	Glucose	Fasting, exercise
		Ketones	Fasting
		VLDL-TAG	Fed state
Skeletal muscle	Glycogen	Lactate	Intense exercise
	Protein[a]	Alanine, glutamine	Fasting

[a]Mobilizable structural proteins.
VLDL, very low-density lipoprotein; TAG, triacylglycerol.

25.3.1   Liver

The liver plays a major role in all aspects of energy metabolism. When glucose is plentiful, the liver utilizes glucose as fuel, stores glycogen, and metabolizes excess glucose to acetyl-CoA. The acetyl-CoA, in turn, is used to synthesize fatty acids and ultimately TAG, which is exported from the liver in the form of VLDL. By contrast, when glucose is required by other cells, the liver switches to utilizing fatty acids to generate energy, mobilizes glycogen stores to maintain plasma glucose levels, and begins synthesizing both glucose and ketones. Utilization of the carbon skeletons of amino acids such as alanine and glutamine for gluconeogenesis is accompanied by conversion of their amino groups to urea.

25.3.2   Adipose Depot

Triacylglycerols are the major fuel stores of the body, and adipocytes are the major site of triacylglycerol storage. In response to hormonal (e.g., glucagon, hydrocortisone) and neuroendocrine (epinephrine) stimulation, free fatty acids are released when needed: for example, during fasting or to meet the increased energy demands of exercise, stress, and trauma. The glycerol generated when TAG is hydrolyzed is available to the liver for gluconeogenesis. By contrast, in the fed state, the body directs dietary fatty acids and glucose into triacylglycerol stores. Lipoprotein lipase in adipose capillaries hydrolyzes the TAG of VLDL: the fatty acids thus released are taken up by adipocytes, incorporated into TAG and stored. In the fed state, adipocytes also oxidize glucose, both to provide the glycerol backbone of TAG and to generate acetyl-CoA for a modest amount of fatty acid synthesis.

25.3.3   Skeletal Muscle

25.3.3.1   *Muscle in the Fed State.*   When glucose (and insulin) levels rise, muscle cells take up glucose via GLUT4 transporters and store that glucose as glycogen. Eating a meal and the subsequent rise in circulating insulin levels also stimulate uptake of amino acids into muscle and promote protein synthesis.

25.3.3.2 *Muscle in the Fasted State.* During an overnight (or longer) fast, skeletal muscle plays a major role in providing fuel to other organs, including the brain. Since muscle lacks glucose 6-phosphatase, muscle glycogen cannot be used to maintain plasma glucose levels. There is, however, considerable catabolism of muscle proteins during a fast. The carbon skeletons of branched-chain amino acids are primarily utilized as fuel by muscle, whereas alanine and glutamine are exported to support gluconeogenesis in liver and kidney, respectively. In the fasted state, muscles also use plasma free fatty acids and ketones to satisfy their fuel needs.

25.3.3.3 *Exercising Muscle.* Physical activity requires muscles to markedly increase the rate of ATP production. The mixture of fuels used by the muscle is dependent on both the intensity and duration of the exercise.

Sprinting. The immediate source of energy for muscles during a rapid sprint is ATP itself, along with the modest intramuscular stores of creatine phosphate, which can sustain a 6-second sprint. Muscle glycogen is also used by a sprinter, and under intense activity, muscle exports lactate into the circulation.

Walking and Similar Moderate Exercise. Fatty acids are the preferred substrates for exercise up to about 50% of $VO_{2\,max}$ (the maximum amount of oxygen the body can use).

Moderate-Intensity Exercise. As the rate of sustained exercise increases from 65% to 85% of $VO_{2\,max}$, the relative contribution of carbohydrate to total metabolism increases, with the ratio of ATP generated from carbohydrate and fat oxidation being in the range 40:60 to 60:40. As muscle glycogen is depleted, there is greater reliance on a mixture of bloodborne fatty acids and bloodborne glucose, with a concomitant drop in RQ from > 0.9 to as low as 0.75. Under these conditions, muscle fatigue may occur if the workload intensity is not decreased. It should be noted that it is only the glycogen stored in the exercising muscles that is depleted; the amount of glycogen in less active muscles (i.e., the arm muscles of a bicyclist) does not decrease.

Adaptations with Athletic Training. Highly fit persons (e.g., triathletes) are able to exercise at greater workloads and sustain their activity for long intervals. Physically fit persons have greater intramuscular stores of both glycogen and TAG than those of relatively inactive persons. They also have an increased $VO_{2\,max}$ values which results in the same level of exercise (i.e., speed of running) occurring at a lower VO_2, thus permitting a greater reliance on fatty acids than glucose to satisfy their energy needs.

25.3.3.4 *Heart Muscle.* Although the heart is never at rest, its metabolism is similar to that of the skeletal muscles of a person at rest, in that when the body is at rest the heart preferentially utilizes free fatty acids as fuel. Cardiac glycogen stores are mobilized for the greater cardiac work that exercise demands.

25.4 INTEGRATION OF ORGAN METABOLISM IN DIFFERENT PHYSIOLOGICAL STATES

Coordination of the metabolic activities of different organs serves to support glucose homeostasis and provide a steady supply of glucose to meet the needs of the brain and erythrocytes, both of which are constantly dependent on glucose as fuel. Integrated metabolism also serves to store fuel efficiently in times of plenty in order to provide for periods of fuel scarcity or times of high energy utilization.

25.4.1 Fasting State

Figure 25-1 shows the role of different organs in the coordinated metabolism of the basal metabolic state, when a person is at rest after an overnight fast. During the night, glucagon stimulates glycogenolysis and the glycogen stores in the liver become depleted. By morning, the major source of plasma glucose is hepatic (and

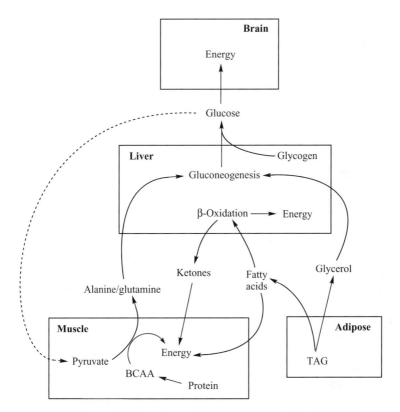

FIGURE 25-1 Major metabolic pathways in the fasted state. The dashed line represents utilization of glucose as a component of the alanine cycle. BCAA, branched-chain amino acids; TAG, triacylglycerols.

to some extent, renal) gluconeogenesis. Substrates for gluconeogenesis are provided by adipocytes (glycerol), muscle (alanine and glutamine), and erythrocytes (lactate). The increase in the rate of protein catabolism in muscle that is associated with gluconeogenesis is accompanied by increased hepatic synthesis of urea. β-Oxidation of fatty acids provides the large amounts of ATP required for both gluconeogenesis and ureagenesis. As a result of diversion of oxaloacetate from the TCA cycle to gluconeogenesis, the liver uses most of the acetyl-CoA from β-oxidation to synthesize ketones.

During a fast, free fatty acids mobilized from TAG in adipocytes are the major fuel supply for organs other than the brain and erythrocytes. As the fast continues, skeletal muscle cells oxidize a mixture of free fatty acids released from adipocytes, ketones produced by the liver, and branched-chain amino acids generated through catabolism of muscle proteins. Muscle cells also require a certain amount of glucose to generate the carbon skeleton of alanine, which is the means by which they export the amino groups released during the catabolism of the branched-chain amino acids.

25.4.1.1 Why Doesn't the Brain Oxidize Fatty Acids? Erythrocytes lack mitochondria and therefore cannot utilize either β-oxidation or the TCA cycle, both of which are mitochondrial processes. By contrast, although neural cells do have mitochondria, they do not use free fatty acids as an energy source during fasting because free fatty acids and other lipophilic substances do not readily pass through the blood–brain barrier. Thus, the mechanism for protecting the brain from a variety of deleterious substances renders the brain strongly dependent on a constant supply of glucose fuel.

25.4.1.2 What Happens During Prolonged Fasting? The changes in fuel utilization that occur with long-term fasting are referred to collectively as the *adaptation to starvation* (Fig. 25-2). Circulating levels of ketones rise markedly during the first few weeks of a prolonged fast and the brain begins to use ketones as well as glucose as fuel; after 2 to 3 weeks of fasting, ketones can satisfy as much as two-thirds of the energy requirement of the brain. Although the brain still does not use free fatty acids directly, neural oxidation of ketones represents, in essence, the brain's ability to utilize some of the energy originally stored in fatty acids. Increased ketone availability to the brain is facilitated by muscle, which stops oxidizing ketones and turns almost entirely to free fatty acids for energy. As this transition occurs, there is less demand for glucose by the brain and a concomitant decrease in the rate of catabolism of muscle proteins to provide gluconeogenic substrate for the liver and kidney. At the same time, relatively more of the amino acid–derived nitrogen excreted in the urine will be in the form of ammonium ions rather than urea. The ammonium ions buffer urinary acetoacetic acid and β-hydroxybutyric acid. Renal production of ammonium ions is directly coupled to an increase in renal use of the carbon skeleton of glutamine for gluconeogenesis.

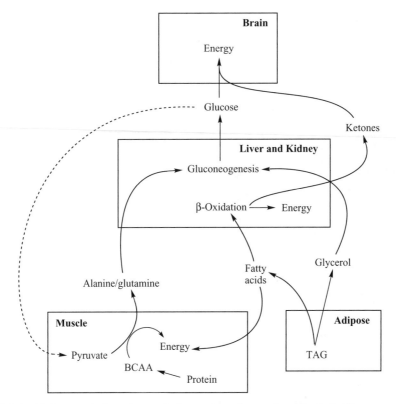

FIGURE 25-2   Metabolic pathways during prolonged starvation. The dashed line represents utilization of glucose as a component of the alanine cycle. BCAA, branched-chain amino acids, TAG, triacylglycerols.

25.4.2 Metabolism in the Fed State

The changes in metabolism in various organs that occur after ingestion of a mixed meal (carbohydrate, fat, and protein) reflect the assimilation of these nutrients and their processing for both immediate utilization and storage (Fig. 25-3).

25.4.2.1 Liver. When the plasma glucose concentration is high, the liver extracts glucose from the blood. Some of that glucose is used for glycogen synthesis; the remainder is oxidized to acetyl-CoA and used primarily for fatty acid synthesis. The resulting long-chain fatty acids are secreted from the liver as VLDL triacylglycerol.

In the fed state, the liver utilizes amino acids primarily for protein synthesis. However, with high protein intakes the excess amino acids are catabolized, with their carbon skeletons being converted to fatty acids and their amino groups utilized for urea synthesis.

25.4.2.2 Adipocytes. In the fed state, the adipose depot synthesizes and stores TAG. Free fatty acids are obtained from the exogenous TAG of chylomicrons and the

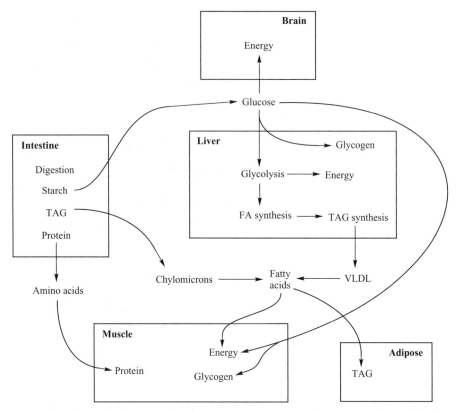

FIGURE 25-3 Major metabolic pathways in the fed state. Substantial amino acid uptake and protein synthesis also occurs in the liver (not shown). FA, fatty acid; TAG, triacylglycerols.

endogenous TAG of VLDL. Glucose is utilized to synthesize the glycerol backbone of TAG as well as the synthesis of long-chain fatty acids.

25.4.2.3 Muscle. When circulating levels of glucose and insulin are increased, muscle extracts glucose from the blood and uses it to synthesize glycogen. Under normal conditions, the synthesis of muscle glycogen functions merely to replenish glycogen stores. However, if carbohydrates are consumed after muscle glycogen has been depleted by strenuous exercise, resynthesis of glycogen may result in even higher glycogen levels than were present prior to the exercise. Athletes commonly refer to this phenomenon as *glycogen loading.*

25.4.3 Metabolism During Moderate Exercise

As discussed above, skeletal muscle can utilize a variety of fuels during exercise. Particularly during short bouts of intense exercise (the 100-meter dash), muscle cells derive energy from creatine phosphate and glycogen stores within the muscle itself.

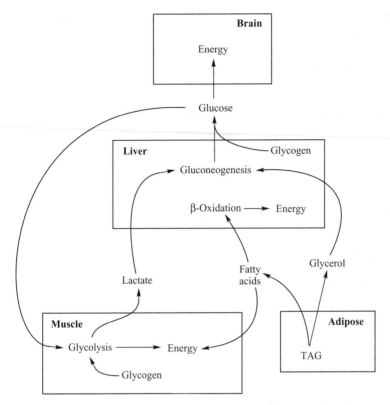

FIGURE 25-4   Major metabolic pathways during exercise. The extent of lactate generation will vary with the intensity of the exercise. TAG, triacylglycerols.

During strenuous exercise, much of the glycolysis that occurs in muscle is anaerobic; the resulting lactate is exported from the muscle and taken up by the liver, where it can either be oxidized further or used as a substrate for gluconeogenesis.

By contrast, moderate exercise relies on circulating free fatty acids and glucose as well as stores within the muscle (Fig. 25-4). TAG stored in adipocytes are the primary source of those free fatty acids, while liver glycogen provides plasma glucose. After 60 to 90 minutes of vigorous running, liver glycogen becomes depleted and hepatic gluconeogenesis is required to maintain plasma glucose levels. Initially, gluconeogenesis utilizes gluconeogenic precursors generated within the liver and glycerol derived from the breakdown of adipose TAG. However, as the exercise period lengthens, there is increased breakdown of muscle proteins to provide alanine and glutamine for gluconeogenesis, with the carbon skeletons of branched-chain amino acids providing an additional fuel source for the muscle. The increased utilization of amino acids for gluconeogenesis is reflected in an increase in hepatic urea synthesis and excretion of urea, primarily in sweat. As expected, people with low carbohydrate stores will excrete more urea during exercise than will those who are carbohydrate replete when they begin exercising.

25.4.3.1 Burning Glucose Versus Fat: Which Is Better for Weight Loss?

It is common for people wishing to lose weight to seek an exercise regime that burns fat rather than glucose. It turns out, however, that the crucial issue in exercise is total caloric expenditure. Most types of exercise promote utilization of a mixture of fuels, although more moderate exercise (brisk walking as opposed to running) utilizes relatively more fatty acid than glucose. To the extent that one utilizes free fatty acids during exercise, one decreases fat stores proportionally. Oxidation of glucose during moderate-duration exercise tends to deplete muscle and liver glycogen stores. When the next meal is consumed, the glycogen stores of the muscles are replenished, and less of the dietary carbohydrate is converted into fat, thereby resulting in lower adipose triacylglycerol stores than if the person had not exercised.

25.5 REGULATION OF METABOLISM

The regulatory affects of hormones on individual metabolic pathways are discussed in other chapters. Nevertheless, it is useful at this point to consider how hormones coordinate the multiorgan integration of metabolism. In general, insulin is anabolic and acts to stimulate the metabolic responses to the fed state, while a number of other hormones (often referred to as *insulin counterregulatory hormones*) oppose the actions of insulin and coordinate fuel mobilization and energy production during fasting or exercise.

25.5.1 Insulin

Insulin acts on many different tissues and has several effects in each, all of which are consistent with the anabolic needs of the body in the fed state. Thus, insulin stimulates translocation of GLUT4 transporters to the plasma membrane and uptake of glucose into both adipocytes and muscle cells. Insulin increases secretion of lipoprotein lipase by adipocytes, thereby increasing the release of free fatty acids from both chylomicrons and VLDL. Within adipocytes, insulin stimulates glycolysis, a modest amount of fatty acid synthesis, and triacylglycerol synthesis. Concurrently, insulin promotes both glycogen synthesis and glycolysis in muscle cells as well as uptake of amino acids into muscle and protein synthesis therein. Not all tissues are regulated by insulin. In particular, the uptake of glucose into neural cells and subsequent glycolysis are glucose-independent.

Although transport of glucose into hepatocytes is insulin-independent, insulin does stimulate the activity of key regulatory enzymes in the pathways that use glucose (glycogen synthesis, glycolysis) while inhibiting pathways that generate glucose (glycogenolysis, gluconeogenesis). Insulin increases the activity of acetyl-CoA carboxylase, thus stimulating hepatic fatty acid synthesis and subsequent VLDL synthesis. In addition, increased acetyl-CoA carboxylase activity generates malonyl-CoA, which prevents concurrent β-oxidation of fatty acids by inhibiting carnitine palmitoyl transferase (CPT-1). Insulin also stimulates synthesis of the sterol response element–binding proteins, SREBP-1 and SREBP-2, thereby increasing gene transcription of the lipogenic enzymes involved in fatty acid and cholesterol synthesis, respectively.

25.5.2 Glucagon

Glucagon stimulates hepatocytes and adipocytes to release glucose and fatty acids, respectively, into the circulation. In the liver, glucagon stimulates both glycogenolysis and gluconeogenesis while inhibiting the pathways of fuel storage (e.g., glycogen and fatty acid synthesis). In adipocytes, glucagon stimulates lipolysis and release of free fatty acids and glycerol into the circulation.

25.5.3 Epinephrine (Adrenaline)

The synthesis of epinephrine from tyrosine in the adrenal medulla is stimulated by stress, endurance exercise, and hypoglycemia. Epinephrine acts through the same G-protein, cAMP-dependent protein kinase signaling pathway as glucagon and has similar effects in both liver and adipocytes. Unlike glucagon, however, epinephrine also acts on muscle cells. Since muscle cells lack glucose 6-phosphatase, the epinephrine-stimulated breakdown of glycogen results in enhanced glycolysis.

25.5.4 Hydrocortisone

Hydrocortisone, a glucocorticoid synthesized by the adrenal cortex, stimulates fuel mobilization from liver, muscle, and adipocytes. However, unlike glucagon and epinephrine, hydrocortisone acts primarily by regulating gene transcription and mediates longer-term metabolic changes during starvation, sepsis, and stress.

25.5.5 Adipocytokines

Adipose tissue is more than a site for triacylglycerol storage; it is also an endocrine organ. The hormones and cytokines secreted by adipocytes include leptin, adiponectin, adipsin, interleukin-6, and tumor necrosis factor α (TNFα). Production of the various adipocytokines varies with a person's energy status (e.g., normal weight or obese) and the anatomical location of the adipose depot (e.g., visceral or subcutaneous).

Leptin signals energy sufficiency and acts to inhibit further lipid storage in adipocytes and stimulate lipolysis of intracellular TAG. It also decreases appetite. Adipsin (acylation-stimulating protein) has effects opposite to those of leptin; adipsin stimulates glucose uptake and increases the activity of diacylglycerol acyltransferase, thus causing adipocytes to retain fatty acids in the form of TAG.

Adiponectin, another adipokine, increases fatty acid oxidation in both muscle and liver. Increased plasma levels of adiponectin are correlated with higher levels of HDL-cholesterol and lower plasma levels of TAG.

25.5.6 Exercise

In addition to the actions of hormones, sustained physical activity exerts major effects on the regulation of energy metabolism. There are many long-term metabolic adaptations to aerobic exercise, including increased muscle mass, resulting in an increase

gy expenditure, and up-regulation of mitochondrial energy metabolism.
ly also increase total energy expenditure by enhancing peroxisomal pro-
and thus oxidation of long-chain fatty acids not directly coupled to ATP
and increasing the expression of uncoupling proteins in mitochondria.

One important regulator of muscle metabolism is AMP-kinase, which is activated when ATP depletion results in increased intracellular levels of AMP. AMP kinase inhibits acetyl-CoA carboxylase and lowers the level of malonyl-CoA in the cytoplasm, thereby stimulating fatty acid oxidation by increasing the activity of carnitine palmitoyl transferase-1. AMP-activated protein kinase also increases glucose uptake into the muscle by insulin-independent recruitment of GLUT4 glucose transporters to the plasma membrane. This phenomenon explains why glucose uptake by skeletal muscle is greatly increased during exercise, when there is no increase in the insulin level. Exercise also renders muscle cells more sensitive to insulin, in part by stimulating intramuscular accumulation of TAG, and removing potentially deleterious fatty acid metabolites.

25.6 CONDITIONS WHERE NORMAL METABOLIC INTEGRATION IS IMPAIRED

25.6.1 Obesity

Obesity is the accumulation of excess adipose tissue. It is commonly assessed by calculating the body mass index (BMI), defined as weight (kg)/height (m)^2. A BMI of 25 or greater is considered overweight, and a BMI in excess of 30 is considered obese. For a 5ft 4in.-tall woman, overweight would therefore be > 145 lb and obesity > 175 lb. BMI is applicable to virtually all adults, except for highly trained athletes such as body builders who have an unusually large muscle mass.

Obesity is the result of a chronic imbalance between energy intake and energy expenditure. One pound of adipose tissue represents approximately 3500 kcal [454 g $\times$ 9 kcal/g (for TAG) $\times$ 0.85 (the fraction of adipose tissue that is triacylglycerol)]. A person who consumes a caloric excess of 100 kcal/day will therefore gain 10 lb/year. Conversely, a person with a 500-kcal/day caloric deficit will lose approximately 1 lb/week. It should be noted that a person who utilizes 2000 kcal/day cannot be expected to lose more than 4 lb of adipose tissue per week even if he or she is fully fasting. More-rapid weight-loss programs actually represent loss of water and some muscle protein rather than the desired loss of adiposity. Most medical recommendations for weight loss suggest maintaining a healthy diet with a deficit of 500 to 1000 kcal/day.

Although there are certainly wide variations in body structure and metabolism between individuals, the current epidemic of obesity in the United States and many other countries is due more to environmental than to genetic factors. The efficient storage of excess calories as triacylglycerol may have been advantageous for our distant ancestors who were physically very active and for whom food scarcity was often the norm. However, in the current context of sedentary lifestyles and in environments where there is an overabundance of calorically dense foods containing large amounts of fats and sugars, efficient fuel storage in the form of fat can result in obesity and its undesirable medical and social sequelae.

25.6.2 Type I Diabetes Mellitus

Type I diabetes is caused by absent or insufficient insulin production. Although the most common cause is autoimmune destruction of the β-cells of the pancreas, it can also result from chronic pancreatitis. The resulting insulin deficiency has been described as "starvation in the midst of plenty," with metabolic pathways active in fasting mode despite high plasma levels of nutrients in the blood. A lack of insulin results in fasting hyperglycemia with both overproduction of glucose by the liver and underutilization of glucose by both muscle and adipocytes. The high rate of gluconeogenesis is often accompanied by extensive ketogenesis and severe ketoacidosis. Chronic hyperglycemia results in damage to many organs, including the eyes, kidneys, blood vessels, and nerves.

Diabetes affects lipid and protein metabolism as well as that of glucose. The high glucagon/insulin ratio stimulates adipose triacylglycerol hydrolysis and increases the plasma free fatty acid concentration; increased uptake and reesterification of fatty acids in the liver results in hypertriglyceridemia. At the same time there is increased catabolism of muscle proteins and hyperaminoacidemia.

Currently, the only effective treatment for type I diabetes mellitus is insulin therapy. The hormone is currently provided either by injection or using an insulin pump. Dosage and timing must be adjusted to a person's food intake and exercise, so as to maintain normoglycemia. Determination of glycosylated hemoglobin (HbA1c) is used to measure the efficacy of glucose control.

25.6.3 Insulin Resistance and Type II Diabetes

Most people with diabetes mellitus have type II diabetes rather than type I, and it is type II diabetes that is now reaching epidemic incidence in many countries. Type II diabetes mellitus is characterized by insulin resistance rather than primary insulin insufficiency. A person has insulin resistance when larger-than-normal amounts of insulin are required to support insulin-dependent metabolic processes. Insulin resistance is also commonly seen in the obese and in those with the metabolic syndrome. In most instances of insulin resistance, insulin secretion is not impaired and insulin receptors are functional. Although people in the early stages of insulin resistance can maintain normal blood glucose concentrations by increasing their insulin secretion, this compensation often becomes inadequate and they eventually progress to hyperglycemia and eventually to type II diabetes.

Like type I diabetes, type II diabetes is characterized by hyperglycemia, hypertriglyceridemia, hyperaminoacidemia, and elevated levels of free fatty acids. The high levels of free fatty acids in the blood are the result of increased TAG lipolysis by adipocytes. Elevated free fatty acid levels, in turn, result in increased TAG synthesis by the liver and export of TAG-rich VLDL particles. Unlike people with type

I diabetes, those with type II diabetes usually do not develop ketoacidosis, in part because – at least in the early stages of the disease – the liver is less insulin resistant than skeletal muscle or adipocytes.

Obesity often leads to insulin resistance in muscle. In people with insulin resistance, the muscle cells do not sufficiently up-regulate the acyltransferases involved in triacylglycerol synthesis to cope with the increased availability of free fatty acids. As a result, a high intracellular concentration of metabolic intermediates inhibits glucose uptake and glycolysis by muscle cells, and higher levels of insulin are required for glucose utilization. Although the mechanisms have not been fully elucidated, the relatively insulin-resistant state associated with obesity may be the result of an imbalance in adipokine production as well as elevated plasma levels of free fatty acids released from the excess adipocyte stores.

Exercise stimulates both TAG and glycogen synthesis in skeletal muscle and improves insulin sensitivity in both normal-weight and obese persons. Among the mechanisms involved are up-regulation of GLUT4 transporters and induced expression of diacylglycerol acyltransferase. In fact, moderate exercise and weight-loss regimes are often sufficient to preclude the need for pharmacological intervention in people with milder forms of type II diabetics. Type II diabetes can also be treated with a variety of drugs that stimulate insulin secretion (e.g., sulfonylureas) or increase insulin sensitivity (e.g, thiazolidinediones), or reduce hepatic gluconeogenesis (e.g., metformin). Many cases of type II diabetes, however, eventually progress to the point of pancreatic β-cell failure and dependence on exogenous insulin.

25.6.4 "Starvation Diets"

Weight-reduction plans usually involve a balanced diet that provides a reduced caloric intake and maintains the body's normal adaptations to the fed and fasted states. Under certain conditions, however, it is preferable to adopt a weight-reduction regimen that resembles starvation.

25.6.4.1 *Protein-Sparing Modified Fasts (PSMF).* Adaptation to starvation involves increased ketone utilization by the brain which decreases—but does not prevent—depletion of muscle protein. PMSF diets are designed to mimic starvation but prevent muscle loss. They usually provide 400 to 800 kcal/day of protein, with essentially no carbohydrates or fat, and are reserved for the morbidly obese (BMI > 40 kg/m^2) who are under medical supervision. The very low caloric intake maximizes weight loss, whereas the provision of exogenous protein provides substrate for gluconeogenesis, thus minimizing muscle wasting.

25.6.4.2 *Diets with Only Minimal Carbohydrate.* A number of popular low-carbohydrate diets utilize a variation of the PSMF in that the person consumes relatively unlimited protein and fat, but carbohydrate intake is strictly limited. This regimen results in an initial rapid loss of water weight; however, subsequent weight loss depends on maintaining an energy deficit. For some people the loss of appetite that accompanies ketosis aids compliance. Very low carbohydrate diets can be helpful

for insulin-resistant type II diabetic persons since there is a decreased demand for insulin; once weight loss is achieved, the person's insulin sensitivity often improves. However, very low carbohydrate diets are not recommended for long-term use because they restrict intake of fruits, vegetables, legumes, and dairy products, all of which provide essential nutrients.

25.6.5 Kwashiorkor and Marasmus

Kwashiorkor is a form of protein–calorie malnutrition that is caused by dietary protein deficiency and is often exacerbated by infection. The classic presentation, particularly in poorer countries, is a young child who has been weaned to an adult diet that lacks sufficient protein to sustain healthy growth. The characteristics of kwashiorkor include growth failure, edema, fatty liver, and "flaky paint" patches of skin. Because of the low protein intake, there is a deficiency of amino acids for synthesis of serum albumin and other plasma proteins, resulting in edema and the characteristic swollen abdomen and limbs. The situation is made worse by the availability of ample dietary carbohydrates, which stimulate insulin secretion and thus inhibit mobilization of amino acids from skeletal muscle. This dietary carbohydrate also provides substrate for fatty acid synthesis, which in the absence of adequate protein synthesis results in fatty liver and hepatomegaly.

The clinical manifestation of a diet deficient in both protein and energy is marasmus, which results in severe muscle wasting and marked growth retardation. Marasmus is the form of malnutrition that occurs when an infant does not receive adequate breast milk or formula. The factors that determine whether an older child develops kwashiorkor or marasmus are complex and not fully elucidated.

25.6.6 Hypercatabolic States

Sepsis, trauma, and burns result in hypercatabolic states, characterized by markedly increased fuel consumption; a negative nitrogen balance in which excretion of nitrogen—mostly as urea—exceeds nitrogen intake; fat mobilization; and marked catabolism of muscle proteins. Hydrocortisone is the main mediator of these changes. Hyperglycemia is a common finding in hypercatabolic states because, even though the insulin level may not be depressed, the metabolic effects of insulin are overcome by increases in the serum levels of the insulin-counterregulatory hormones.

INDEX

Medical Biochemistry: Human Metabolism in Health and Disease   By Miriam D. Rosenthal and Robert H. Glew
Copyright © 2009 John Wiley & Sons, Inc.

411